Solar Energy: Photovoltaic Technology and Systems

Solar Energy: Photovoltaic Technology and Systems

Editor: Catherine Waltz

R CALLISTO REFERENCE

www.callistoreference.com

Callisto Reference,
118-35 Queens Blvd., Suite 400,
Forest Hills, NY 11375, USA

Visit us on the World Wide Web at:
www.callistoreference.com

ISBN: 978-1-64116-117-6 (Hardback)

Cataloging-in-Publication Data

Solar energy : photovoltaic technology and systems / edited by Catherine Waltz.
p. cm.
Includes bibliographical references and index.
ISBN 978-1-64116-117-6
1. Solar energy. 2. Photovoltaic power generation. 3. Photovoltaic power systems.
4. Solar power plants. I. Waltz, Catherine.
TJ810 .S65 2019
621.47--dc23

Table of Contents

Preface

Solar energy is a renewable source of energy. The field of photovoltaics (PV) studies the conversion of sunlight into electricity using semiconducting materials. This field explores the phenomenon of photovoltaic effect for energy conversion. PV technology is used to power orbiting satellites, spacecraft and to generate power in power grids. Research in photovoltaic technology strives to increase efficiency and reduce costs, while also exploring newer photovoltaic structures and technologies like perovskites, organic and polymer photovoltaic, etc. This book presents recent innovations in photovoltaic technologies in a comprehensive manner. It will provide innovative insights into the field of photovoltaic technology and systems. It will benefit engineers, experts and students working in this domain.

This book is a result of research of several months to collate the most relevant data in the field.

When I was approached with the idea of this book and the proposal to edit it, I was overwhelmed. It gave me an opportunity to reach out to all those who share a common interest with me in this field. I had 3 main parameters for editing this text:

1. Accuracy – The data and information provided in this book should be up-to-date and valuable to the readers.

2. Structure – The data must be presented in a structured format for easy understanding and better grasping of the readers.

3. Universal Approach – This book not only targets students but also experts and innovators in the field, thus my aim was to present topics which are of use to all.

Thus, it took me a couple of months to finish the editing of this book.

I would like to make a special mention of my publisher who considered me worthy of this opportunity and also supported me throughout the editing process. I would also like to thank the editing team at the back-end who extended their help whenever required.

Editor

Fe-Cu metastable material as a mesoporous layer for dye-sensitized solar cells

Abdul Hai Alami[1,2], Jehad Abed[2], Meera Almheiri[2], Afra Alketbi[2] &
Camilia Aokal[2]

[1]Center for Advanced Materials Research, University of Sharjah, PO Box 27272 Sharjah, United Arab Emirates
[2]Sustainable and Renewable Energy Engineering Department, University of Sharjah, PO Box 27272 Sharjah, United Arab Emirates

Keywords
Dye-sensitized solar cells, mechanical alloying, mesoporous materials, metastable compounds

Correspondence
Abdul Hai Alami, Center for Advanced Materials Research, University of Sharjah, PO Box 27272 Sharjah, United Arab Emirates.
E-mail: aalalami@sharjah.ac.ae

Funding Information
No funding information provided.

Abstract

This study investigates the performance of dye-sensitized solar cells constructed with a Fe-Cu metastable material as the mesoporous layer on which a natural organic dye is applied. The synthesis of the Fe-Cu material is done via a high throughput process that produces nanosized particles from elemental metallic powders. Xanthophyll is singled out as the organic natural dye of choice among other dyes that were extracted, as it exhibited wider spectral absorptivity in terms of wavelength range and magnitude. Two compact solar cells were constructed and tested; one is a reference cell with a TiO_2 working electrode and the other with a Fe-Cu working electrode. The results show a better power conversion efficiency for the Fe-Cu-based solar cell 0.943% compared to 0.638% for the TiO_2, and the number of carriers in the former is found to be orders of magnitude higher than the latter (10^{19} vs. 10^{32}, respectively). A thorough optical, electrical, and thermal analysis of the Fe-Cu material is conducted and used to explain the obtained results.

Introduction

The dye-sensitized solar cells (DSSC) are formidable competitors to the industry-standard silicon-based solar cells due to their low cost, easy fabrication, and relatively high efficiencies, reaching up to 13% in porphyrin dye and cobalt (II/III) DSSC as reported by Grätzel group and others [1, 2]. Sensitizers are key cell components, critical to the widespread adoption of this technology. They possess excellent radiation absorbing properties in the visible wavelength range with their ability to mimic the light harvesting strategies found in nature to generate the required excitons. Many different sensitizers have been investigated for DSSCs applications, most of which use their carboxyl groups to enhance their attachment to the semiconductor [3]. The main types of sensitizers used nowadays are the metal complexes that use the metal to ligand (MTL) charge transfer phenomenon to increase photovoltaic performance, such as ruthenium complex sensitizers. It has been noted that DSSCs sensitized using pure organic sensitizers have higher absorption coefficients due to intramolecular π–π^* transitions. Their redox potentials, LUMO, and HUMO, are easily controlled for better performance and their ease of purification increases the chances of their acceptance in the market. Importantly, organic dyes have the availability aspect that metal complexes lack [4, 5]. Critical to achieving the reported efficiency levels, a proper choice of the nanostructured mesoporous substrate in order to achieve the desired dye molecules adsorption. This substrate is conventionally a metal oxide, usually a thin TiO_2 film, applied to the working electrode to maximize the surface area available for dye adsorption, optimize incident light harvesting, and enhance electrolyte diffusivity. The available molecular structures of TiO_2-containing mesoporous layers include: highly ordered nanorods with PCE of 2.9% [6, 7], spheres [8, 9], rice grain shapes and hollow fibers [10]; all pertaining to nanoscale dimensions. Some references in the literature have reported on the utilization of different materials to replace or be used in conjunction with the mesoporous semiconducting layer, for example, graphene has been used as the transparent conductive photoelectrode to great thermal and chemical stability [11, 12]. The use of organic perovskite electrodes is also a promising third generation solar cell technology employing lead or tin halide-based materials as the light harvesting electrodes, with reported efficiencies of around 20.1% in 2015, according to the National Renewable Energy Laboratory (NREL) [13].

The Fe-Cu intermetallic phase is attractive for many applications due to its high strength and traditionally attractive thermal and electrical properties [14–19]. But similar to other metastable intermetallic systems, for example, Ni-Ag, Cu-V, and Co-Cu, its synthesis suffers from the main drawback of limited immiscibility of its components as solid solutions due to their positive energy of mixing (around 13 kJ/mol for the Fe-Cu system) [20, 21]. Thus they will not form intermetallic compounds and will have negligible mutual solid solubility in equilibrium at temperatures below 700°C [22–26]. One effective and easy method to synthesize the Fe-Cu system with no conventional energy requirement in the form of applied heat or voltage is mechanical alloying (MA), which has the advantages of low-temperature processing, easy control of compositions, the production of relatively large amount of samples [20] and results in a significant extension of mutual solubility of the elements relative to the equilibrium values, which can be observed through X-ray diffraction patterns. This method involves ball-milling powders of the pure constituents to obtain the sought solid solutions. In this process, the coherent lattices of the pure metals undergo simultaneous shear induced deformation and thermal interdiffusion, with the resulting composition being determined by the equilibrium between the mechanically driven alloying and the diffusion-controlled decomposition [27–30]. It was found that low-energy ball milling of FCC and BCC metals leads to a refinement of the crystallite size to the nanometer scale [31–34]. With a work difference (ΔE_w) of around 0.43 eV, the Fe-Cu alloy system has a high intrinsic absorption coefficient that further optimizes its optical absorptance [35]. Also, any observed roughness of the microstructure allows the interreflection of incident irradiation in the UV–Vis range that reduces reflection and scattering losses [36], and decreases the impedance between space and the absorber, which also leads to better absorbance properties [37].

In this work an alternative to TiO_2 as a mesoporous substrate of the working electrode of a DSSC in the form of a Fe_{50}-Cu_{50} binary intermetallic phase is proposed and tested. The structural and optical properties of the obtained alloy are compared with those of TiO_2, and in addition, two DSSCs are constructed to evaluate the power, overall PCE, and fill factor (FF) for tested cells. The natural organic dye sensitizers (chlorophyll, anthocyanin, and xanthophyll) are extracted and processed locally and their absorptivities compared to choose the one with the widest spectral absorptivity in the UV–Vis wavelength. The Fe-Cu metastable alloy is produced via mechanical alloying, which is a facile and economic process that utilizes high-energy ball milling. It is an old but effective technique with high throughput, which will enhance the power-to-price ratio, increasing the appeal of DSSC in general as a third generation photovoltaic option. A comprehensive thermal, optical, and electrical investigation of the properties of the proposed Fe-Cu alloy is presented.

Experimental

Synthesis

The synthesis by MA takes place in a Retsch PM 100 planetary ball mill in a 25 mL stainless steel grinding bowl to mechanically alloy a starting amount of 9 g of high-purity copper (<425 μm, 99.5%) and iron (≥99%) powders, used as received from the supplier (Sigma-Aldrich, http://www.sigmaaldrich.com/united-states.html [United States of America]). A target composition 50:50 of Fe-Cu (% wt.) is used at a controlled milling speed of 600 rpm. Six 10-mm stainless-steel balls are used, making the filling ratio within the bowl 5:1. Milling is carried out for an hour at a time, pausing afterwards to cool the equipment and take a few milligrams of the powder for further characterization and testing. The run is to be terminated once microstructural changes become small, which in the present case took place after 6 h.

Microstructural analysis via SEM-EDX and XRD

The powder X-ray diffraction (XRD) patterns, plotted for five powder samples collected at a 2-h interval for 8 h, provide an insight into extent and progress of crystallization and the composition and grain structure of the developing solid solution. The X-ray patterns are recorded in the 2θ geometry between 40 and 90° at 0.02° 2θ/sec with a Bruker D8 Advance DaVinci multipurpose X-ray diffractometer with Cu Kα radiation operating at λ = 1.5406 Å, 40 kV tube voltage, and 40 mA current. The microstructural results are used to calculate important quantities such as the lattice parameter from Cohen's method, grain size by the Full-width half-max (FWHM) analysis with the Hall–Williamson method and Bragg's formula. Fused pieces of the material collected after 6 h milling time are examined under a scanning electron microscope (SEM) and the coupled energy dispersive X-ray spectrometer (EDX). The SEM is a VEGA3 XM by TESCAN, operating at 5 kV, whereas the EDX analysis is conducted with both map and point modes at the same operating voltage; the former was acquired during 3 min, whereas the latter was from four different spots of the sample during 30 sec live time.

Thermal analysis

Differential scanning calorimetry (DSC) is performed on the milled powders to provide insight on the energy of formation and mixing of the resulting compounds. The stability of these materials is due to the balance between the effectiveness of the MA process and thermal decomposition at high temperatures. For an endothermic reaction, heat flow indicates the phase shift within the solid solution (a peritectoid reaction). The calorimeter used is a Q20 from TA Instruments, running on 120 V_{ac}, 47–63 Hz, 500 W (4.5 amps) and equipped with a liquid nitrogen cooling system (LNCS) that allows automatic and continuous temperature control within a full range of −180°C to 550°C. A few milligrams of as-is and 2, 4, 6, and 8 h powders is encapsulated in an aluminum pan, and an empty reference pan sit on a thermoelectric disk surrounded by a furnace. As the temperature of the furnace is changed, heat is transferred to the sample and reference through the thermoelectric disk. The differential heat flow to the sample and reference is measured by area thermocouples. The phase formation, total enthalpy and heat flow through 0–500°C temperatures are examined for the current test.

Optical (spectroscopic) analysis

Spectral measurements of absorption in the ultraviolet, visible, and near-infrared (UV–Vis–NIR) regions were carried out on all the powders (as-is and 2, 4, 6, and 8 h milling time) with an Ocean Optics HR2000 high-resolution spectrometer. The HR2000 has a 300 lines per mm diffraction grating, 10 μm entrance slit, a Sony ILX511 2048-Pixel element linear CCD array detector, and is operating in the effective wavelengths range 300–1100 nm. The spectrometer is connected to a fiber optic reflection probe R200-7-SR, 2-m long, and of a 200-μm-core diameter. The reflection probe consists of a tight bundle of seven optical fibers in a stainless steel ferrule with six illuminating fibers around one axial read fiber, fixed at ~4 mm from the sample where losses due to scattering is assumed to be negligible, and all diffuse reflectance is collected at the probe. The source end of the reflection probe is connected to a tungsten halogen light source (Ocean Optics LS-1-LL). A reference surface in the form of a reflection standard (B0071519) is used to store baseline absorptance (0%) spectra to facilitate comparison between the various compositions. The integration time

was set to 30 msec to contain the intensity of the highest acquired peak. The recorded time-resolved spectra were averaged over 10 readings to increase the signal-to-noise ratio.

Electrical impedance and Mott–Schottky analysis

The impedance of the resulting Fe-Cu material is investigated and compared with values obtained for TiO_2 by virtue of the four-probe method impedance spectroscopy. The four probes, arranged as seen in Figure 1A, cover an area of 2 cm^2 and are connected to a Biologic

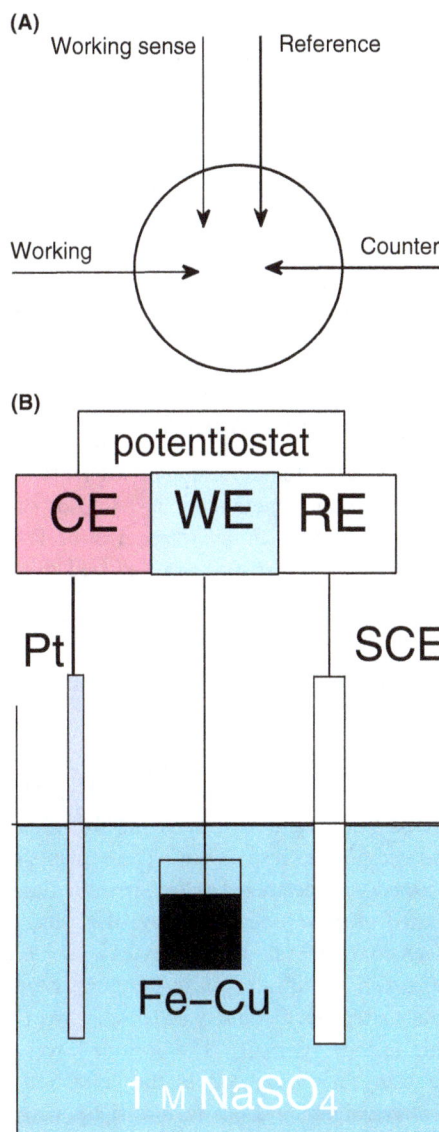

Figure 1. (A) Four-electrode arrangement for impedance spectroscopy and (B) Mott–Schottky cell setup.

VSP-300 potentiostat, providing impedance spectroscopy (10 μHz to 7 MHz) $\pm$ 12 V compliance, an automatic current range from 1 μA up to 500 mA (seven decades) and a potential resolution down to 750 nV. Impedance parameters were determined by fitting of impedance spectra using Z-view software and generating the pertinent Nyquist plot. The Fe-Cu material and the TiO$_2$ were deposited on respective glass substrates by spreading an emulsion of each in glycerol ($C_3H_8O_3$) and subsequently sintering the setup at 450°C for 30 min. It is noted that no phase change was observed afterwards in the postsintered XRD results. To investigate the interaction between the Fe-Cu layer and the electrolyte, a Mott–Schottky analysis was performed with the mesoporous FeCu deposited on ITO glass, which acts as the working electrode and immersed in an aqueous solution of 1 mol/L Na$_2$SO$_4$ (pH 8.5) with respect to a standard calomel (SCE) reference electrode and a Pt counter electrode [38]. The three-electrode arrangement for the electrochemical cell is shown in Figure 1B and is connected to a Biologic VSP-300 potentiostat providing an applied voltage range between −1 V and 1 V vs. SCE (20 potential steps recorded every 0.1 sec with waiting interval of 5 sec/step) and a potential resolution of 50 μV. The frequency range was from 0.1 Hz to 200 kHz (10 points per decade) with sinusoidal amplitude of 25 mV.

Surface area determination

The surface area of the resulting Fe-Cu material is determined by the BET method in an Autosorb iQ from Quantachrome USA machine, where 1 g of Fe-Cu and 1 g of TiO$_2$ (for comparison) are degassed for 16 h at 300°C then tested with adsorbent nitrogen at 77.4 K.

Natural organic dye extraction and absorptivity testing

The most commonly used natural organic dyes in literature are chlorophyll, anthocyanin, and xanthophyll. These dyes are extracted from plants rich in them, like spinach, cabbage, and red leaves, respectively. These plants are collected, washed with deionized water, dried by tissue paper and trimmed using scissors to remove the stalk, mid rib, and veins. A container full of sliced leaves is cooled downed by placement in liquid nitrogen to minimize degradation. Immediately after, leaves are soaked completely in a beaker filled with acetone (C_3H_6O). The acetone-leaves mixture is stirred using a stirring rod as the extraction process carries on overnight for 24 h in dim light. Later, liquid dye extracts are filtered through a funnel equipped with small pore filter paper and placed in a centrifuge at 10,000 rpm for 10 min. In the case of chlorophyll, the

beaker is covered with punched aluminum foil to allow natural vaporization of the acetone carrier for 48 h in dim light. On the other hand, each of the anthocyanin and xanthophyll are placed in a rotary evaporator to rotate and heat the filtrates up to the boiling point of extraction solution. The final products – chlorophyll, anthocyanin and xanthophyll extracts – are stored in a refrigerator to minimize any light-induced degradation [39–41].

In order to perform the absorptivity test, the dyes are diluted by distilled water, which is taken as the absorptivity baseline. The absorptivity measurement test is conducted using a PerkinElmer EZ301 spectrometer in ultraviolet–visible (UV-VIS) range (250 and 750 nm).

Solar cell fabrication and testing

To compare the effectiveness of utilizing the Fe-Cu material as a mesoporous material in DSSCs, two cells were constructed, one having TiO$_2$ as the photoelectrode, whereas the other had Fe-Cu. For depositing the former, the process starts with the preparation of TiO$_2$ paste by adding 1.5 mL of weak acetic acid (CH3COOH) to 1 g of titanium dioxide nanopowder (brookite nanopowder, <100 nm, 99.99%, used as received from Sigma-Aldrich) in a mortar while vigorously grinding by the pestle and then glycerol is added as a surfactant. The prepared homogenous paste is spread as a thin layer by employing the doctor blading technique on ITO-coated glass (15–25 Ω/sq., used as received from Sigma-Aldrich). The photoelectrode is then sintered by heating in a furnace at 450°C in static air for 30 min to remove solvents, surfactants, and any other organic materials.

As for the Fe-Cu photoelectrode, an identical process is applied, but instead of using acetic acid, glycerol is used to evenly spread the homogeneous layer on the ITO glass. The assembled cell is shown in Figure 2.

The electrolyte used is the classic I^-/I_3^- redox electrolyte prepared from 127 mg iodine crystals, 830 mg potassium iodide, and 10 mL ethylene glycol are mixed thoroughly until they are completely dissolved. This electrolyte will be the diffusion medium for ionic species between the working photoelectrode prepared above and the counter electrode made up of pure copper with carbon deposits on an area of 2.5 × 2.5 cm to enhance the surface area and consequently the conductivity. The operation mechanism (injection of electrons and diffusion of ionic species) of DSSCs is well explained and documented, especially for the I^-/I_3^- redox electrolyte [40].

The testing of the solar cell to measure its power conversion efficiency, fill factor, and IV characteristics are conducted using a solar simulator Xenon Arc Lamps setup, capable of providing irradiance between 0.1 and 1 suns (up to 1367 W/m^2) is used to provide a constant

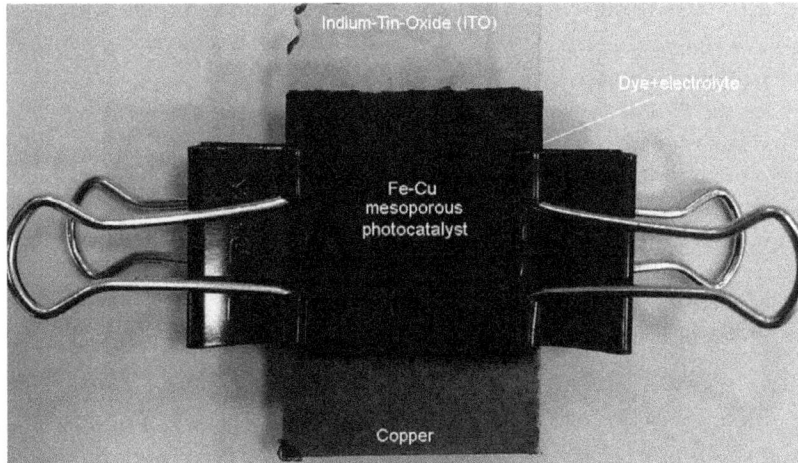

Figure 2. Assembled Fe-Cu sensitized solar cell.

irradiation on the cells, and was kept at a vertical distance of 30 cm at 1.5 AM at 25°C. The cell characterization is conducted using VSP-300 potentiostat from Biologic with a potential resolution of 1 μV and a control voltage of ±10 V up to ±48 V. The voltage was varied at a rate of 10 mV/sec and the current recorded at each point until the measured voltage reached the open circuit voltage, V_{OC}. A PV cell analysis software built into the potentiostat is also used to determine the power, efficiency, fill-factor (FF), and also the Nyquist plots of impedance for each cell configuration using the Z-fit postprocessor.

Results and Discussion

SEM and EDS

The SEM micrograph of the Fe-Cu system is shown in Figure 3A for a few milligrams removed after 6 h of milling. There is obvious presence of the FCC phase caused by the interdiffusion of Fe and Cu, resulting in different grain size distribution from a few nanometers to around 100 nm. Upon further magnification, Figure 3B shows a homogeneous phase with the apparent shearing effect from milling, which produces a lamellar structure [24]. An EDS analysis at two points (1) and (2) indicated on the figure shows a composition of 53% Fe and 47% Cu (%wt.) as evidence of successful interdiffusion and the formation of the new single phase. This is in contrast to other synthesis methods involving casting [21], deposition or electron-beam forming that reported severely segregated phases when the material is cooled below 700°C.

The lattice strain energy ΔU_{strain} can be calculated from the relation $\Delta U_{strain} = (\bar{E}\delta)(2\bar{r})^2\bar{d}$ where $\bar{E}$ is the average Young modulus for Cu and Fe, $\bar{d}$ the average

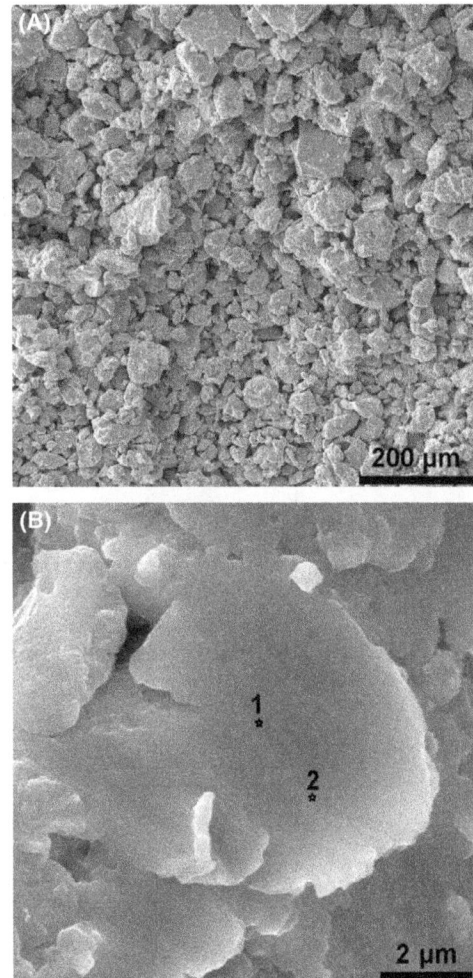

Figure 3. SEM photomicrographs of the resulting microstructure at (A) 250 × magnification, showing the granular microstructure, and (B) 20 k × magnification, showing homogeneous phases of Fe and Cu, with EDS test locations indicated at 1 and 2.

displacement of atoms $(\bar{d}=\delta\bar{r})$ and lattice distortion, δ, is calculated by using the equation $\delta=\sqrt{\sum_{i=1}^{n}X_i(1-\frac{r_i}{\bar{r}})^2}$, where X_i is the fraction of the ith component (n here equals 2) and r_i and $\bar{r}$ are the ith and the average atom radii, respectively [42]. The lattice distortion energy is found to be 8.61 kJ/mol. Although the relations are non-linear and highly dependent on the Cu concentration in the Fe-Cu material, this value is still orders of magnitudes higher compared to the martensitic transformation in iron system (around 751 J/mol) that will form a homogenous single martensite phase with a high density of lattice defects [43].

XRD

The result of XRD analysis is shown in Figure 4 for the as-is and 2, 4, 6, and 8 h milling time, to follow the evolution of the crystal structure. The as-is patterns show intense sharp peaks of elemental iron and copper are slightly shifted towards lower angles, which are consistent with the similar atomic sizes of Fe and Cu present in the same mixture [24], and the recorded signals are strong with little noise. It is interesting to observe these peaks widening and shifting further as milling proceeds due to the interdiffusion of Fe and Cu atoms caused by the shearing action and the friction-induced temperature increase, promoting larger grain sizes. After 6 h, a single phase is observed in which Fe atoms are diffused in the Cu matrix, or an FCC Cu(Fe) solid solution, as corroborated by the results from SEM analysis in Figure 4B. These

results are also consistent with previous reports [24, 44] and indicate the successful formation of the desired homogenous phase.

The grain size change with milling time is depicted in Figure 5 calculated from Bragg's law and Scherrer formula. It is noted from the figure that the grain size decreases rapidly as milling time increases, then it almost plateaus around 4–8 h, which is consistent with trends reported elsewhere [42, 44, 45].

Thermal analysis results

The results obtained from the DSC are shown in Figure 6 for all milling times and also for the as-is specimen. While the latter shows no phase reaction and its thermal response line remains flat, the rest of the samples show a trend that indicates more thermal stability as milling time increases. The results for 6 and 8 h show the least difference between the obtained peaks, indicating a limiting time for the milling process, which can be correlated with the XRD patterns of Figure 4. The results are inverted to show a positive scale and are in line with those found in previous work [18]. The peaks also occur at successively higher temperatures, indicating better phase stability.

Spectroscopy

The evolution of the microstructure with milling has impacted the optical absorptivity of the material, as can be

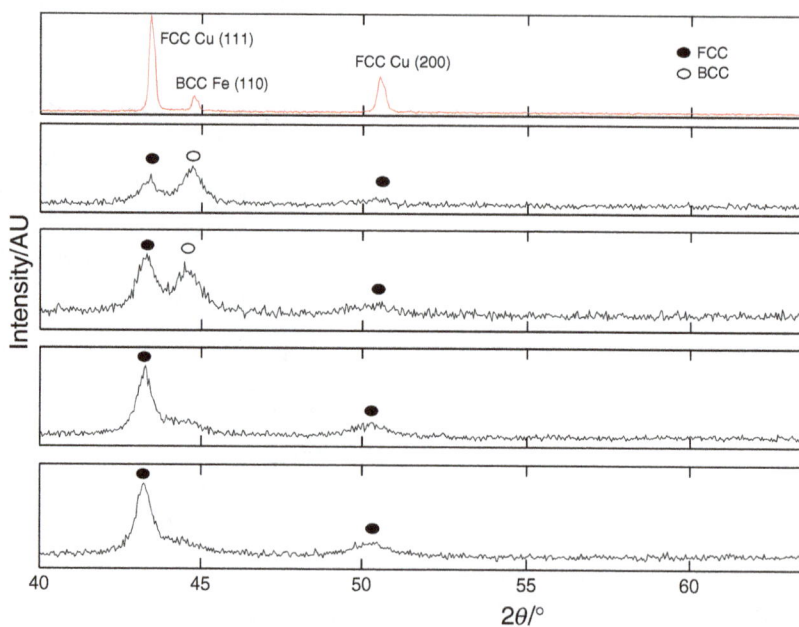

Figure 4. XRD plots for progression of milling times.

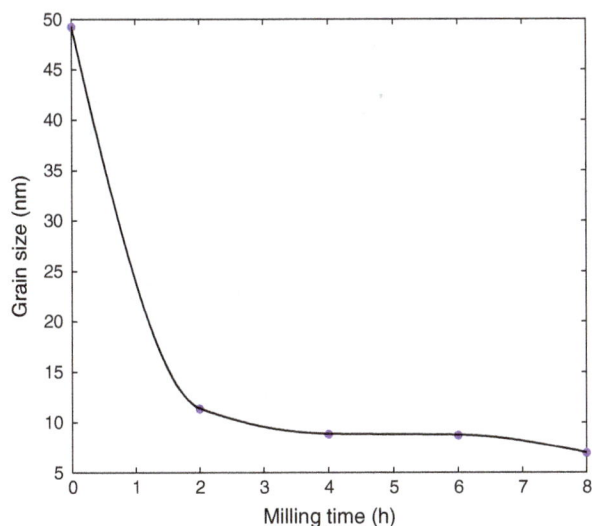

Figure 5. Grain size variation with milling time.

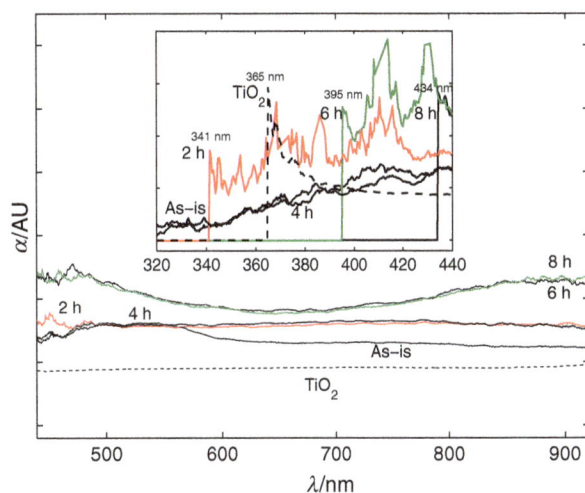

Figure 7. Absorptivity results at different milling times with TiO_2 superimposed in dotted line.

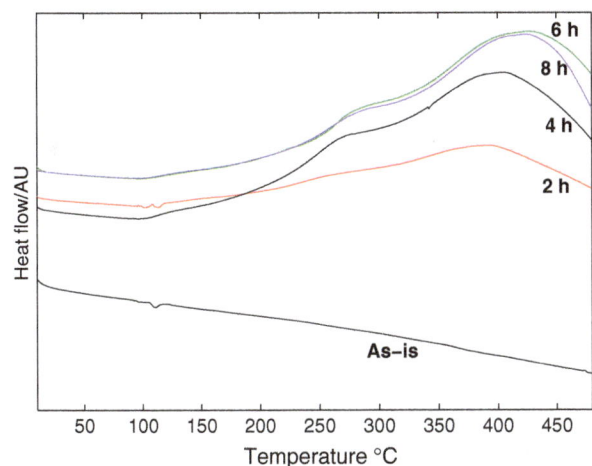

Figure 6. Differential scanning calorimetry heat flow plots versus milling times.

seen in Figure 7. The original powder shows a slight increase in absorptivity with respect to the mirror reference up to 500 nm, at which it plateaus until 550 nm, after which absorptivity decreases asymptotically towards 900 nm. As for milling times of 2, 6, and 8 h, the onset of absorptivity is seen to be an abrupt step increase at progressive wavelengths of 341, 396, and 434 nm, respectively. Another interesting observation from the curves can be made in conjunction with the XRD plots of Figure 3 for the 2–4 h on one hand and 6–8 h materials on the other. The former set shows a virtually constant absorptivity across the tested wavelength of Vis–NIR (450–900 nm) due to the existence of both FCC and BCC structures, whereas the latter start with a high onset value,

then assumes a flat parabolic shape with an apex at 650 nm, thought to be due to an all-FCC microstructure. Longer milling times of 6 and 8 h, exhibit a 63% and 81%, enhanced absorptivity, respectively, compared to as-is powder if the areas under the absorptivity curves are integrated from 450 to 900 nm. This enhancement is due in part to the roughness in microstructure that augments the probability of absorbing the diffracted and trapped light beams by adjacent grains. Roos et al. [46–48] have reported similar trends of optical properties of copper oxide thin films prepared by thermal and chemical oxidations and plotted against processing time, and the results are in general agreement of what is reported in this work.

It is also useful to compare the obtained optical properties of the prepared materials against TiO_2, a known photo catalyst that is used extensively in DSSCs. Thus, the optical absorptivity in the UV–Vis–NIR (280–900 nm) range was measured for an as-received TiO_2 powder (Sigma-Aldrich) and shown as a dashed line in Figure 6. The curve shows standard optical characteristics of the TiO_2 powder as reported in literature [49], where negligible absorptivity below 365 nm is replaced with a sharp increase in absorptivity followed by an exponential decay right before the 400 nm mark where it levels off as Figure 6 depicts. The advantage of using the suggested mesoporous Fe-Cu material is thus twofold, one is in the tuning of absorption response to higher wavelengths that can be well defended in situations where target solar cells are to be operated indoors or at different incident radiation intensities, and the other benefit is the obvious enhanced absorptivity magnitudes over the rest of the Vis–NIR spectrum which has many benefits for both solar thermal and solar photovoltaic applications.

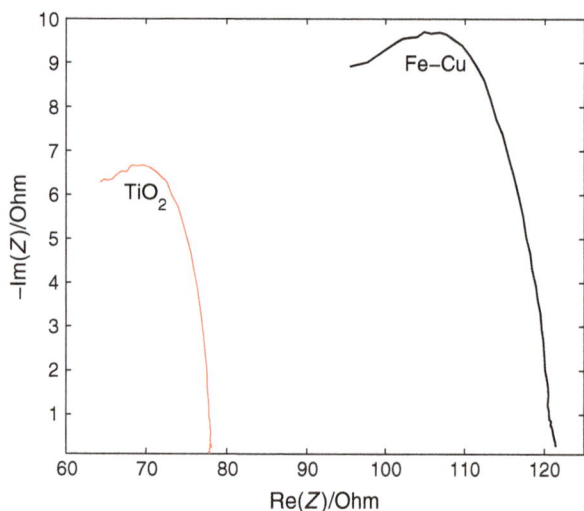

Figure 8. Impedance spectroscopy of 8 h milled Fe-Cu versus TiO$_2$.

Figure 9. (A) SEM of the Fe-Cu photoanode on ITO with the EDX map for (B) Fe, (C) Cu, and (D) oxygen; being only 11% wt. of the present components.

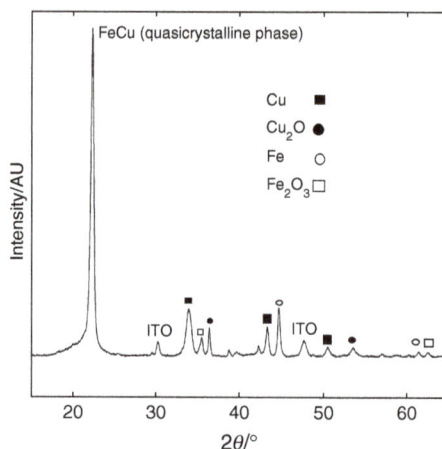

Figure 10. XRD of the Fe-Cu material deposited on ITO glass after solar testing.

Electrical impedance

The electrical impedance of a thin film of an 8 h milled Fe-Cu versus TiO$_2$, each deposited on a glass substrate is shown in the Nyquist plot of Figure 8.

Compared with the near zero real impedance of the as-is starting Fe-Cu powder, the 8 h milling time displayed a large impedance value (around 120 Ω), larger than the one measured for titania (around 78 Ω) as seen in the figure, suggesting a fundamental change in the electric behavior of the new Fe-Cu composite compared to the as-is starting mix, akin to the thermal conductivity of an alloy being substantially less than that of its pure components. This result is important, especially that the traces of oxygen observed in the EDX analysis of the components indicates that no oxides were formed from either iron or copper as seen in the mapping of Figure 9, showing that oxygen as only 11% of the elements present, with Fe, Cu, and I as 42%, 43%, and 4%, respectively.

An XRD testing is also performed on the sintered Fe-Cu on ITO, and the results show a significant peak belonging to the FeCu FCC phase that is sharper than the one seen in the original XRD due to sintering on glass at 400°C and a shift towards $2\theta = 21°$, with minor peaks belonging mostly to elemental Fe and Cu that have been segregated from the unsintered mixture. This has resulted in using up the available oxygen in the form of Cu$_2$O and Fe$_2$O$_3$, with the former being more pronounced and is an intermediate phase of copper oxide that is less stable chemically and structurally than CuO [50]. From the XRD plot of Figure 10, the Fe-Cu materials shows good stability after the 50 cycles of solar cell testing while being in contact with the dye and electrolyte, exhibiting what is believed to be mild oxidation of the surface of the material occupied mostly by the Cu

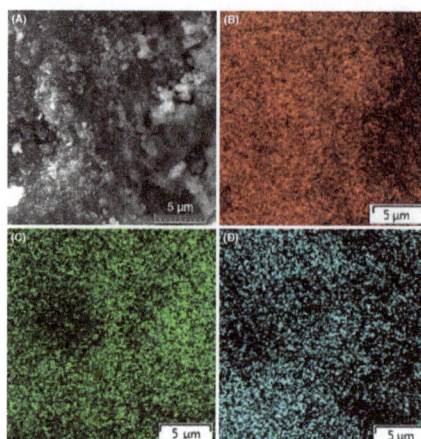

molecules due to the presence of the hydroxyl (OH$^-$) species that caused the following pathway to take place [51]:

$$Cu + OH^- \rightarrow CuOH \tag{1}$$

$$2CuOH \rightarrow Cu_2O + H_2O \tag{2}$$

where the copper oxide film is gradually formed on the outer surface of the Fe-Cu material by the adsorption of OH$^-$, followed by the dehydration of copper hydroxide into Cu$_2$O. This is another indication of the formation of the FCC structure with Fe atoms being forced into the structure of the Cu grains, whereas the Fe$_2$O$_3$ oxidation takes place on the Fe atoms that diffused out and were segregated from the Fe-Cu material due to the annealing temperature [51–56].

The semiconductive behavior of the resulting material paves the way for using it in place of the more expensive titania that is apparently an inferior absorber in the useful solar spectrum as clearly seen in Figure 7. Since the Fe-Cu material exhibits this semiconductive behavior, an estimation of the resulting bandgap is shown in Figure 11. The band gap is determined from the traditional Tauc relationship; $\alpha.h\nu = A.(h.\nu-E_g)^n$, by plotting $(\alpha.h.\nu)^{1/n}$ versus the photoenergy h.ν, where α is the absorption coefficient, A is the edge width parameter, E_g is the optical band gap value, and n is a constant dependent on the nature of the transition ($n = 1/2$ for a direct allowed transition and 2 for an indirect transition) [47, 48]. An extrapolation of the linear region of the plot for when the ordinate equals to zero, gives an absorption edge energy that corresponds to the value of the optical band gap E_g. With n being 1/2 for best fitting, the results in Figure 11 indicate direct transitions in Fe-Cu material with milling time. According to the figure, band gap results start to appear after 6 h of milling as elements diffusion becomes pronounced and the micrsostructure shifts to metastability, which is also inferred from the XRD plots of Figure 4. The obtained optical band gap values is 1.8 eV.

The semiconductive behavior of quasicrystalline materials has been observed by other researchers; some correlating the change in behavior with the increased role of disorder in the aperiodicity that has a profound effect on the metal-insulator transition [57], and others have reported a 2.2 eV bandgap value for the Fe-Cu solid solution and attributed it to a contribution similar to d transitions in the noble metal and follows the direct transitions from a virtual bound state to the Fermi level [58, 59]. In the particular case of quasicrystalline materials produced by mechanical alloying, the semiconductive behavior is generally attributed to the enhanced interdiffusion of elemental components that creates supersaturated structures that

leads to a metallic-covalent bonding conversion, and indeed a bandgap is reported by Takagiwa and Kimura [60] for an aluminum-based quasicrystalline material, and by our group [56].

Mesoporous media surface area

The results obtained from the BET testing show that the surface area of the Fe-Cu material is 0.78 m^2/g, whereas that of TiO$_2$ equals 41.1 m^2/g. While the fit for the volume adsorbed versus the differential pressure applied is strictly linear for TiO$_2$ ($R = 0.9995$), the variation in grains sizes for the Fe-Cu specimen as seen in the SEM of Figure 9 has caused inconsistencies in its adsorption as shown in Figure 12 and the fitted data shows a deviation from the linear behavior ($R = 0.867$).

Dye absorptivity results

The results obtained from testing chlorophyll, anthocyanin, and xanthophyll are shown in Figure 13. The spectral absorptivity in the UV–VIS range (250–750 nm) for each dye shows a high onset of absorption at the UV spectral range, compared to the distilled water reference. The three curves show a step fall after around 350, 400, and 450 nm for anthocyanin, chlorophyll, and xanthophyll, respectively. Xanthophyll exhibits superior absorptivity behavior, especially in the visible range (400–700 nm) where the drop in absorptivity is seen to be less abrupt than its two counterparts, never reaching zero even at the extreme end of the sampling spectrum (750 nm). This behavior, along with the broad spectral range of absorptivity of xanthophyll makes it an ideal choice for sensitizing the mesoporous materials of the experimental cells.

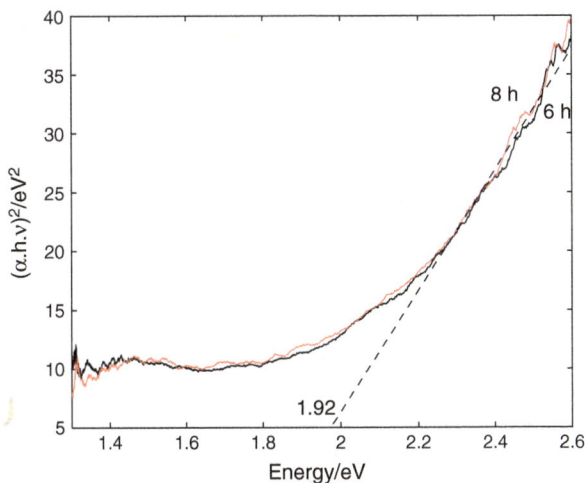

Figure 11. Energy bandgap of the resulting Fe-Cu material.

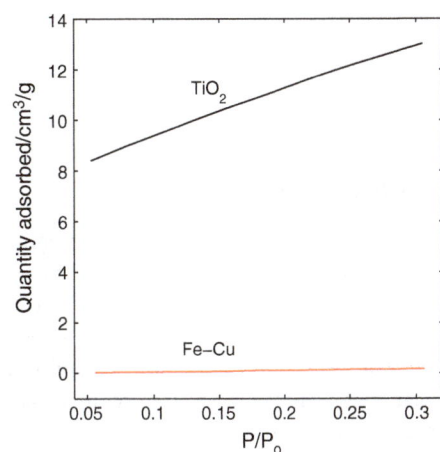

Figure 12. BET surface area for Fe-Cu versus TiO$_2$.

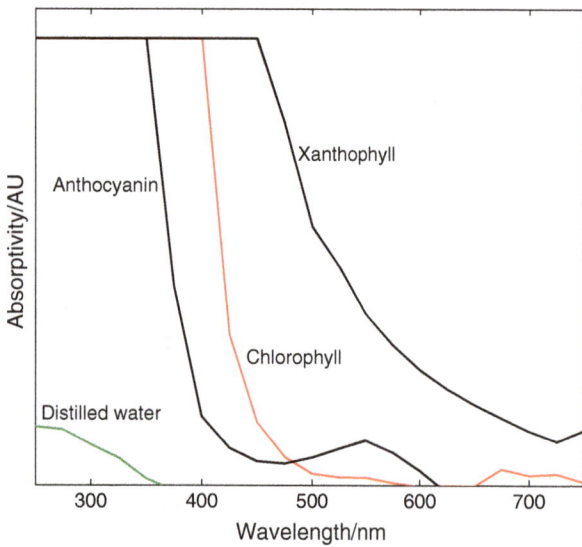

Figure 13. Absorptivity test of various natural organic dyes against distilled water.

Table 1. Comparison of parameters for the two DSSCs experimental cells.

Parameter	TiO$_2$-based cell	Fe-Cu-based cell
I_{sc}, [mA]	0.173	0.53
E_{oc}, [V]	0.248	0.269
FF (P_{max}/P_{Theo})[%]	48.4	25.0
η, [%]	0.638	0.943

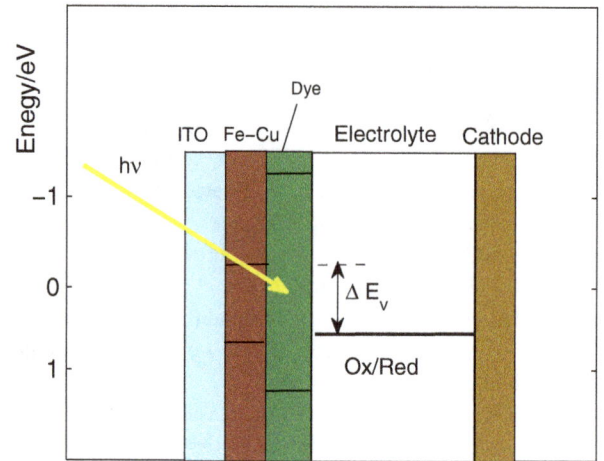

Figure 15. Illustration of energy levels diagram of a DSSC based on Fe-Cu mesoporous material.

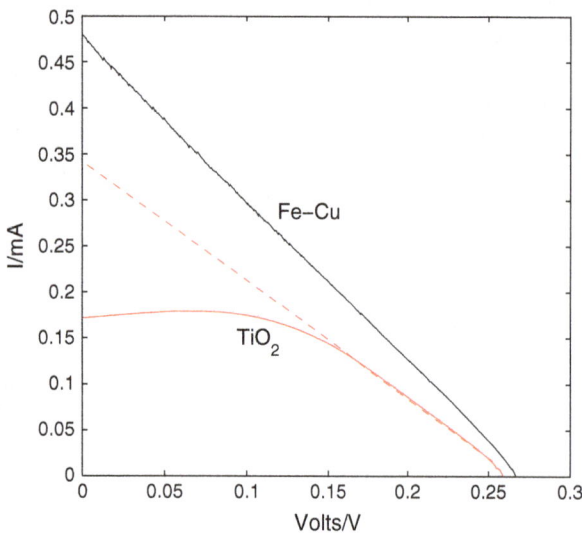

Figure 14. IV characteristic curves for TiO$_2$ and Fe-Cu xanthophyll-sensitized solar cells.

Solar cell test results

The results of the characteristics test for the manufactured DSSC are shown in Figure 14. It is noted that there is a linear decrease in the voltage of the Fe-Cu cell in proportion to the generated current, which is usually an indication of a large series resistance in multicrystalline silicon solar cells. The series resistance in solar cells is mostly due to contact resistance, and hence the recommendations to mitigate its effects are numerous, like changing the concentration of the electrolyte, the counter

electrode or the clips that connect the cell to the load. It is also noted from the curve that the series resistance in both have similar values (as shown from the line tangent to the TiO$_2$ IV curve in Figure 14).

In general, the Fe-Cu-based solar cell exhibits better performance when other characteristics are calculated, such as the fill factor, power, and efficiency. A summary of these parameters is shown in Table 1.

An explanation of the measured enhancement over the TiO$_2$-based cell is presented with reference to the energy levels diagram in Figure 10 that schematically shows the Fe-Cu bandgap to be around 1.8 eV (±0.9 eV at each side of the 0 eV mark), where the open circuit voltage, measured as ΔE_v is the potential difference between the dye/electrolyte interface and the upper end of the mesoporous material energy gap. The conduction band for the Fe-Cu has to be above 0 V versus normal hydrogen electrode (NHE) for the photovoltaic effect to take place (Fig. 15).

To determine the cell potential behavior at the Fe-Cu/electrolyte interface, a Mott–Schottky (MS) analysis is conducted using the potentiostat at frequencies 63, 79, and 100 Hz. The plots are shown in Figure 16. The MS analysis probes the depletion capacitance at a Schottky or *p-n* junction which is determined by the width of the

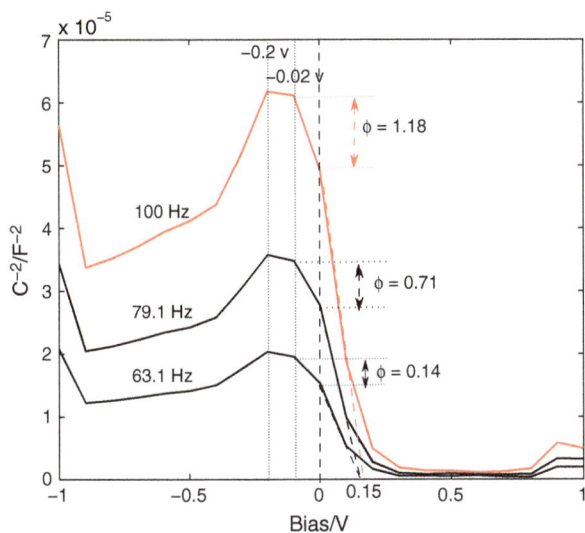

Figure 16. Mott–Schottky diagrams for Fe-Cu-based photoelectrode.

bias-dependent depletion region, hence the depletion capacitance, C, is also bias dependent and can be expressed as follows:

$$\frac{1}{C^2} = \frac{2(V_{bi} - V)}{A^2 q \varepsilon \varepsilon_0 N} \qquad (3)$$

where V is the applied bias, A is the device area, q is the elementary charge, ε is the compound dielectric constant, and ε_0 is the permittivity of free space. The built-in bias and doping density are then found by fitting equation 1 to the linear portion of the C^{-2} versus bias voltage plot, as shown by the converging dotted lines in Figure 16. The intersection of the line equals the flat band potential, whereas the slope of the line is used to estimate the carrier density, N, as the slope equals $2/(\varepsilon \cdot \varepsilon_0 \cdot q \cdot N)$. The Helmholtz capacitance on the electrolyte side of the interface is large enough that it is negligible since the observed capacitance, C, is given by $1/C = 1/C_H + 1/C_{SC}$, where C_H, and C_{SC}, are the capacitances of the Helmholtz layer and the semiconductor, respectively.

The MS plots reveal that the Fe-Cu behaves like a p-type semiconductor (SC), contrary to the n-type TiO_2. When the p-SC is in contact with an electrolyte solution, the electrons in the SC are transported to the vacant level of the electrolyte solution, which makes the SC positively charged and the electrolyte solution negatively charged. Upon irradiation, at the interface between the Fe-Cu layer and the electrolyte solution, the energy of which is larger than the bandgap (E_g) of the semiconductor (for Fe-Cu $E_g \sim 0.1.8$ eV), excitation of electrons from the valence band to the conduction takes place leaving holes in the valence band. The electron and the hole form the short-lived excitons. The space charge

layer, φ, which is measured as the difference between the C^{-2} value at the inflection point before crossing the zero-voltage bias (see Figure 16) for the three frequency values of 63 Hz, 79 Hz, and 100 Hz is measured as 0.14, 0.71, and 1.18 F^{-2}, respectively. The physical significance of the space charge layer is that a hole can migrate towards the SC interface, and the electron can migrate toward the inside of the SC bulk so that the chances of their recombination are smaller. When an electron donor is present in the contacted liquid phase, the holes can oxidize the donor in the electrolyte, and the electrons are transported first to the ITO glass conductive layer through Fe-Cu grain boundaries and then to the counter electrode, reducing electron acceptance there. From Figure 14, the flat band potential of the Fe-Cu layer is found to be 0.15 V compared with that of TiO_2 at −0.16 V [52]. To calculate the carrier density, whereas the relative permittivity of Fe-Cu is chosen to be in the order of magnitude of 10^3 [53], ε_0 the vacuum permittivity = 8.854×10^{-12} F/m, and q the elementary electric charge = 1.602×10^{-19} C; then the carrier density per meter cube is estimated to be $4.27 \times 10^{32}/cm^3$. This value is higher than the value reported for TiO_2 ($6.96 \times 10^{19}/cm^3$) in literature [51], proving that the quasicrystalline Fe-Cu thin film can function as an excellent electron conducting material under irradiation conditions when a strong electron donor is present in the electrolyte. This explains the bigger current values in the IV curves of Figure 12 when Fe-Cu is used as a mesoporous layer instead of TiO_2, and with the facile and cheap synthesis procedure of Fe-Cu, the latter is considered a real competitor in the manufacture of dye-sensitized solar cells to drive the cost-per-watt even lower [54–56].

Conclusion

In this study, an economical and facile method for the production of a mesoporous material made from Fe-Cu metastable alloy system is investigated for DSSC photoelectrodes. High-energy ball milling, a mechanical alloying technique, was used to synthesize the alloy and the resulting microstructure was examined using SEM-EDS and XRD at 2 h intervals starting from the as-is powder up to the final duration of 8 h. The resulting alloy is a homogenous, FCC single phase of Fe diffused in Cu with no segregation of elements. The optical absorptivity in the Vis–NIR ranges for the alloy was measured for the same durations and the results show a strong correlation between the microstructural morphology and the optical behavior and enhancement of spectral absorptivity between 400 and 900 nm. The spectral absorptivity is enhanced up to 81% for the 8 h milled alloy compared to the original powder, and around 95% higher than TiO_2, which is the mesoporous material of choice for DSSCs. The increase in the impedance up to 120 Ω is an indication of the fundamental change of the electrical properties of the resulting phase after 8 h of milling, which has created the reported bandgap. Two DSSCs sensitized by the natural organic dye xanthophyll were constructed, one with Fe-Cu and the other with TiO_2 as the mesoporous layer, and the former had better performance in terms of power (0.078 vs. 0.022 mW, respectively) and efficiency (0.638% compared to 0.943%, respectively) for a 6.25 cm^2 cell. The carrier density for cells was also calculated using Mott–Schottky tests and Fe-Cu had orders of magnitude more carriers (4.27×1032 vs. $6.69 \times 1019/cm^3$) for the TiO_2. The Fe-Cu intermetallic phase is a promising mesoporous material due to its cost-effectiveness, mass-production capable synthesis technique, excellent spectral absorptivity in the UV–Vis range, recyclability, and good thermal and mechanical properties.

Conflict of Interest

None declared.

References

1. Green, M. A., K. Emery, Y. Hishikawa, W. Warta, and E. D. Dunlop. 2015. Solar cell efficiency tables (Version 45). Prog. Photovoltaics Res. Appl. 23:1–9.

2. Mathew, S., A. Yella, P. Gao, R. Humphry-Baker, B. F. E. Curchod, N. Ashari-Astani, et al. 2014. Dye-sensitized solar cells with 13% efficiency achieved through the molecular engineering of porphyrin sensitizers. Nat. Chem. 6:242–247.

3. Puxeu, L. 2014.Exploring novel dye concepts in dye sensitized solar cell, Thesis.

4. Chen, Z., F. Li, and C. Huang. 2007. Organic D-π-A dyes for dye-sensitized solar cell. Curr. Org. Chem. 11:1241–1258.

5. Hara, K., and N. Koumura. 2009. Organic dyes for efficient and stable dye-sensitized solar cells. Mater. Matters 92:4.4.

6. Mor, G. K., K. Shankar, M. Paulose, O. K. Varghese, and C. A. Grimes. 2006. Use of highly-ordered TiO_2 nanotube arrays in dye-sensitized solar cells. Nano Lett. 6(2):215–218.

7. Zhu, K., N. R. Neale, A. Miedaner, and A. J. Frank. 2007. Enhanced charge-collection efficiencies and light scattering in dye-sensitized solar cells using oriented TiO_2 nanotubes arrays. Nano Lett. 7(1):69–74.

8. Kim, Y. J., M. H. Lee, H. J. Kim, G. Lim, Y. S. Choi, N.-G. Park, et al. 2009. Formation of highly efficient dye-sensitized solar cells by hierarchical pore generation with nanoporous TiO_2 spheres. Adv. Mater. 21:3668–3673.

9. Park, Y.-C., Y.-J. Chang, B.-G. Kum, E.-H. Kong, J. Y. Son, Y. S. Kwon, et al. 2011. Size-tunable mesoporous spherical TiO_2 as a scattering overlayer in high-performance dye-sensitized solar cells. J. Mater. Chem. 21:9582–9586.

10. Chen, J., J. Wang, F. Bai, L. Hao, Q. Pan, and H. Zhang. 2013. Connection style and spectroscopic properties: theoretical understanding of the interface between N749 and TiO_2 in DSSCs. Dyes Pigm. 99:201–208.

11. Wang, X., L. Zhi, and K. Müllen. 2008. Transparent, conductive graphene electrodes for dye-sensitized solar cells. Nano Lett. 8:323–327.

12. Sun, S., L. Gao, and Y. Liu. 2010. Enhanced dye-sensitized solar cell using graphene-TiO_2 photoanode prepared by heterogeneous coagulation. Appl. Phys. Lett. 96:083113.

13. NREL. 2015. www.nrel.gov, last accessed 31 December 2015.

14. Liu, J. Z., A. van de Walle, G. Ghosh, and M. Asta. 2005. Structure, energetics, and mechanical stability of Fe-Cu bcc alloys from first-principles calculations. Phys. Rev. B 72:144109.

15. Hasebe, M. 1981. T. Nishizawa and Further study on phase diagram of the iron-copper system. Calphad 5:105–108.

16. Huang, X., and T. Mashimo. 1999. Metastable BCC and FCC alloy bulk bodies in Fe–Cu system prepared by mechanical alloying and shock compression. J. Alloy. Compd. 288:299–305.

17. Weeber, A. W. 1987. Application of the Miedema model to formation enthalpies and crystallisation temperatures of amorphous alloys. J. Phys. F: Met. Phys. 17:809.

18. He, J., J. Z. Zhao, and L. Ratke. 2006. Solidification microstructure and dynamics of metastable phase transformation in undercooled liquid Cu–Fe alloys. Acta Mater. 54:1749–1757.

19. Mazzone, G., and M. V. Antisari. 1996. Structural and thermodynamic factors of suppressed interdiffusion kinetics in multi-component high-entropy materials. Phys. Rev. B 54:441–446.

20. Ma, E., M. Atzmon, and F. E. Pinkerton. 1993. Thermodynamic and magnetic properties of metastable FexCu100− x solid solutions formed by mechanical alloying. J. Appl. Phys. 74:955–962.

21. Xu, J., U. Herr, T. Klassen, and R. S. Averback. 1996. Formation of supersaturated solid solutions in the immiscible Ni–Ag system by mechanical alloying. J. Appl. Phys. 79:3935–3945.

22. Hansen, M., and K. Anderko. 1965. Constitution of binary alloys, Issue 1, 2nd ed. McGraw-Hill, New York.

23. Lyasotsky, I., N. Dyakonova, D. Dyakonov, and E. Vlasova. 2008. Metastable phases and nanostructuring of Fe-Nb-Si-B base rapidly quenched alloys. Rev. Adv. Mater. Sci. 18:695–702.

24. Fu, G., Z. Hu, L. Xie, X. Jin, Y. Xie, Y. Wang, et al. 2009. Electrodeposition of nickel hydroxide films on nickel foil and its electrochemical performances for supercapacitor. Int. J. Electrochem. Sci. 4:1052–1062.

25. Gupta, R., N. Sukiman, M. Cavanaugh, B. Hinton, C. Hutchinson, and N. Birbilis. 2012. Metastable pitting characteristics of aluminium alloys measured using current transients during potentiostatic polarisation. Electrochim. Acta 66:245–254.

26. Li, Q. 2009. Formation of bulk ferromagnetic nanostructured $Fe_{40}Ni_{40}P_{14}B_6$ alloys by metastable liquid spinodal decomposition. SP Sci. China Press 1:1919–1922.

27. Das, N., J. Mittra, B. Murty, S. Pabi, U. Kulkarni, and G. Dey. 2013. Miedema model based methodology to predict amorphous-forming-composition range in binary and ternary systems. J. Alloy. Compd. 550:483–495.

28. Greenfield, M., and C. Pierce. 1973. Postweld aging of a metastable beta titanium alloy. Weld. J. 52:524–528.

29. Ravi, C., C. Wolverton, and V. Ozoliņš. 2006. Predicting metastable phase boundaries in Al–Cu alloys from first-principles calculations of free energies: the role of atomic vibrations. Europhys. Lett. 73:719.

30. Xu, J., G. S. Collins, L. S. J. Peng, and M. Atzmon. 1999. Deformation-assisted decomposition of unstable Fe50Cu50 solid solution during low-energy ball milling. Acta Mater. 47:1241–1253.

31. Oleszak, D., and P. Shingu. 1996. Nanocrystalline metals prepared by low energy ball milling. J. Appl. Phys. 79:2975–2980.

32. Jartych, E., J. K. Żurawicz, D. Oleszak, and M. Pękała. 2000. X-ray diffraction, magnetization and Mössbauer studies of nanocrystalline Fe–Ni alloys prepared by low- and high-energy ball milling. J. Magn. Magn. Mater. 208:221–230.

33. Roeder, J., J. Sculac, and M. Notis. 1984. The precipitation of iron in early smelted copper from Timna. Microbeam Anal. 243:243–246.

34. Ying-Yu, C., S. Rainer, and C. Y. Austin. 1984. Thermodynamic analysis of the iron-copper system I: the stable and metastable phase equilibria. Metall. Trans. A 15:1921–1930.

35. Kennedy, C. E. 2002. Review of mid- to high-temperature solar selective absorber materials. Technical Report NREL/TP-520-31267. National Renewable Energy Laboratory.

36. Alami, A. H., A. Allagui, and H. Alawadhi. 2015. Synthesis and optical properties of electrodeposited crystalline Cu2O in the Vis–NIR range for solar selective absorbers. Renewable Energy 82:21–25.

37. Rephaeli, E., and F. S. 2008. Tungsten black absorber for solar light with wide angular operation range. Appl. Phys. Lett. 92:211107–1–3.

38. Geldermon, K., L. Lee, and W. Donne. 2007. Falt-Band potential of a semiconductor: using the Mott-Schottky equation. J. Chem. Educ. 84:685–688.

39. Hoerner, L. 2013. Photosynthetic solar cells using chlorophyll and the applications towards energy sustainability. Thesis. University of South Florida, St. Petersburg.

40. Chien, C., and B. Hsu. 2013. Optimization of the dye-sensitized solar cell with anthocyanin as photosensitizer. Sol. Energy 98:203–211.

41. Patil, G., M. Madhusudhan, B. R. Babu, and K. Raghavarao. 2009. Extraction, dealcoholization and concentration of anthocyanin from red radish. Chem. Eng. Process. 48:364–369.

42. Zhang, Y., Y. J. Zhou, J. P. Lin, G. L. Chen, and P. K. Liaw. 2008. Solid-solution phase formation rules for multi-component alloys. Adv. Eng. Mater. 10:534–538.

43. Takaki, S., K. Fukunaga, J. Syarif, and T. Tsuchiyama. 2004. Effect of grain refinement on thermal stability of metastable austenitic steel. Mater. Trans. 45:2245–2251.

44. Gaffet, E., M. Harmelin, and F. Faudot. 1993. Far-from-equilibrium phase transition induced by mechanical alloying in the Cu-Fe system. J. Alloy. Compd. 194:23–30.

45. Lucas, F., B. Trindade, B. Costa, and G. Le Caër. 2003. The influence of pre-milling on the microstructural evolution during mechanical alloying of a $Fe_{50}Cu_{50}$ Alloy. J. Metastable Nanocrystalline Mater. 18:49–56.

46. Roos, A., T. Chibuye, and B. Karlsson. 1983. Properties of oxidized copper surfaces for solar applications I. Solar Energy Mater. 7:453–465.

47. Tauc, J., R. Grigorovici, and A. Vancu. 1966. Optical properties and electronic structure of amorphous germanium. Phys. Status Solidi B 15:627–637.

48. Erdogan, I. Y., and O. Gullu. 2010. Optical and structural properties of CuO nanofilm: its diode application. J. Alloy. Compd. 492:378–383.

49. Kormann, C., D. W. Bahnemann, and M. R. Hoffmann. 1988. Preparation and characterization of quantum-size titanium dioxide. J. Phys. Chem. 92:5196–5201.

50. Nakano, Y., S. Saeki, and T. Morikawa. 2009. Optical bandgap widening of p-type Cu_2O films by nitrogen doping. Appl. Phys. Lett. 94:022111.

51. Pike, J., S. W. Chan, F. Zhang, X. Wang, and J. Hanson. 2006. Formation of stable Cu_2O from reduction of CuO nanoparticles. Appl. Catal. A 303:273–277.

52. Kaneko, M., H. Ueno and J. Nemoto. 2011. Schottky junction/ohmic contact behavior of a nanoporous TiO_2

thin film photoanode in contact with redox electrolyte solutions. Beilstein J. Nanotechnol. 2:127–134.

53. Brian, Cantor (Book editor). 2005. Novel nanocrystalline alloys and magnetic nanomaterials, Brian Cantor. Institute of Physics Publishing, Sussex, UK.

54. Alami, A. H., A. Alketbi, M. Almheiri, and J. Abed. 2015. The Fe-Cu metastable nano-scale compound for enhanced absorption in the UV-Vis and NIR ranges. Metall. Mater. Trans. E Mater. Energy Syst. 2:229–235.

55. Alami, A. H., A. Alketbi, M. Almheiri, and J. Abed. 2016. Assessment of Al-Cu-Fe compound for enhanced solar absorption. Int. J. Energy Res. doi: 10.1002/er.3468 [Epub ahead of press].

56. Alami, A. H., A. Alketbi, and M. Almheiri. 2015. Synthesis and microstructural and optical characterization of Fe-Cu metastable alloys for enhanced solar thermal absorption. Energy Procedia 75 C:410–416.

57. Axel, F., F. Denoyer, and J. P. Gazeau. 2000. From quasicrystals to more complex systems, 13, 1, pp. 76. Springer-Verlag, Berlin Heidelberg.

58. Korn, D., H. Pfeifle, and G. Zibold. 1979. Optical properties of metastable CuFe solid solutions. J. Phys. F: Metal Phys. 9:1709–1–7.

59. Liu, B. X., C. H. Shang, L. J. Huang, and H.-D. Li. 1990. Fe-Cu icosahedral phase and its thermal and magnetic properties. J. Non-Cryst. Solids 117:785–788.

60. Takagiwa, Y., and K. Kimur. 2014. Metallic–covalent bonding conversion and thermoelectric properties of Al-based icosahedral quasicrystals and approximants, a review. Sci. Technol. Adv. Mater. 15:044802, (12 pp).

Spectral response measurements of multijunction solar cells with low shunt resistance and breakdown voltages

Juan P. Babaro[1], Kevin G. West[2] & Behrang H. Hamadani[3]

[1]Instituto Nacional de Tecnología Industrial- Física y Metrología, San Martín, Buenos Aires, Argentina
[2]SensorMetrix, San Diego, California, 92126
[3]National Institute of Standards and Technology, Engineering Laboratory, Gaithersburg, Maryland 20899

Keywords
multijunction solar cells, spectral response, shunt resistance, breakdown voltage

Correspondence
Behrang H. Hamadani, National Institute of Standards and Technology, Engineering Laboratory, Gaithersburg, MD 20899.
E-mail: behrang.hamadani@nist.gov

Funding Information
No funding information provided.

Abstract

Spectral response measurements of germanium-based triple-junction solar cells were performed under a variety of light and voltage bias conditions. Two of the three junctions exhibited voltage and light bias-dependent artifacts in their measured responses, complicating the true spectral response of these junctions. To obtain more insight into the observed phenomena, a set of current-voltage measurement combinations were also performed on the solar cells under identical illumination conditions, and the data were used in the context of a diode-based analytical model to calculate and predict the spectral response behavior of each junction as a function of voltage. The analysis revealed that both low shunt resistance and low breakdown voltages in two of the three junctions influenced the measured quantum efficiency of all three junctions. The data and the modeling suggest that combination of current-voltage measurements under various light bias sources can reveal important information about the spectral response behavior in multijunction solar cells.

Introduction

With significant advances in high-efficiency multijunction solar cell (MJSC) technologies, the number of subcells within each structure has steadily increased over the years with the goal of maximizing the operating voltage while preserving the other performance parameters such as the short circuit current (I_{sc}) and the fill factor [1–6]. Recently, five-junction solar cells based on semiconductor bonded technology have demonstrated confirmed non-concentrator terrestrial air mass 1.5 global spectrum power conversion efficiency of $\approx$ 39% [7] and concentrator 4-junction cell efficiencies of 46% [8]. With increased complexities in design and architecture of these types of cells, the task of performing electrical characterization and spectral response (SR) measurements of the cells will also become more involved. Although there are well-established procedures for performing current–voltage (I–V) or spectral response measurements in MJSCs [9, 10], it has been shown previously that certain factors such as low shunt resistance [11–17], low reverse breakdown voltage [10, 18] or luminescence coupling [12, 19–25] can cause artifacts

in the spectral response or the quantum efficiency (QE) of the device.

Luminescence coupling is usually a factor between two adjacent junctions and its effects generally appear as an increased QE signal in the wavelength range corresponding to energies above the bandgap of the current-limited junction [24–26]. It has also been reported that luminescence coupling has a light bias intensity dependence, resulting in a decrease in QE of the junction of interest with an increase in the light bias intensity of the junction above it [12]. This reduction factor is substantially dependent on the LC coupling coefficients between the junctions. When considering artifacts originating mostly from shunt and breakdown voltage issues, most of the work reported to-date in QE measurements of MJSCs, particularly 3-junction cells based on a germanium (Ge) bottom cell (BC), has been focused on cases where only a single junction has been found to behave nonideally. If the nonideality is due to a low breakdown voltage only, then an applied voltage bias of enough magnitude is able to fully recover the QE of the junction of interest without much dependence on the light bias intensity [18].

If the junction of interest suffers from a low shunt resistance, increasing the light bias intensity of the nonlimiting junctions and application of a voltage bias is generally needed to achieve a good measurement of the QE signal [14]. There are even reported cases where a combination of luminescence coupling and low shunt resistance affect the QE response of the Ge junction [12].

Although generalized opto-electronic models have been proposed that combine and explain a variety of simultaneous effects in MJSCs measurements [21], it is still very illuminating to consider simpler analytical models to explain the influence of the most severe artifacts in certain devices. Our aim has particularly been focused on cells where two junctions revealed both low shunt resistance and low breakdown voltage effects and when LC has not been a major factor in causing artifacts in the QE response of the junctions. Even though several previous works have been dedicated to this subject and some variations of it [10, 12, 14–18], we found that significant insight can be obtained into the behavior of each junction when we first investigated the I–V response of one of these types of cells under several combinations of light sources where each light source is a high-power light emitting diode (LED) with a unique wavelength emission peak suitable to light bias a single junction. For a triple junction cell, this approach involves eight combinations of measurements (3 single-LED illuminations, 3 double LED combinations, one triple LED illumination, and one with no illumination). These data can then be used within the context of a straight-forward two-diode equivalent circuit (EC) model to determine a set of parameters that will fit all eight data combinations at once. This technique eliminates the need for separate iso-cell measurements for parameter estimates because there are sufficient data to allow for obtaining good fits to the data. After the EC model parameters are determined, the results are used to predict SR and QE of each junction with similar levels of light bias as a function of voltage and compare the findings to the experimental results. We generally find very good agreement between the predicted model results and the experimental data. When more than one junction suffers from nonideal shunt and breakdown voltage effects generally caused during manufacturing or nonoptimal material processing, the ideal diode-like behavior expected from a well-fabricated p-n junction is nonexistent. In these cases, the QE measurement of the intended junction will reveal artifacts originating from all of the nonideal junctions, in a combined way. Furthermore, in almost all cases studied, the magnitude of the measured QE itself, as measured across the two terminals of the entire stack, will be less than the individual junctions' actual QE response, even under the best voltage bias conditions.

Experimental

For this study, the samples were monolithic, two-terminal triple junction GaInP/GaAs/Ge solar cells of commercial grade with a nominal area of 2.3 cm^2 and air mass (AM) 1.5 G 1-sun efficiency of 27% at 25°C [4]. The current-voltage and the spectral response data reported here are representative of multiple solar cells tested within the batch received. Combinations of current-voltage measurements were performed with a source-meter electrometer and three high-power LEDs with projector optics illuminating the entire area of the cell with bright illumination beams of good uniformity. The spectral irradiance of all three LEDs at the incident plane of illumination, was measured by a calibrated spectroradiometer. The irradiance has a Gaussian form, with the three-center wavelengths at 520 nm, 741 nm, and 1077 nm. All three LEDs were operated with battery sources close to their rated source current values for better control over signal-to-noise ratios when performing differential QE measurements. The reported irradiance values for each LED were obtained by integrating the spectral irradiance curves over wavelength. For QE, a dual-light source monochromator set up was used, utilizing the principle of differential spectral response (DSR) measurement as described previously [27] with calibrated monitor detectors from 300 nm to 1900 nm, allowing for simultaneous measurements of the cell photocurrent and the incident monochromatic beam power. The monochromator beam was chopped by a mechanical chopper, hence generating a small AC photocurrent that is collected via a transimpedance preamplifier with the ability to separate AC and DC components of the signal and built-in voltage bias capability. The AC signal of the cell was measured by a lock-in amplifier. The voltage bias was applied in the forward bias direction of the cell operation from 0 V to values slightly higher than the open circuit voltage, V_{oc}, of the cell under the illumination conditions to verify modeling results. Although application of higher voltage amounts is possible, it can lead to significant instability in the recorded AC lock-in signal. The monochromator measurement uncertainty for spectral response measurements under low voltage bias is generally better than 1%. This includes the uncertainty of the reference photodiode and the transfer calibration onto the monitor diode, the repeatability and reproducibility measurements and the standard deviation of the collected data. However, at bias voltages approaching the V_{oc}, it can reach 5% or more, mostly due to signal instability.

The Modeling Framework

In this section, we first describe the mathematical model used to fit the composite I–V curve data under each unique lighting condition. Then, by use of a simplified analysis, we demonstrate how the true value of the spectral response of a given junction is affected by various artifacts.

Mathematical model for each junction

We consider a typical two-diode model for each junction [28], with its corresponding shunt and series resistances. The implicit current density versus voltage $J-V$ curves for each junction $i = 1,2,3$ can be written in the forward bias direction:

$$
\begin{aligned}
J_i(V) = &-J_{i_{ph}} + J_{i_{01}} \left(\exp\left(\frac{e\left(V - R_{i_s} A J_i(V)\right)}{kT} \right) - 1 \right) \\
&+ J_{i_{02}} \left(\exp\left(\frac{e\left(V - R_{i_s} A J_i(V)\right)}{n_i kT} \right) - 1 \right) \\
&+ \frac{V - R_{i_s} A J_i(V)}{R_{i_{sh}} A},
\end{aligned} \tag{1}
$$

and in reverse bias direction:

$$
\begin{aligned}
J_i(V) = &\left(-J_{i_{ph}} + J_{i_{01}} \left(\exp\left(\frac{e\left(V - R_{i_s} A J_i(V)\right)}{kT} \right) - 1 \right) \right. \\
&+ J_{i_{02}} \left(\exp\left(\frac{e\left(V - R_{i_s} A J_i(V)\right)}{n_i kT} \right) - 1 \right) \\
&\left. + \frac{V - R_{i_s} A J_i(V)}{R_{i_{sh}} A} \right) \times \frac{1}{1 - \left(\frac{V}{V_{i_{bd}}} \right)^{n_{i_m}}},
\end{aligned} \tag{2}
$$

where $J_{i_{ph}}$ is the photocurrent density of the ith junction, $J_{i_{01}}$ is the saturation current density, $J_{i_{02}}$ is the saturation current density for the depletion region, A is the device area, n_i is the ideality factor, $R_{i_{sh}}$ is the shunt resistance, R_{i_s} is the series resistance, n_{i_m} is the Miller index and $V_{i_{bd}}$ is the breakdown voltage. In equation (2), a multiplicative factor is added to account for the cell's avalanche breakdown in the reverse bias [28]. We define the differential spectral responsivity of each junction as the difference in its short circuit current between the two chopper states of *closed* and *open*, r, divided by the power of the incident monochromatic beam. The difference between these two current densities is simply:

$$
J_i^{\text{chopper open}}(0) - J_i^{\text{chopper closed}}(0) = r_i. \tag{3}
$$

Note that in reality, in a MJSC, only the voltage across the entire cell stack can be fixed, say at 0 V, leaving each junction under a different operating voltage. However, this definition is consistent with single-cell measurements where the short circuit current is the quantity of interest for

spectral response measurements. For a very small value of r, which is the case for differential monochromator-based measurements, this change is equivalent to a small displacement of the entire curve, $J_i(V)$ in an amount equal to r.

To calculate the composite $J(V)$ curve for the entire cell, which constitutes the series sum of the individual junction $J_i(V)$, first the best fit parameters for each implicit equation corresponding to each junction are selected and the J_i versus V functional form is calculated. Inverting these functions and obtaining $V(J_i)$ curves then allow for summation of the voltages at fixed current values. Finally, this composite curve can be inverted back to the $J(V)$ form. As it will be described in Discussion and Results, the objective of this modeling was to obtain the best fit parameters from as many combination $I-V$ curves as possible so that a robust estimate of the voltage bias dependence of the spectral responsivity of each junction can be made. Here, by best fit, we mean the best visual fit of the model to the experimental data, that is, no least square calculations were employed.

Factors limiting the spectral response measurements

The mathematical model given by equations (1) and (2) allows for determination of the relationship between the measured spectral response quantity, which is influenced by the whole composite curve and measured at the terminals, and the real increase r of the zero-voltage current for the junction of interest. We have observed that in cases where one or more junctions have nonideal diode properties, application of a voltage bias or light bias will not fully recover the entire value r, as desired. Although this concept can be mathematically explained by use of equations (1) and (2), here we present a simplified case that demonstrates this point well.

Figure 1 shows an example of an MJSC case with a well-behaved GaAs junction but with faulty Ge and GaInP junctions with low shunt resistances and low breakdown voltages. The plotted $J-V$ curves are representative of the sort of behavior we have seen in our measurement and modeling work. This plot represents a case where the intention is to extract the SR of the GaInP junction by light biasing the Ge junction (with the 1077 nm LED) and the GaAs junction (with the 741 nm LED), while light modulating the GaInP junction with the much weaker 520 nm monochromatic light. For the sake of clarity, the $J-V$ curve corresponding to the open chopper state is drawn in an expanded view with an exaggerated r value, but in reality, r is many orders of magnitude smaller than the other two junctions' light bias-induced short circuit current values. Given this, the composite $J-V$ curve can be constructed for both chopper states, showing that points

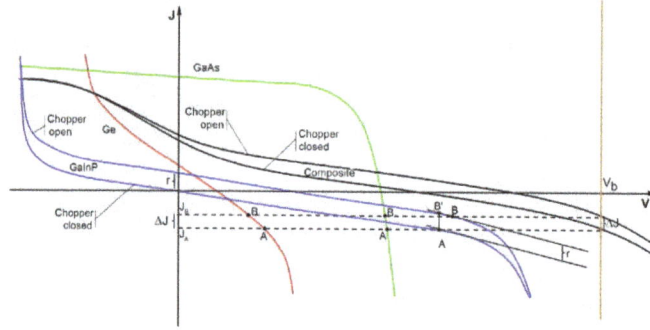

Figure 1. *J* versus *V* curves for each junction of a triple junction solar cell presenting a good diode-like GaAs junction behavior in contrast to Ge and GaInP junctions exhibiting low shunt resistances and low breakdown voltages. While Ge and GaAs junctions are light biased, the GaInP junction is only light modulated with a chopped monochromatic beam. Its short circuit current density rise *r* is very small, but is exaggerated for demonstration purposes.

A and B represent the actual operating points for each curve as the chopper turns. Points A and B on each *J* versus *V* curve are close enough so that the segments of the curves joining these points are parts of straight lines. Point B' on the chopper-open GaInP *J–V* curve is a point that shares the same voltage with point A of the chopper-closed GaInP *J–V* curve. The effect of opening the chopper can be approximated as an upward shift of the *J–V* curve equal to *r*. However, it can be seen from the graphics that the actual operating point for the chopper-open curve is point B for the given V_b. We now can apply equations of straight lines with their corresponding slope *m* and origin intercept *b* to this region. In the low voltage bias range, we can think of the GaInP *J* versus *V* curve (blue curve) for the open chopper state to be equal to the *J* versus *V* curve with the closed chopper case, but being displaced upwards by a value *r*, making the tangents to the points A and B' parallel. The zero-intercept value of the BB' line is equal to the line tangent at A plus *r*. We are not interested in B' but as *r* is so small, we can consider that B and B' share the same tangent line. So, the following four linear equations for the *J* versus *V* graphics near the points A and B for all 4 curves are evident:

$$J_{GaAs}(V) = -m_{GaAs} V + b_{GaAs} \rightarrow V_{GaAs}(J) = \frac{b_{GaAs} - J}{m_{GaAs}}, \quad (4)$$

$$J_{Ge}(V) = -m_{Ge} V + b_{Ge} \rightarrow V_{Ge}(J) = \frac{b_{Ge} - J}{m_{Ge}}, \quad (5)$$

$$J_{GaInP}^{chopper\,closed}(V) = -m_{GaInP} V + b_{GaInP} \rightarrow V_{GaInP}^{chopper\,closed}(J)$$
$$= \frac{b_{GaInP} - J}{m_{GaInP}}, \quad (6)$$

$$J_{GaInP}^{chopper\,open}(V) = -m_{GaInP} V + b_{GaInP} + r \rightarrow V_{GaInP}^{chopper\,open}(J)$$
$$= \frac{b_{GaInP} + r - J}{m_{GaInP}}, \quad (7)$$

Since the three junctions are in series, the current density through them is the same and we sum the voltages to obtain the *V* versus *J* relations, with the chopper closed and open:

$$V^{chopper\,closed}(J) = V_{GaInP}^{chopper\,closed}(J) + V_{Ge}(J) + V_{GaAs}(J)$$
$$= -J\left(\frac{1}{m_{GaInP}} + \frac{1}{m_{Ge}} + \frac{1}{m_{GaAs}}\right)$$
$$+ \frac{b_{GaInP}}{m_{GaInP}} + \frac{b_{Ge}}{m_{Ge}} + \frac{b_{GaAs}}{m_{GaAs}}, \quad (8)$$

$$V^{chopper\,open}(J) = V_{GaInP}^{chopper\,open}(J) + V_{Ge}(J) + V_{GaAs}(J)$$
$$= -J\left(\frac{1}{m_{GaInP}} + \frac{1}{m_{Ge}} + \frac{1}{m_{GaAs}}\right)$$
$$+ \frac{b_{GaInP} + r}{m_{GaInP}} + \frac{b_{Ge}}{m_{Ge}} + \frac{b_{GaAs}}{m_{GaAs}}, \quad (9)$$

From this set of equations, we calculate the current density that passes through the cell when the chopper is closed and open and when a certain voltage bias V_b applied:

$$V^{chopper\,closed}(J_A) = V_b \rightarrow J_A = \frac{\frac{b_{GaInP}}{m_{GaInP}} + \frac{b_{Ge}}{m_{Ge}} + \frac{b_{GaAs}}{m_{GaAs}} + V_b}{\frac{1}{m_{GaInP}} + \frac{1}{m_{Ge}} + \frac{1}{m_{GaAs}}}, \quad (10)$$

$$V^{chopper\,open}(J_B) = V_b \rightarrow J_B = \frac{\frac{b_{GaInP} + r}{m_{GaInP}} + \frac{b_{Ge}}{m_{Ge}} + \frac{b_{GaAs}}{m_{GaAs}} + V_b}{\frac{1}{m_{GaInP}} + \frac{1}{m_{Ge}} + \frac{1}{m_{GaAs}}}. \quad (11)$$

The difference Δ*J* between these two current densities is the amplitude of the AC current density that is actually being measured:

$$\Delta J = J_B - J_A = \frac{\frac{r}{m_{GaInP}}}{\frac{1}{m_{GaInP}} + \frac{1}{m_{Ge}} + \frac{1}{m_{GaAs}}} = \frac{r}{1 + \frac{m_{GaInP}}{m_{Ge}} + \frac{m_{GaInP}}{m_{GaAs}}}. \quad (12)$$

ΔJ is always lower than the change r in the short circuit current density of the junction being measured, but gets close to it only as we tend to certain values for the slopes m of the J versus V curves. This behavior can be easily understood by observing the following limit:

$$\lim_{\substack{\frac{m_{GaInP}}{m_{Ge}} \to 0,\ \frac{m_{GaInP}}{m_{GaAs}} \to 0}} \Delta J = r \qquad (13)$$

Equation (12) indicates that the measurement of the spectral response of one junction, such as the GaInP junction, can be affected by the J–V characteristics of the other junctions as well. Furthermore, it can be seen from equation (13) that having a higher intensity light bias on the Ge junction would make m_{Ge} larger, therefore increasing the ΔJ closer to the r limit, in essence increasing the measured spectral response of the GaInP junction. The section on voltage bias dependence of spectral response discusses this issue in detail.

The responsivity of a junction is defined as:

$$R = \frac{r}{P}, \qquad (14)$$

where P is the monochromatic beam power. However, the measurement gives the following value:

$$R_m = \frac{\Delta J}{P}. \qquad (15)$$

Therefore, the ratio of these two quantities, R_m/R describes a factor that effectively reduces the spectral response from its maximum expected value:

$$\frac{R_m}{R} = \frac{\frac{\Delta J}{P}}{\frac{r}{P}} = \frac{\Delta J}{r}. \qquad (16)$$

In the section on the voltage bias dependence of spectral response, it will be shown that this ratio is close to 1 for well-behaved junctions and lower than 1 for bad cases of low shunt resistance or breakdown voltages. This ratio can be extracted from combinations of I–V measurements at a representative wavelength for each junction, *prior to* performing any spectral response measurements.

Discussion and Results

Performing a set of I–V measurements of the cell under all possible combinations of the three LED light sources *prior* to performing QE measurements can significantly help in determining: (1) which junctions are likely to be affected by artifacts, (2) what level of light and voltage bias should be used to obtain the best measurement results, and finally (3) how the best QE measurement results compare with model predictions for the maximum QE that each junction is able to produce internally.

Combinations of *I*–*V* measurements

In order to use the mathematical model described in Mathematical model for each junction to predict the SR behavior, the values of all the parameters in equations (1) and (2) need to be determined as accurately as possible. This was accomplished by performing eight combinations of I–V measurements on the solar cell, with three unique LEDs appropriate for generating a photocurrent signal in each of the junctions. These were the 520 nm LED for excitation of the GaInP junction, the 741 nm LED for the GaAs junction, and the 1077 nm LED for the Ge junction. The best fit parameters for each junction are chosen so that the calculated composite I–V curve gives the best fit to the measured cell's I–V curve for each of the eight unique measurements. Figure 2(aa'–hh') show all the eight different combinations of I–V measurements possible, with the modeled and measured composite I–V curves on the left panel and the individual junction I–Vs under the given illumination conditions on the right panel. Note that if the cell has either one or no nonideal junctions, it will suffice to perform just one I–V curve measurement with all three LEDs on to obtain a good estimate of the parameters. However, in a case such as this where two junctions have poor diode qualities, data from additional combinations of I–V measurements significantly improve the estimate of the parameters. The modeling results here indicate that both the GaInP junction and the Ge junction suffer from low shunt resistance and low breakdown voltages [14, 18].

The two-diode model seems to provide a reasonably good fit to the various composite I–V curves. However, even a single-diode model may be sufficient for obtaining reasonable fits to the data. The data fitting, especially in the forward bias direction, could potentially be improved further with additional circuit elements and fit parameters, but those additions will add complexities to the simple model presented here.

Table 1 reports the values of the fit parameters for each junction. As expected, the shunt and breakdown voltage values for the GaInP and the Ge junctions are lower than what has typically been reported for solar cells [14] The photocurrent density values J_{ph} correspond to cases where the junctions are illuminated by their respective LEDs, that is, J_{ph} would be zero for an LED off-state for a particular junction. The Miller indices are set to 3, as suggested elsewhere [28].

Voltage bias dependence of spectral response

Once the I–V curve parameters are determined, the measured spectral response behavior as a function of voltage

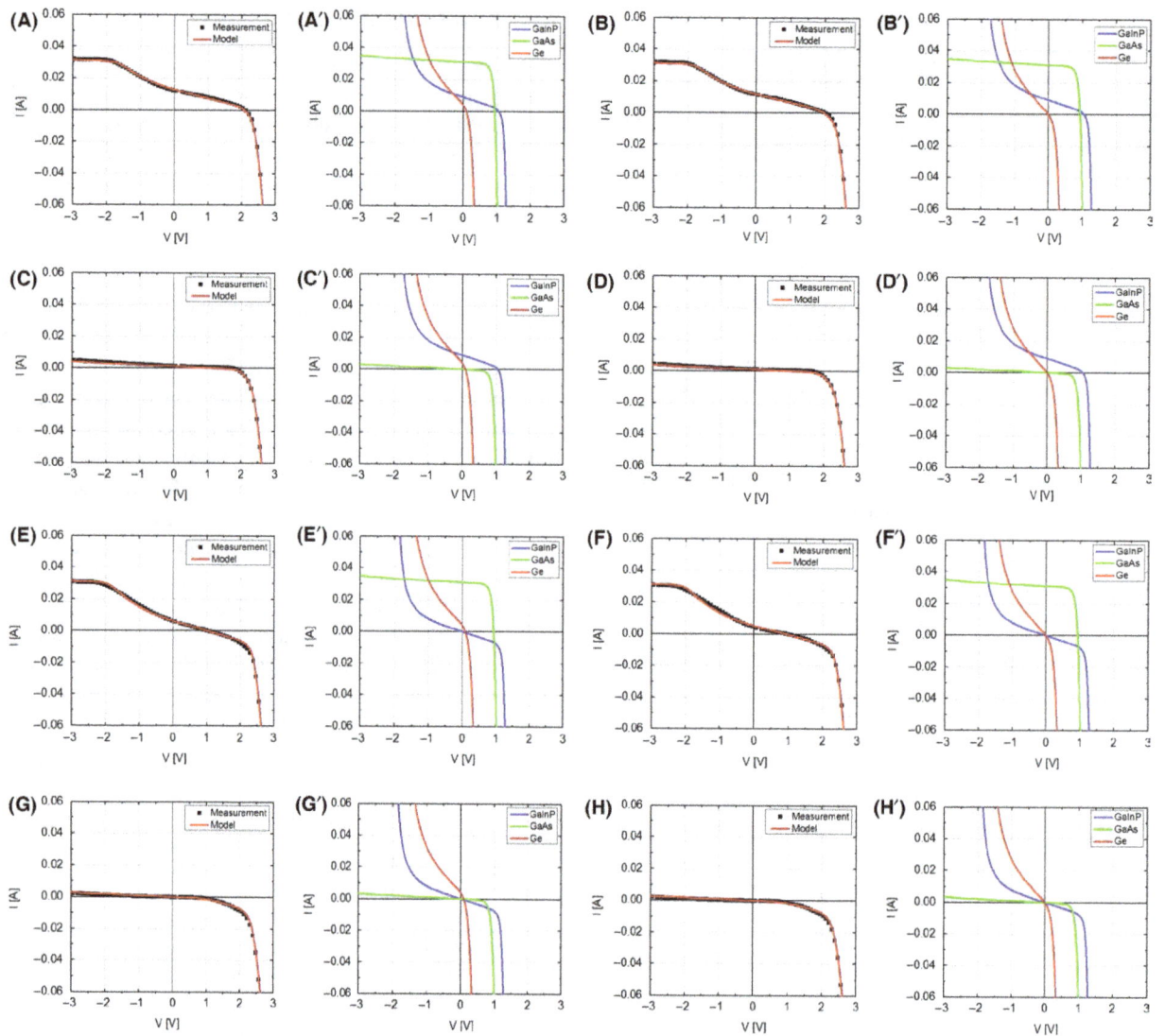

Figure 2. *I* versus *V* curves measured with eight combinations of the three different light biases. For each case, we show the measurement and the model's best fit on the left panel and the modeled *I–V* curves for each junction on the right panel. The eight cases are as follows: (A) All three LEDs on. (B) 520 and 741 nm LEDs on. (C) 520 and 1077 nm LEDs on. (D) 520 nm LED on. (E) 741 and 1077 nm LEDs on. (F) 741 nm LED on. (G) 1077 nm LED on. (H) No LED on.

can be calculated and compared to the measured data. This calculation is based on the modeled junction *I–V*s and not based on equation (12) which was derived with a simpler explanation of the observed effects. Furthermore, the important ratio R_m/R can also be plotted as a function of V_b to see the discrepancy between the measured SR of a junction versus the predicted response from the modeled individual *I–V* curve. Since the SR is a differential measurement, the mechanical motion of the chopper results in a pair of *I–V* composite curves, one for when the chopper blocks the monochromatic light and one for when it unblocks it. The SR calculations are straightforward. Taking the GaInP junction as an example, the individual *I–V*

curves for GaAs and Ge junctions are simulated under the 741 nm and the 1077 nm LEDs 'on', and the *I–V* curve for GaInP junction is simulated under dark (520 nm LED 'off'). A second simulation of this case is run again, with the singular difference that a small photocurrent value is picked for the GaInP junction to simulate its response under the open chopper scenario. This value is many orders of magnitude lower than the other two photocurrent values, just as is the case in actual measurements. From this result, the change due to photogenerated current in the GaInP junction at 0 V, *r*, is calculated from the simulated subcell *J–V* curve. This of course is proportional to the real responsivity of the junction. However, the actual

Table 1. The model fit parameters obtained from fitting the measured composite I–V curves to the model based on equations (1) and (2). Parameter n is higher than the typical value of 2, most likely due to Ohmic or recombination current distributions [28].

Junction	R_s [Ω]	R_{sh} [Ω]	J_{ph} [A/m²]	J_{01}[A/m²]	J_{02} [A/m²]	n	n_m	V_{bd} [V]
GaInP	0.5	145	39.1	2.2×10^{-19}	3×10^{-5}	3.3	3	−2
GaAs	0.5	1000	134.8	4.3×10^{-15}	2.2×10^{-4}	2.8	3	−10
Ge	0.5	45	17.4	4.3×10^{-8}	2.2	2.5	3	−1.8

quantity that is measured, R_m, is proportional to ΔJ, the difference in the composite current between the two chopper states of on and off, as previously shown in Figure 1. ΔJ is also obtained from the simulation as a function of bias voltage. Plotting the calculated ratio $\Delta J/r$ as a function of voltage is very revealing; for example, it can show whether a true, artifact-free spectral response for a given junction can be obtained from a measurement.

Figures 3A–C show the calculated and measured spectral response as a function of V_b for the GaInP junction with the monochromator (MC) wavelength at 520 nm and the following light bias scenarios: (A) both 741 and 1077 nm LEDs, (B) only the 741 nm LED, and (C) only the 1077 nm LED. The left Y-axis represents the measured and calculated responsivities and the right Y-axis simultaneously shows the ratio R_m/R, obtained purely from the calculations. Remembering that a $R_m/R = 1$ ratio means that the measured responsivity is equal to the actual junction responsivity, that is, photocurrent generated in the junction is the photocurrent measured at the terminals, two important observations can be made here: First, the diode-based model predicts the V_b dependence of the responsivity relatively well, predicting a peak response that can be achieved by an application of $V_b \approx 1.8$ V in all three cases with the exception that both combinations of the 1077 and 741 nm LEDs are needed to obtain the maximum response. Note

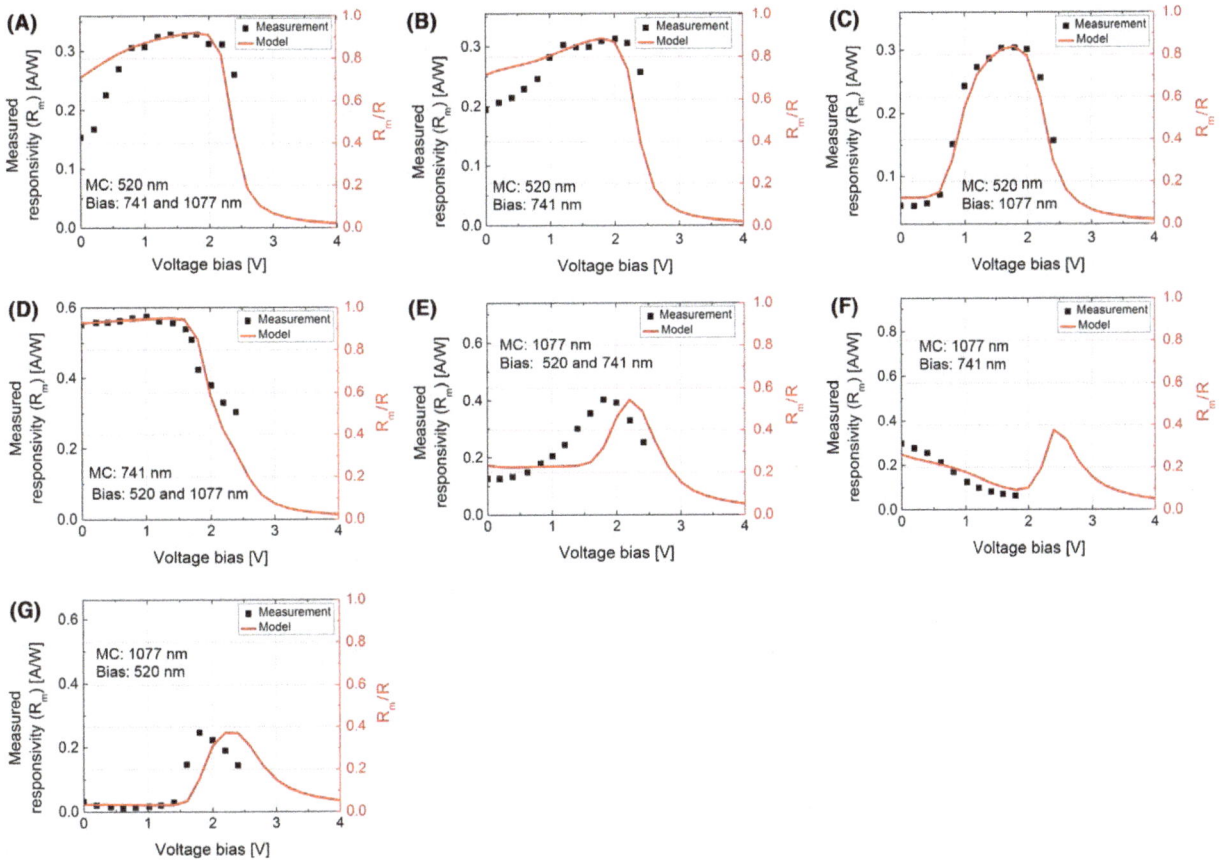

Figure 3. Measurements of the spectral response versus voltage bias for fixed wavelengths corresponding to a region of responsivity of a given junction: (A–C) GaInP junction response with ac monochromatic (MC) wavelength of 520 nm and dc bias lights as labeled. (D) GaAs junction with ac monochromatic wavelength of 741 nm and dc bias wavelengths of 520 and 1077 nm. (E and F) Ge junction response with ac monochromatic wavelength of 1077 nm and dc bias lights as labeled.

that beyond the peak value, the responsivity should drop to 0, as the data also suggest, though the DSR measurement becomes very unstable at higher V_b. The second observation is that in Figure 3A, R_m/R is reaching a maximum value at 0.92 for the GaInP junction, meaning that no matter what voltage bias is used, 8% of the GaInP-generated photocurrent can never be recovered at the external terminals.

Figure 3D shows the V_b dependence of the responsivities for the GaAs junction, with the monochromator wavelength at 741 nm, and both the light bias 520 and 1077 nm LEDs on. The data and the model are in good agreement here, showing no significant V_b dependence until the responsivity drops at higher voltages. This behavior is consistent with the high breakdown voltage and high shunt resistance of the GaAs junction. Yet, the R_m/R ratio for this junction is at 0.96, indicating that the poor behavior of the other two junctions slightly affects the maximum measured responsivity of an otherwise ideal junction. The other two light bias cases for this junction were similar and are therefore not shown.

Finally, Figure 3E–G show the V_b dependence of the responsivities for the Ge junction, with the monochromator wavelength at 1077 nm and the light bias scenarios: (E) both 520 and 741 nm LEDs on, (F) with only 741 nm LED on, and (G) with only 520 nm LED on. Although there are small discrepancies between the modeled and measured results, the findings are similar to the GaInP case in that a peak response is achieved at a voltage bias higher than 0 V, typically around 2 V. One notable exception is that unlike the GaInP junction measurements, the maximum R_m/R for the Ge junction is 0.55, indicating that a much smaller photogenerated current value is extracted at the terminals for this junction. A portion of this 45% photocurrent loss can be recovered for the Ge junction if the 520 nm LED intensity were increased.

We believe that the discrepancies observed between the model and the data, particularly at the low voltage values for Figure 3A and B and the shifts in the peak positions for Figure 3E and F, originate from imperfections of the double-diode I–V model in capturing all the relevant physics inside these imperfect devices. Since we are using the parameters strictly obtained from I–V curve fits, we do not make any alterations or adjustments to the individual R_m/R ratio plots afterwards. Actual spectral response measurements are differential lock-in based measurements and furthermore, the monochromatic light intensity is very low and hence it is possible that there are nonlinearity or second-order effects present between the monochromator "on" and "off" states that we have not captured in our simulations; that is, it is possible the shape of the I–V curve for the chopper-on and chopper-off states change

slightly. There is no way for us to know this, because our instruments are not sensitive to this while there is a large light bias signal present on the other junctions. The purpose of this section is to show that it is possible to understand the voltage bias behavior but the exact source of discrepancy between the model and the data is beyond the scope of this work.

Complete wavelength measurements

Now that the voltage dependence of the SR has been described at fixed wavelengths within the regions of responsivity of each junction, we now turn our attention to the full wavelength measurements for all three junctions. Figure 4A shows the measured spectral responses of all three junctions under the best voltage bias conditions and light bias conditions corresponding to LED light intensities of 116 W/m^2, 228 W/m^2 and 31 W/m^2 for the 520 nm, 741 nm, and the 1077 nm LEDs, respectively. Notice that the nonzero photocurrent response for the GaInP top cell in the region $\lambda > 700$ nm is likely due to some leaked photocurrent from the GaAs middle cell due to the low shunt resistance values of the Ge and the GaInP subcells. However, the nonzero response (0.03 A/W) for the Ge junction for $\lambda < 860$ nm is related to LC from the middle junction. The calculated R_m/R ratios for each junction from the previous section are used as correction factors for each corresponding curve. The corrected spectral response results are plotted as in Figure 4B. The largest correction factor of 1/0.55 for the Ge junction substantially increases the responsivity of that junction, to values expected from a well-behaved Ge junction. In Figure 4C, the corrected QE curves are also shown for reference.

Furthermore, the analytical model allows one to predict the magnitude of the light bias needed to obtain the best QE measurement for the affected junctions. Focusing on the Ge junction's response, we know that the poor diode behavior of the GaInP junction is partially to blame for Ge's low SR values. Applying a stronger light bias to the GaInP junction should shift the operating voltage point for that junction farther to the right towards a higher shunt resistance region, that is, $m_{GaInP} \to \infty$, causing $\Delta J \to r_{Ge}$, hence approaching the corrected SR curve. To show this trend, the 520 nm LED intensity was progressively increased from the initial value of 116 W/m^2 to 147 W/m^2 and 243 W/m^2, while the 741 nm LED was kept fixed at 228 W/m^2, partially to maintain the effect of LC between the GaAs and the Ge junctions at a fixed level. To calculate the model predicted R_m/R ratios, a linear relationship was assumed between the 520 nm LED light intensity and the photogenerated current in the GaInP junction as a result. The predicted R_m/R ratios for the Ge junction should rise from 0.55 to 0.63 and finally to 0.81 for the three 520 nm

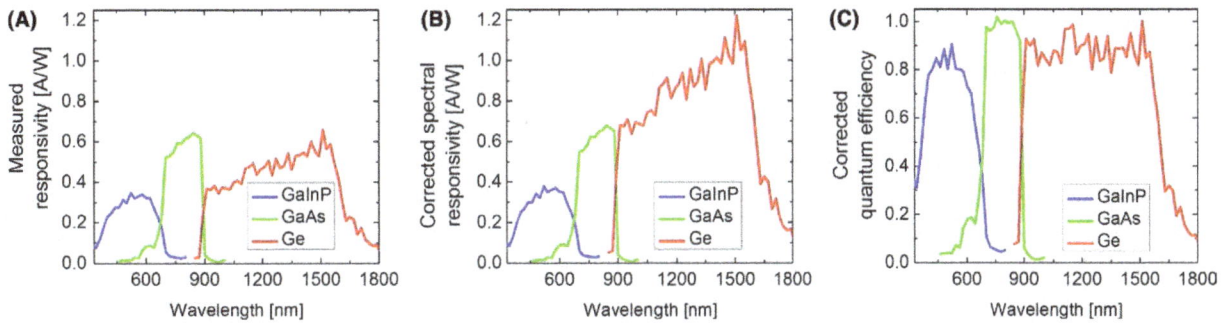

Figure 4. The complete spectral response measurements of all three junctions. (A) Best SR measurements under a given set of bias light and voltage conditions. (B) The corrected spectral responsivity according to the model. GaInP, GaAs and the Ge measured responsivities in (A) are multiplied by the correction factors 1/0.92, 1/0.96, and 1/0.55, respectively. (C) Quantum efficiency obtained after the corrections in (B).

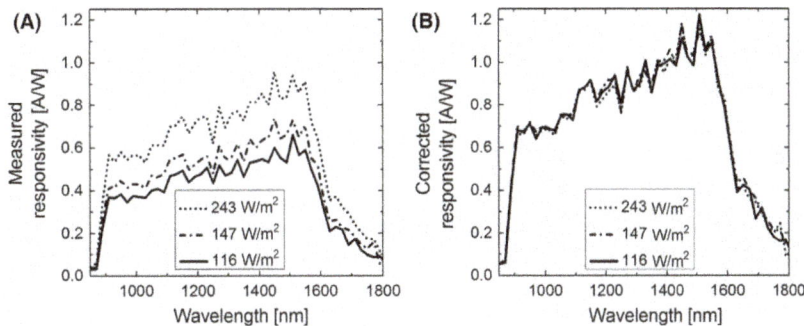

Figure 5. (A) Spectral responsivity of the Ge junction with different intensities of the 520 nm LED light bias, while keeping the 741 nm LED light intensity constant and $V_b = 2$ V. (B) Agreement between the measurements after multiplying each curve by the correction factors predicted by our model.

LED light intensity settings. Figure 5A and B show the results of these three measurements for the Ge junction, and the corrected Ge responsivities based on these calculated ratios. The plots indicate good agreement between the model and the measurements and show that all three distinct curves can be collapsed onto one curve, within the context of the simple analytical model. Since the 741 nm bias light was kept constant for all three measurements and only the 520 nm light was changed, the luminescence coupling between the middle and the bottom junctions were mostly unchanged as evidenced by the plateau below 870 nm. So the shunt resistance effect is the dominant mechanism affecting the QE response of these devices.

Conclusions

Combinations of I–V measurements under various types of light bias sources were used within the context of an analytical two-diode I–V curve model to predict the voltage bias and light bias dependence of the differential spectral response measurements in a triple junction solar cell. The solar cells studied for this work were Ge based, with both the GaInP and the Ge junctions presenting severe cases of low shunt resistance and low reverse bias breakdown voltages. We have shown that these conditions can significantly affect the spectral response measurements of all three junctions, with the Ge junction affected the most. Our work suggests that it is valuable to perform these sets of current voltage measurements and subsequent analysis *prior* to any spectral response measurements so that one can choose a priori how to set up, or what to expect from, the DSR measurements.

Acknowledgments

B. H. Hamadani thanks H. Yoon of NIST for useful discussions and providing reference detectors for calibrations of the spectral response system, J. Roller for writing the computer code needed to run the system, and Matthew Lumb of the Naval Research Laboratory for useful discussions. J. P. Babaro expresses his gratitude to NIST for the hospitality during his stay, while this work was done.

Conflict of Interest

None declared.

References

1. Cotal, H., C. Fetzer, J. Boisvert, G. Kinsey, R. King, P. Hebert et al. 2009. III–V multijunction solar cells for concentrating photovoltaics. Energy Environ. Sci. 2:174.

2. Kurtz, S., and J. Geisz. 2010. Multijunction solar cells for conversion of concentrated sunlight to electricity. Opt. Express 18:A73–A78.

3. Yamaguchi, M. 2003. III–V compound multi-junction solar cells: present and future. Sol. Energy Mater. Sol. Cells 75:261–269.

4. Yoon, H., J. E. Granata, P. Hebert, R. R. King, C. M. Fetzer, P. C. Colter et al. 2005. Recent advances in high-efficiency III-V multi-junction solar cells for space applications: ultra triple junction qualification. Prog. Photovoltaics Res. Appl. 13:133–139.

5. Boisvert, J., D. Law, R. King, D. Bhusari, X. Liu, A. Zakaria et al. 2010. Development of advanced space solar cells at Spectrolab. *35th IEEE Photovolt. Spec. Conf.*, 123–127.

6. King, R. R., C. M. Fetzer, D. C. Law, K. M. Edmondson, H. Yoon, G. S. Kinsey et al. 2007. Advanced III-V multijunction cells for space. *Conf. Rec. 2006 IEEE 4th World Conf. Photovolt. Energy Conversion, WCPEC-4*, 2:1757–1762.

7. Chiu, P. T., D. C. Law, R. L. Woo, S. B. Singer, D. Bhusari, W. D. Hong et al. 2014. 35.8% space and 38.8% terrestrial 5J direct bonded cells. *IEEE 40th Photovolt. Spec. Conf.*

8. 2014. Press Release, Fraunhofer Institute for Solar Energy Systems, 1 December 2014.

9. ASTM Standard E2236-10: Standard test methods for measurement of electrical performance and spectral response of non-concentrator multijunction photovoltaic cells and modules.

10. King, D. L., B. R. Hansen, J. M. Moore, and D. J. Aiken. 2000. New methods for measuring performance of monolithic multi-junction solar cells. *Conf. Rec. Twenty-Eighth IEEE Photovolt. Spec. Conf. - 2000 (Cat. No.00CH37036)*, 1197–1201.

11. Li, J.-J., S. H. Lim, and Y.-H. Zhang. 2012. A novel method to eliminate the measurement artifacts of external quantum efficiency of multi-junction solar cells caused by the shunt effect. Proc. SPIE, 8256:825616–825623.

12. Li, J.-J., S. H. Lim, C. R. Allen, D. Ding, and Y.-H. Zhang. 2011. Combined effects of shunt and luminescence coupling on external quantum efficiency measurements of multijunction solar cells. IEEE J. Photovolt. 1:225–230.

13. Li, J.-J., and Y.-H. Zhang. 2013. Elimination of artifacts in external quantum efficiency measurements for multijunction solar cells using a pulsed light bias. IEEE J. Photovolt. 3:364–369.

14. Meusel, M., C. Baur, G. Letay, A. W. Bett, W. Warta, and E. Fernandez. 2003. Spectral response measurements of monolithic GaInP/Ga(In)As/Ge triple-junction solar cells: measurement artifacts and their explanation. Prog. Photovoltaics Res. Appl. 11:499–514.

15. Paraskeva, V., M. Hadjipanayi, M. Norton, M. Pravettoni, G. E. Georghiou. 2013. Voltage and light bias dependent quantum efficiency measurements of GaInP/GaInAs/Ge triple junction devices. Sol. Energy Mater. Sol. Cells 116:55–60.

16. Pravettoni, M., R. Galleano, A. Virtuani, H. Müllejans, and E. D. Dunlop. 2011. Spectral response measurement of double-junction thin-film photovoltaic devices: the impact of shunt resistance and bias voltage. Meas. Sci. Technol. 22:045902.

17. Siefer, G., C. Baur, and A. W. Bett. 2010. External quantum efficiency measurements of Germanium bottom subcells: Measurement artifacts and correction procedures. *Conf. Rec. IEEE Photovolt. Spec. Conf.*, 704–707.

18. Barrigón, E., P. Espinet-González, Y. Contreras, and I. Rey-Stolle. 2015. Implications of low breakdown voltage of component subcells on external quantum efficiency measurements of multijunction solar cells. Prog. Photovolt. Res. Appl. 23:1597–1607.

19. Allen, C. R., S. H. Lim, J.-J. Li, and Y.-H. Zhang. 2011. Simple method for determining luminescence coupling in multi-junction solar cells. *2011 37th IEEE Photovolt. Spec. Conf.*, 000452–000453.

20. Derkacs, D., D. T. Bilir, and V. A. Sabnis. 2013. Luminescent coupling in GaAs/GaInNAsSb multijunction solar cells. IEEE J. Photovolt. 3:520–527.

21. Geisz, J. F., M. A. Steiner, I. García, R. M. France, W. E. McMahon, C. R. Osterwald, and D. J. Friedman. 2015. Generalized optoelectronic model of series-connected multijunction solar cells. IEEE J. Photovolt. 5:1827–1839.

22. Steiner, M. A., S. R. Kurtz, J. F. Geisz, W. E. McMahon, and J. M. Olson. 2012. Using phase effects to understand measurements of the quantum efficiency and related luminescent coupling in a multijunction solar cell. IEEE J. Photovolt. 2:424–433.

23. Steiner, M. A., and J. F. Geisz. 2012. Non-linear luminescent coupling in series-connected multijunction solar cells Non-linear luminescent coupling in series-connected multijunction solar cells. Appl. Phys. Lett. 100:251106.

24. Yoon, H., R. R. King, G. S. Kinsey, S. Kurtz, and D. D. Krut. 2003. Radiative coupling effects in GaInP/GaAs/Ge multijunction solar cells. *3rd World Conf. PV Energy Conv.* 745–748.

25. Baur, C., M. Hermle, F. Dimroth, and A. W. Bett. 2007. Effects of optical coupling in III-V multilayer systems. Appl. Phys. Lett. 90:13–16.

26. Geisz, J. F., S. Kurtz, M. W. Wanlass, J. S. Ward, A. Duda, D. J. Friedman et al. 2007. High-efficiency

GaInP/GaAs/InGaAs triple-junction solar cells grown inverted with a metamorphic bottom junction. Appl. Phys. Lett. 91:12–15.

27. Hamadani, B. H., J. Roller, B. Dougherty, F. Persaud, and H. W. Yoon. 2013. Absolute spectral responsivity measurements of solar cells by a hybrid optical technique. Appl. Opt. 52:5184–5193.

28. Breitenstein, O. 2013. The physics of industrial crystalline silicon solar cells. Adv. Photovolt. 89:1–75.

3

Automatized analysis of IR-images of photovoltaic modules and its use for quality control of solar cells

Johannes Hepp[1,2], Florian Machui[1], Hans-J. Egelhaaf[1], Christoph J. Brabec[1,3] & Andreas Vetter[1,3]

[1]Bavarian Center for Applied Energy Research (ZAE Bayern), Haberstraße 2a, 91058 Erlangen, Germany

[2]Erlangen Graduate School in Advanced Optical Technologies (SAOT), Friedrich Alexander University Erlangen-Nuremberg (FAU), Paul-Gordan-Str. 6, 91052 Erlangen, Germany

[3]Materials for Electronics and Energy Technology (iMEET), Friedrich Alexander University Erlangen-Nuremberg (FAU), Energie Campus Nürnberg (EnCN), 90429 Nürnberg, Germany

Keywords

Imaging, IR-thermography, PV, quality control, segmentation, solar cell

Correspondence

Andreas Vetter, Materials for Electronics and Energy Technology (iMEET), Friedrich Alexander University Erlangen-Nuremberg (FAU), Energie Campus Nürnberg (EnCN), 90429 Nürnberg, Germany.
E-mail: andreas.vetter@fau.de

Funding Information

German Ministry of Economy and Energy (Grant / Award Number: 'OptiCIGS, 0325724C') State of Bavaria (Grant / Award Number: 'Bavaria on the move') German Research Foundation (Grant / Award Number: 'Entwicklung von bildgebenden Verfahren zur Defekte').

Abstract

It is well known that the performance of solar cells may significantly suffer from local electric defects. Accordingly, infrared thermography (i.p. lock-in thermography) has been intensely applied to identify such defects as hot spots. As an imaging method, this is a fast way of module characterization. However, imaging leads to a huge amount of data, which needs to be investigated. An automatized image analysis would be a very beneficial tool but has not been suggested so far for lock-in thermography images. In this manuscript, we describe such an automatized analysis of solar cells. We first established a robust algorithm for segmentation (or recognition) for both, the PV-module and the defects (hot spots). With this information, we then calculated a parameter from the IR-images, which could be well correlated with the maximal power (P_{mpp}) of the modules. The proposed automatized method serves as a very useful foundation for faster and more thorough analyses of IR-images and stimulates the further development of quality control on solar modules.

Introduction

Energy supply by renewable sources such as solar modules (PV) will be a key issue for societies for the next decades. Common solar cells of the "first generation" (based on Silicon) contribute significantly to the electricity generation in various countries already today [1]. The success story of PV was heavily promoted by decreasing silicon-PV prices. However, solar cells based on thin film absorbers, such as CIGS, CdTe, or organic photovoltaics (OPV), start to gain larger parts of the market share. For illustration, about 10% of the installed modules today are based on thin film technology [2]. This is very promising as thin film solar modules have a strong potential for further substantial decrease in price, such enabling a further increase in green electricity production.

Solar cells based on organic compounds are definitely one of the most thrilling options when aiming for a huge decrease in production costs. One key aspect here is the possibility to print organic solar cells in large scale, which would decrease strongly the price of OPV. While the production of silicon PV is a rather mature process, processing of organic solar cells still exhibits room for improvement. One factor, which decreases the efficiency of organic solar cells, is electrical defects introduced during production. We want to point out that electrical defects also may and do occur when producing inorganic solar cells. The detrimental effect of "hot spots" has been reported in a various number of publications for different solar cell types (for example for silicon [3–10], CIGS [11–16], CdTe [17–19] and for OPV [20–23]). In any case, an automatized recognition and analysis of the defects is a highly desirable tool.

Electrical defects lead to local short circuit currents with heat being dissipated at the hot spot thereby reducing the performance of the solar cell [13, 24]. Origin of such local short circuits may be bad edge isolation or small electrical conductive contamination connecting locally front and back contact. Identification of such defects has been found to be an important task and, accordingly, the problem has been tackled in particular by fast imaging via IR cameras. The localization of even very small hot spots (showing only a minor temperature gradient) may be realized by applying lock-in thermography [25]. In this method, a pulsed excitation and a phase-sensitive detection increase the sensitivity vastly. The method is named dark lock-in thermography (DLIT) when applying electric current for excitation of the samples. By applying a voltage sweep (reverse and forward bias), DLIT enables a detailed defect characteristics [26]. Accordingly, DLIT is an important and commonly used tool in R&D labs of solar cell manufactures.

Processing conditions generally do vary over time when producing solar cells. These fluctuations most likely affect the composition, and the morphology of the module, as well as the number and types of defects. Hence, a large number of modules are studied and, in particular true for imaging methods, a large data set needs to be analyzed. IR-Images of the modules contain the foreground (the actual module) and a background. Furthermore, there may be hot spots on the foreground, which reduce the performance of the module. An algorithm, which automatically recognizes or detects both, the foreground and the defects, would be of great help to thoroughly analyze the huge amount of data obtained in R&D labs when aiming for a detailed characterization of defects on solar cells. This is, because an automatized analysis of the influence of defects on the module performance may be carried out with this information.

In this study, we describe an automatized analysis of lock-in thermography images and provide a proof-of-principle of its applicability for solar cell quality analysis. To do so, we establish an algorithm, which allows for an automatized segmentation of the module and an automatized segmentation of the defects. With segmentation we mean the recognition or detection of the according pixels of the digital image, that is, the pixels belonging to the solar module and the pixels belonging to hot spots. We then post-process the images and correlate the calculated image parameter with the crucial electrical module parameter for quality control, the maximum power of the module (P_{mpp}). The maximum power of a module is the key parameter for the price of module. Next to P_{mpp}, important parameters are the open circuit voltage V_{oc} and the short circuit current J_{sc}. We focus our work on organic solar cells; however, the described method is not restricted at all to OPV.

Materials and Methods

We led the proof-of principle with innovative semitransparent test OPV modules produced in our lab (Fig. 1). The module substrate size was 16.5 cm × 16.5 cm consisting of 30 individual cells on an active area of 197 cm^2. We processed four modules with the same processing parameters as described below.

Inverted structure OPV devices were processed on fluorine-doped tin oxide (FTO) coated glass with the layer sequence ZnO nanoparticles/PBTZT-stat-BDTT-8: phenyl-C61-butyric acid methyl ester (PCBM)/poly(3,4-ethylenedioxythiophene)/polystyrene sulfonate (PEDOT:PSS)/silver nanowires (AgNW). FTO substrates were laser patterned to achieve P1 with a fluence of 0.41 J/cm^2, 50% overlap and 0.9 J/cm^2, 98% overlap, respectively. All layers were processed via slot-die coating with a 20 cm wide heatable slot-die head. ZnO

Figure 1. Visible image of the investigated semitransparent modules.

nanoparticles dispersed in isopropyl alcohol (Nanograde AG, Zurich, Switzerland) were coated via slot-die coating optimizing coating and drying conditions to achieve a film thickness of 50 nm. Afterwards, the layer was dried at 80°C for 5 min. For the photoactive layer (PAL), PBTZT-stat-BDTT-8 was purchased from Merck Chemicals GmbH, Darmstadt, Germany and PCBM (technical grade 99%) from Solenne BV, Groningen, the Netherlands, and dissolved at a concentration of 35 mg/mL in a weight ratio 1:2 in o-xylene: tetrahydronaphthalene (9:1) and stirred for 12 h at 80°C before coating. The PAL was then slot die coated aiming at a dried film thickness of 290 nm. PEDOT:PSS (Clevios FHC) from Heraeus was diluted in deionized water (1:1 volume ratio) and then coated via slot-die coating aiming at a dried film thickness of 100 nm. The substrates were then annealed at 120°C for 5 min and afterwards patterned by laser ablation to achieve P2 (fluence of 0.08 J/cm^2, 94% overlap, 3 times). The final wet film application completing the devices was done by slot-die coating AgNWs (Cambrios Advanced Materials Corp, Sunnyvale, CA, USA.) from aqueous solution which were afterwards annealed at 120°C for 5 min and laser patterned to achieve electrical separation (P3 – fluence of 0.08 J/cm^2 and 94% overlap). Laser patterning was achieved with an LS 7xxP setup built by LS Laser Systems GmbH (München, Germany). The heart of the system is a femtoREGENTM UC- 1040–8000 fsec Yb SHG from High Q Laser GmbH (Rankweil, Austria) emitting at 1040 nm (fundamental wavelength) and 520 nm (first harmonic wavelength) with a pulse duration of <350 fsec at repetition rates up to 960 kHz. The devices were encapsulated in glass with DELO Katiobond LP655. Current–voltage (IV) characteristics were measured with a source measurement unit from Keysight (Santa Rosa, CA, USA) and a Tracer software (ReRa Tracer 3, ReRa Solutions, Nijmegen, The Netherlands). Illumination was provided by an LOT quantum device solar simulator Class A, AM1.5G spectra at 1000 W/m^2.

After production, characterization of the samples was carried out by DLIT (dark lock-in thermography). In order to acquire lock-in IR-images, the IR-camera (Equus 327k; IRcam GmbH, Erlangen, Germany) sends a modulation signal to an electrical excitation. This way both devices are synchronized, which allows the application of a correlation function. The software of the camera then integrates (weighted average) over all captured images, in order to reduce the noise of the final output. The result is a phase-sensitive average of IR-images, commonly known lock-in thermography. When the measurement is done in the absence of illumination (electrical excitation instead), it is called dark lock-in thermography (DLIT). It is called illuminated lock-in thermography (ILIT) when pulsed light is used for excitation.

The measurements were performed at an excitation alternating between 0 V and 35 V. The applied voltage was in the low excitation regime to enhance shunt localization. The Stirling cooled InSb-detector of the camera is sensitive between 1 and 5 μm. The detector size is 640 × 512 pixels. Further settings of the camera were a lock-in frequency of 1 Hz, a frame integration time of 1 msec, and an image acquisition time of 15 min. The thickness of glass substrate was 3.2 mm and, at a frequency of 1 Hz, the captured signal dominantly stemmed from the active layer [27]. Amplitude and phase images were exported as raw data for further segmentation and image analysis.

Algorithm

Automatized recognition of diverse objects has been successfully implemented in machine vision, for example, in food industry [28] or in printed electronics [29]. Rather large noise makes object segmentation more difficult when applying infrared thermography [30, 31]. With segmentation, we mean the recognition of pixels belonging to a certain object in an image. Thus, the foreground may be separated from the background of an image, for example. Yet, a variety of algorithms has been proposed for standard thermography depending on the application, such as IR-surveillance [32–34] or target detection [35]. Unfortunately, the problem of low signal to noise ratio generally becomes much more challenging when applying highly sensitive lock-in thermography.

In a recent publication, we proposed an algorithm for automatized segmentation of PV-modules characterized by highly sensitive lock-in thermography [36]. This method was stimulated by algorithms developed to detect the background and foreground in images obtained by commercial scanners [37–39]. The method proved to work even under extremely low signal to noise ratio of 1.09 (for more details see [36], i.p. Fig. 3 of that reference). In short, the algorithm worked by identifying the four edges of the solar module individually by searching for the edge pixels. The edge pixels then represent the outer frame of the PV-module (this means all pixels inside the frame belong to the module or foreground). Searching was performed by a "moving standard deviation" process. The aim of this process is, simply speaking, to minimize the (relative) standard deviations of the two regions (background and foreground). The edge pixels are calculated by minimizing the standard deviation. For more details, the reader is referred to the original publication [36]. In the current paper, we apply the same algorithm to segment the foreground or active area of the solar modules. However, instead of using the amplitude images (as in the original paper), we use the phase image this time as the phase images exhibit a better contrast regarding the edges.

Next to the segmentation of the module (i.e., the identification of the foreground pixels), the segmentation of the defects (hot spots) is required for a successive quality analysis. The physical meaning of the identified hot spots is a large local short circuit current. Standard segmentation algorithms such as Otsu's method [40] do not work properly as there is no bimodal distribution of the intensities. Figure 2 shows exemplarily a histogram of a DLIT image taken in this study (logarithmic scale). The peak corresponds to the foreground (PV module), while values significantly lower than the peak correspond to the background. As a remark, one can also directly see from this figure that such a bimodal global segmentation algorithm would not work to identify the foreground from the background.

Instead of a second peak, the hot spots are aligned in a very long and low tail in the histogram. Accordingly, two characteristics of the defects will be exploited for hot spot segmentation. Defects (1) show a large intensity in the image, and (2) they are rather sparsely distributed over the module. This leads obviously to the conclusion to search for outstandingly hot pixels and, at the same time, values, which have a low probability of occurrence.

The algorithm for defect identification is described in the following. We use the histogram information of the amplitude images for hot spot detection (see Fig. 2 as an example). P is the probability (of a certain intensity), which can be calculated by eq. 1, with N as number of total pixels of the image and n as numbers of pixels, or counts, in the respective bin (see Fig. 2, y-axis). The peak intensity is called I_{max} (eq. 2). In our case, we apply the amplitude images, this means, the intensity is the DLIT amplitude. Only two parameters are required for the segmentation algorithm: the numbers of bins, $nBins$, typically

256, and a factor f (remark: in Fig. 2, we used only 60 bins for better visualization). The latter parameter characterizes the sensitivity (by defining the maximal threshold probability P_{th}, see eq. 3) of the hot spot detection and was set to 0.1. The idea of "outstandingly hot pixels with low probability" is mathematically defined by eq. 4, the core of the rather simple but robust algorithm. BW_{def} denotes the binary image containing the defect pixels (hot spots) with x and y as coordinates of the image and I as the original intensity image (DLIT, amplitude). This means, pixels in the image BW_{def} with value 1 belong to a hot spot while pixels with value 0 do not belong to a hot spot. In summary, we use an additional probability threshold compared to conventional intensity thresholding.

$$P = \frac{n(I)}{N} \quad (1)$$

$$I_{max} = \arg\max(P(I)) \quad (2)$$

$$P_{th} = \left(\frac{1}{nBins}\right) \times f \quad (3)$$

$$BW_{def}(x,y) = \begin{cases} 1 & \text{if } I(x,y) > I_{max} \ \& \ P(I(x,y)) < P_{th} \\ 0 & \text{else} \end{cases} \quad (4)$$

Results

We examined four OPV modules produced in our lab. Figure 3 shows the phase images of the DLIT imaging on the left side and the amplitude images on the right side in gray values. First of all, we tested the segmentation of the foreground by the algorithm. The best results for this segmentation are found when applying the phase images. In doing so, all four modules were segmented correctly. The automatically segmented edges are marked as yellow rectangles in Figure 3. All pixels inside the yellow rectangle belong to the module and their intensities are referred to as I_{module} in this paragraph.

Next, we tested the algorithm for segmentation of the hot spots. Rather few hot spots were detected as already mentioned in the section "algorithm", see also the histogram in Figure 2. The defects were identified in the amplitude images, see the red indicated pixels in Figure 3, right. The pixels inside the red boundary lines recognized as hot spots and their intensities are referred to as I_{def}. The edge of the module, indicated in yellow, may be copied from the phase image as both images are calculated from the same phase-sensitive transient lock-in measurement. Accordingly, the module cannot be moved unintentionally in between recording both images. Therefore, the yellow frame has exactly the same position in the amplitude image and phase

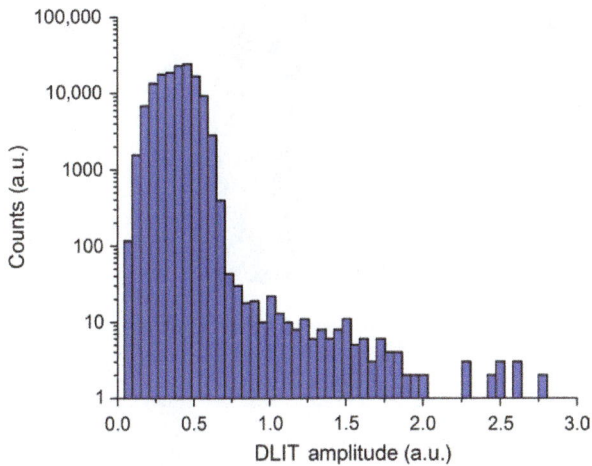

Figure 2. Example of an intensity histogram of a dark lock-in thermography image (amplitude) recorded in this study (in a histogram, the x-axis resembles an intensity ranges or bins and the y-axis the number of pixels in the according bin). The histogram was calculated for 60 bins (rather low value for better visualization).

(A) Phase · (B) Amplitude

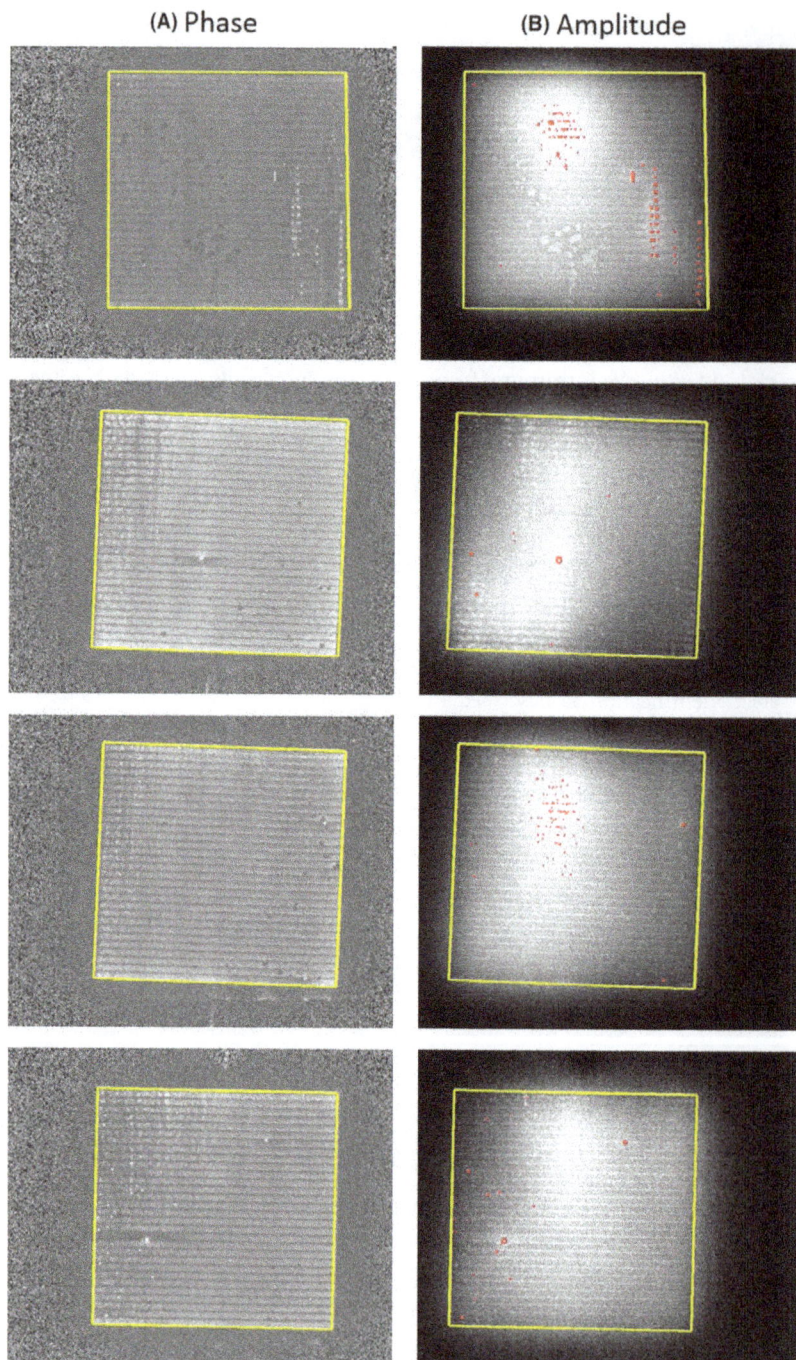

Figure 3. Dark lock-in thermography images of the four different modules. On the left side the phase images are depicted, on the right side the amplitude images. Yellow lines indicate the segmentation of the PV-module (foreground) which was carried out on the phase images. Module segmentation may be applied also to the amplitude image (both images are recorded simultaneously). The red lines indicate the segmented hot spots (or defects), which was carried out with the amplitude images.

image. The number of detected "hot pixels" varied between 86 (sample number 2) and 411 (sample number 1). Sample 3 exhibited 131 defect pixels and sample 4 showed 134 defect pixels. The defect pixels are hard to distinguish by eye when looking at the grey scale images in Figure 3,

right. To verify that the identified pixels are hot spots, Figure 4 compares the DLIT-amplitude image of one sample displayed in a colored intensity scale (Fig. 4A) and in grey scale values (Fig. 4B). The hot spots can more easily be located by eye in the colored image and were correctly

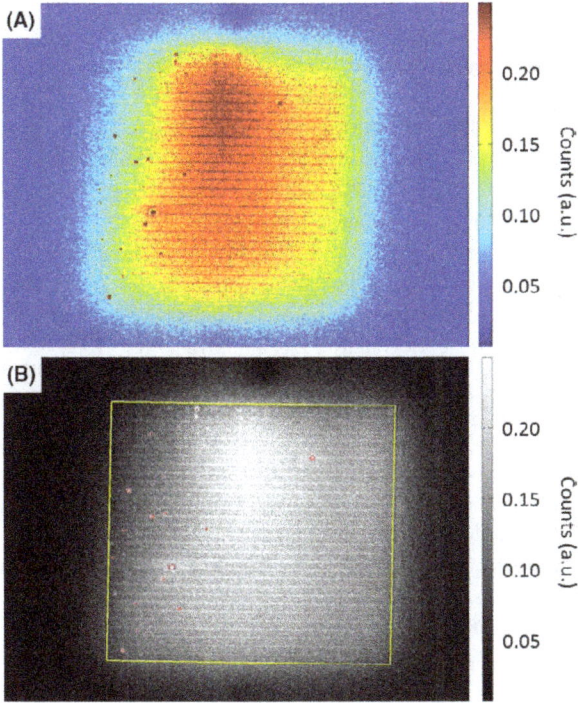

Figure 4. (A) Dark lock-in thermography (amplitude) image of sample 4 in color-coded scale and (B) in grey scale. Yellow lines indicate the PV-module edge and the right lines indicate the detected hot spots by the algorithm.

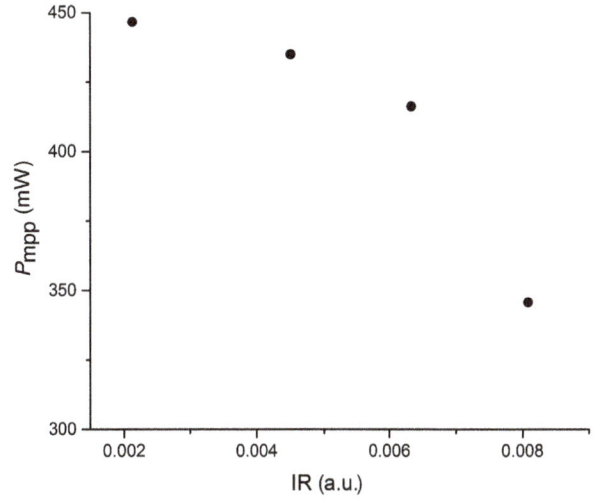

Figure 5. Maximal power (P_{mpp}) of the modules depending on the IR-parameter (calculated according to eq. 5) derived from the dark lock-in thermography amplitude images. Segmentation of the PV-module and the hot spots was carried out prior to this analysis.

$$IR = \frac{\sum_i I_{def}(i) \cdot A_{def}}{\sum_i I_{module}(i) \cdot A_{module}} \tag{5}$$

identified by the algorithm, compare Figure 4B. The colored image, though, has the disadvantage of a more difficult identification of the module edges by eye. This example illustrates the large gradients in temperature and the difficulty of setting the "correct" contrast. Also, by looking at the "freckles" in the images in Figure 4, one can get an impression of the noise present in the highly sensitive lock-in thermography images.

After recognition of the foreground (active PV module area) and the hot spots, a parameter (scalar value) describing or "summarizing" the DLIT image may be determined. The aim is to correlate this parameter with electrical parameters for quality analysis. One of the most crucial parameters in terms of quality control is the maximum power, P_{mpp}. Previous work [13, 25] showed that a promising IR-parameter candidate is the ratio of the intensity of the hot spots and the intensity of the module (as a kind of base signal). The parameter is calculated according to eq. 5. The IR-parameter basically quantifies a contrast between the hot spots and the module normal active area signal. Here, I_{def} denotes the intensity vector of the hot pixels, I_{module} the intensity vector of the module (foreground), A_{def} the area (or number of pixels) of the defect vector, and A_{module} the area (or number of pixels) of the module vector.

We correlate this parameter with P_{mpp} measured by the JV-curve at standard measurement conditions (STM). Figure 5 shows the result with the IR parameter on the x-axis and P_{mpp} on the y-axis. A clear dependency between the IR-parameter and P_{mpp} is observed. In our proof of principle, we found a highly nonlinear decrease in P_{mpp} with increasing value of the IR-parameter. This behavior depends on various influences, such as the thickness of the encapsulation glass, IR-camera, camera calibration or, of course, the choice of the IR-parameter. Accordingly, different relations between P_{mmp} may be found when applying the proposed analysis to a different PV-module types and/or using a different imaging setup. In any case, our results strongly illustrate the suitability of this analysis method for automatized quality control. In the current work, we proofed that the automatized analysis works even for encapsulated modules. Segmentation of IR-images of nonencapsulated samples (taken for example directly in the production line) is much easier due to larger thermal gradients (as glass is a thermal insulator). Previous work based on a manual analysis proved also the applicability of ILIT (illuminated lock-in thermography) as a potential contactless measurement tool for quality control [11]. Accordingly, a transfer of this analysis to ILIT images, and therefore a contactless quality control tool, is straight forward.

Conclusions

We present a combined approach of an automatized segmentation of a PV-module and the defects (hot spots) on the module. Both, segmentation of the module and the hot spots, were carried out successfully for all investigated four encapsulated OPV modules. In previous work [36], module detection worked also for all 10 investigated CIGS modules (samples without front cover glass). We applied the segmentation algorithm to 10 CIGS mini-module, and all defects (hot spots) and the boundaries of the modules were determined correctly (by using the same values of the parameters of the algorithm, f and N_{bins}). Accordingly, we believe that the algorithm is robust and may be applied to any kind of thin film solar cells. The transfer of the method is straight forward, though, in some cases an adaptation of the algorithm parameters might be necessary. The automatized segmentation is an important step toward a thorough analysis of IR-images (and also potentially for luminescence images). We successfully correlated an IR-parameter calculated from the lock-in thermography images with the maximum power (P_{mpp}) of the modules as a proof-of-principle.

The presented approach may be utilized as foundation of adapted (and if necessary more sophisticated) automatized evaluation of large data sets obtained by imaging of PV. The presented (or similar) algorithm facilitates a thorough statistical analysis of a large number of samples also under different working conditions. This strongly helps to improve tools for quality control and also helps to better understand the photo-physical impact of defects on solar modules. While the effect of single defects on the solar module performance have been successfully investigated [13, 41, 42], many open questions remain when studying whole modules with several defects.

Acknowledgments

We gratefully acknowledge the German Ministry of Economy and Energy (OptiCIGS, 0325724C) for funding. Andreas Vetter received funding through the "Bavaria on the move initiative" (Energie Campus Nürnberg) by the State of Bavaria. We thank Viktor Antlitz and Andre Karl for helping with the experimental work. Andre Karl gratefully acknowledges funding by the German Research Foundation (DFG, "Entwicklung von bildgebenden Verfahren zur Defekterkennung in Tandem Solarzellen").

Conflict of Interest

None declared.

References

1. Cumulative installed solar photovoltaics capacity in leading countries and the world, 2000–2013. Earth Policy Institute, Washington, DC. 2015

2. Forstner, H., S. Bandil, M. Zweegers, R. Bollen, G. Coletti, W. Sinke. 2013 Results. International Technology Roadmap for Photovoltaic (ITRPV). VDMA Photovoltaic Equipment, Frankfurt, Germany.

3. Breitenstein, O., J. P. Rakotoniaina, and M. H. Al Rifai. 2003. Quantitative evaluation of shunts in solar cells by lock-in thermography. Prog. Photovolt. Res. Appl. 11:515–526.

4. Breitenstein, O., J. P. Rakotoniaina, M. H. Al Rifai, and M. Werner. 2004. Shunt types in crystalline silicon solar cells. Prog. Photovolt. 12:529–538.

5. Michl, B., M. Padilla, I. Geisemeyer, S. T. Haag, F. Schindler, M. C. Schubert et al. 2014. Imaging techniques for quantitative silicon material ans solar cell analysis. IEEE J. Photovolt. 4:2156–3381.

6. Johnston, S., H. Guthrey, F. Yan, K. Zaunbrecher, M. Al-Jassim, P. Rakotoniaina et al. 2014. Correlating multicrystalline silicon defect types using photoluminescence, defect-band emission, and lock-in thermography imaging techniques. IEEE J. Photovolt. 4:348–354.

7. Rißland, S., T. M. Pletzer, H. Windgassen, O. Breitenstein, and A. S. Preparation. 2013. Local thermographic efficiency analysis of multicrystalline and cast-mono silicon solar cells. IEEE J. Photovolt. 3:1192–1199.

8. Geisemeyer, I., F. Fertig, W. Warta, S. Rein, and M. C. Schubert. 2014. Prediction of silicon PV module temperature for hot spots and worst case partial shading situations using spatially resolved lock-in thermography. Sol. Energy Mater. Sol. Cells 120:259–269.

9. Shen, C., K. Wang, and M. A. Green. 2014. Fast separation of front and bulk defects via photoluminescence on silicon solar cells. Sol. Energy Mater. Sol. Cells 128:260–263.

10. Augarten, Y., T. Trupke, M. Lenio, J. Bauer, J. W. Weber, M. Juhl et al. 2013. Calculation of quantitative shunt values using photoluminescence imaging. Prog. Photovolt. Res. Appl. 21:933–941.

11. Vetter, A., F. Fecher, J. Adams, R. Schaeffler, J.-P. Theisen, C. J. Brabec et al. 2013. Lock-in thermography as a tool for quality control of photovoltaic modules. Energy Sci. Eng. 1:12–17.

12. Adams, J., A. Vetter, F. Hoga, F. Fecher, J. P. Theisen, C. J. Brabec et al. 2014. The influence of defects on the cellular open circuit voltage in CuInGaSe2 thin film solar modules-An illuminated lock-in thermography study. Sol. Energy Mater. Sol. Cells 123:159–165.

13. Vetter, A., F. S. Babbe, B. Hofbeck, P. Kubis, M. Richter, S. J. Heise et al. in press. Visualizing the performance loss of solar cells by IR-thermography - an evaluation study on CIGS with artificially induced defects. Prog. Photovolt. Res. Appl. 24:1001–1008.

14. Misic, B., B. E. Pieters, U. Schweitzer, A. Gerber, and U. Rau. 2015. Thermography and electroluminescence imaging of scribing failures in Cu(In, Ga)Se-2 thin film solar modules. Phys. Status Solidi. A Appl. Mater. Sci. 212:2877–2888.

15. Johnston, S., T. Unold, I. Repins, A. Kanevce, K. Zaunbrecher, F. Yan et al. 2012. Correlations of Cu(In, Ga)Se-2 imaging with device performance, defects, and microstructural properties. J. Vac. Sci. Technol., A 30:111–114.

16. Gerber, A., V. Huhn, T. M. H. Tran, M. Siegloch, Y. Augarten, B. E. Pieters et al. 2015. Advanced large area characterization of thin-film solar modules by electroluminescence and thermography imaging techniques. Sol. Energy Mater. Sol. Cells 135:35–42.

17. McMahon, T. J., T. J. Berniard, and D. S. Albin. 2005. Nonlinear shunt paths in thin-film CdTe solar cells. J. Appl. Phys. 97:054503.

18. Johnston, S., T. Unold, I. Repins, R. Sundaramoorthy, K. M. Jones, B. To et al. 2010. Imaging characterization techniques applied to Cu(In,Ga)Se[sub 2] solar cells. J. Vac. Sci. Technol. 28:665.

19. Katz, N., M. Patterson, K. Zaunbrecher, S. Johnston, and J. Hudgings. 2013. High-resolution imaging of defects in CdTe solar cells using thermoreflectance. Electron. Lett. 49:1559–1560.

20. Hoppe, H., J. Bachmann, B. Muhsin, K.-H. Druee, I. Riedel, G. Gobsch et al. 2010. Quality control of polymer solar modules by lock-in thermography. J. Appl. Phys. 107:014505.

21. Bachmann, J., C. Buerhop-Lutz, R. Steim, P. Schilinsky, J. A. Hauch, E. Zeira et al. 2012. Highly sensitive non-contact shunt detection of organic photovoltaic modules. Sol. Energy Mater. Sol. Cells 101:176–179.

22. Roesch, R., D. M. Tanenbaum, M. Jorgensen, M. Seeland, M. Baerenklau, M. Hermenau et al. 2012. Investigation of the degradation mechanisms of a variety of organic photovoltaic devices by combination of imaging techniques-the ISOS-3 inter-laboratory collaboration. Energy Environ. Sci. 5:6521–6540.

23. Roesch, R., F. C. Krebs, D. M. Tanenbaum, and H. Hoppe. 2012. Quality control of roll-to-roll processed polymer solar modules by complementary imaging methods. Sol. Energy Mater. Sol. Cells 97:176–180.

24. Straube, H., J.-M. Wagner, J. Schneider, and O. Breitenstein. 2011. Quantitative evaluation of loss mechanisms in thin film solar cells using lock-in thermography. J. Appl. Phys. 110:084513.

25. Breitenstein, O., W. Warta, and M. Langenkamp. 2010. Lock-in thermography, basics and use for evaluating electronic devices and materials, 2nd ed. Springer, Berlin.

26. Breitenstein, O.. 2012. Local efficiency analysis of solar cells based on lock-in thermography. Sol. Energy Mater. Sol. Cells 107:381–389.

27. Straube, H., and O. Breitenstein. 2011. Infrared lock-in thermography through glass substrates. Sol. Energy Mater. Sol. Cells 95:2768–2771.

28. Brosnan, T., and D.-W. Sun. 2004. Improving quality inspection of food products by computer vision – a review. J. Food Eng. 61:3–16.

29. Wu, W. Y., M. J. J. Wang, and C. M. Liu. 1996. Automated inspection of printed circuit boards through machine vision. Comput. Ind. 28:103–111.

30. Lin, C.-L., C.-W. Kuo, C.-C. Lai, M.-D. Tsai, Y.-C. Chang, and H.-Y. Cheng. 2011. A novel approach to fast noise reduction of infrared image. Infrared Phys. Technol. 54:1–9.

31. Zhu, B., X. Fan, Z. D. Cheng, D. Wang, Y. Q. Fang, and X. S. Chen. 2015. An IR strong clutter background suppression algorithm based on sparse kernel method. Tien Tzu Hsueh Pao/Acta Electron. Sin. 43:716–721.

32. Elguebaly, T., and N. Bouguil. 2013. Finite asymmetric generalized Gaussian mixture models learning for infrared object detection. Comput. Vis. Image Underst. 117:1659–1671.

33. Bankman, D. J., and T. M. Neighoff. 2008. Pattern recognition for detection of human heads in infrared images. Opt. Eng. 47:046404.

34. Han, J., A. Gaszczak, R. Maciol, S. E. Barnes, and T. P. Breckon. 2013. Human pose classification within the context of near-IR imagery tracking. Proc. SPIE Int. Soc. Opt. Eng. 8901:89010E.

35. Wang, X., G. Lv, and L. Xu. 2012. Infrared dim target detection based on visual attention. Infrared Phys. Technol. 55:513–521.

36. Vetter, A., J. Hepp, and C. J. Brabec. 2016. Automatized segmentation of photovoltaic modules in IR-images with extreme noise. Infrared Phys. Technol. 76:439–443.

37. Herley, C.. 2004. Efficient inscribing of noisy rectangular objects in scanned images. Proc. Int. Conf. Image Process. ICIP, 4:2399–2402.

38. Guerzhoy, M., and Z. Hui. 2008. Segmentation of rectangular objects lying on an unknown background in a small preview scan image. *Proceedings of the 5th Canadian Conference on Computer and Robot Vision, CRV 2008*, 369–375.

39. Kurilin, I. V., I. V. Safonov, M. N. Rychagov, H. Lee, and S. H. Kim. 2014. High-performance automatic cropping and deskew of multiple objects

on scanned images. Proc. SPIE Int. Soc. Opt. Eng., 9016:90160B.

40. Otsu, N. 1979. A threshold selection method from gray-level histograms. IEEE Trans. Syst. Man. Cybern. 9:62–66.

41. Konovalov, I. E., O. Breitenstein, and K. Iwig. 1997. Local current-voltage curves measured thermally (LIVT):

a new technique of characterizing PV cells. Sol. Energy Mater. Sol. Cells 48:53–60.

42. Fecher, F. W., A. P. Romero, C. J. Brabec, and C. Buerhop-Lutz. 2014. Influence of a shunt on the electrical behavior in thin film photovoltaic modules – a 2D finite element simulation study. Sol. Energy 105:494–504.

4

Parametric analysis and systems design of dynamic photovoltaic shading modules

Johannes Hofer[1], Abel Groenewolt[2], Prageeth Jayathissa[1], Zoltan Nagy[1] & Arno Schlueter[1]

[1]Architecture and Building Systems, Institute of Technology in Architecture, ETH Zurich, John-von-Neumann Weg 9, 8093 Zürich, Switzerland
[2]Institute for Computational Design, University of Stuttgart, Keplerstrasse 11, 70174 Stuttgart, Germany

Keywords
Building integrated photovoltaics, energy efficiency, partial shading, solar tracking, system design, thin film PV

Correspondence
Johannes Hofer, Architecture and Building Systems, Institute of Technology in Architecture, ETH Zurich, John-von-Neumann Weg 9, 8093 Zürich, Switzerland. E-mail: hofer@arch.ethz.ch

Funding Information
This research has been financially supported by CTI within the SCCER FEEB&D (CTI.2014.0119) and by the Building Technologies Accelerator program of Climate-KIC.

Abstract

Shading systems improve building energy performance and occupant comfort by controlling glare, natural lighting, and solar gain. Integrating PV (photovoltaics) in shading systems opens new opportunities for BIPV (building integrated photovoltaics) on façades. A key problem of such systems is mutual shading among PV modules as it can lead to electrical mismatch losses and overheating effects. In this work, we present a new modeling framework, which couples parametric 3D with high-resolution electrical modeling of thin-film PV modules to simulate electric energy yield of geometrically complex PV applications. The developed method is able to predict the shading pattern for individual PV modules with high spatio-temporal resolution, which is of great importance for electrical system design. The methodology is applied to evaluate the performance of different dynamic BIPV shading system configurations, as well as its sensitivity to façade orientation and module arrangement. The analysis shows, that there is a trade-off between tracking performance and mutual shading of modules. Distance between modules is a critical parameter influencing the amount of mutual shading and hence limiting solar irradiation and electricity generation of PV shading systems using solar tracking. Planning of module string configuration, PV cell orientation, and location of bypass diodes according to partial shading conditions, reduces electrical mismatch losses and results in significantly higher electricity generation. The integration of parametric 3D and electrical modeling opens new possibilities for PV system design and dynamic control optimization. Though the analysis focuses on BIPV, the method is useful for the planning and operation of solar tracking systems in general.

Introduction

In order to reduce fossil primary energy demand and its effects on the global climate, energy strategies aim at increasing efficiency along with the use of renewable sources. As the building sector contributes significantly to energy use and greenhouse gas emissions, the EU aims for all newly constructed buildings to have close to zero net energy consumption by 2020. The realization of this goal requires local energy harvesting, and the integration of PV (photovoltaic) systems in buildings is an obvious choice for many climatic regions. Due to strong cost reductions and technical developments of PV systems in recent years, there is a growing opportunity for integrating PV elements into buildings. Implementations of BIPV (building integrated photovoltaics) include rooftop installations, external building walls, or semitransparent façades. Even though façade PV systems receive less irradiation than rooftop and ground installations, they open new application possibilities for BIPV in the urban context and can substantially contribute to electricity generation, with higher production in winter and potentially lower diurnal and seasonal variations. Moreover, façade BIPV systems can replace conventional building materials, such as the shading system, thus avoiding related costs and environmental impacts [1].

Building shading systems are an important architectural element, simultaneously influencing building appearance, its energy performance, and occupant comfort. Shading systems can control glare, natural lighting and solar gain, thereby offering reductions in energy demand for heating and cooling. Architectural implementations range from static elements and traditional blinds to automated shading systems. Dynamic shading systems that are capable of reacting to their external environment can result in high-energy savings when compared to static shading systems [2]. The mechanics required to actuate such systems couples seamlessly with the mechanics required for façade integrated PV solar tracking. With the increased use in large window openings and curtain walls in today's architecture, there is a growing opportunity to integrate PV elements into such shading systems [3–7].

A key problem with many PV installations, in particular for tracking systems and tightly integrated building applications, is shading among PV modules and by objects in the environment. Module shading can lead to electrical mismatch losses and overheating effects, thereby affecting power output and system reliability [8, 9]. Possible proposed solutions to reduce these adverse effects include optimized PV module distribution and tracker control to circumvent shading [10–13] or electrical configurations of a PV array to minimize power losses, such as the inclusion of bypass diodes or series/parallel circuit designs [8, 14–16].

Several approaches to model the effect of shading on PV systems power output have been proposed in the literature [8, 17–28]. Only few of the developed models have a resolution high enough to capture the electrical effects of small area shadows, such as those from antennas, chimneys, or other PV modules [28]. In addition, state-of-the-art modeling tools for PV systems simulation consider shading only for simplified conditions, that is, flat, rectangular PV modules, uniform module positioning, and regular building geometries and environments. Some recent studies use architectural 3D modeling software such as AutoCAD or SketchUp for the calculation of PV module shading and combine it with PV electrical modeling at cell level resolution [28–30]. Such a 3D modeling framework provides much higher flexibility than conventional tools in defining arbitrarily shaped objects, realistic building environments, and elevation profiles. The approach is very promising for integrated architectural and technical planning of BIPV systems and provides an opportunity for the development of innovative BIPV solutions.

In this work, we combine for the first time a high-resolution shading analysis performed with the Rhinoceros 3D software [31] and the parametric Grasshopper plugin [32] with electrical modeling of thin-film PV modules at subcell level resolution. We employ the model to analyze mutual shading, solar insolation, and electric energy yield of dynamically actuated PV modules in a building

environment. The method we have developed can accurately predict the shading pattern for individual modules with high spatial resolution. The geometrical simulation is coupled with irradiation data and applied to assess solar insolation for various dynamic shading system configurations, including one-axis actuated horizontal and vertical louver blinds, as well as more complex geometries, such as two-axis actuated rectangular modules in a diamond pattern. The irradiation and shading simulation is coupled with a high-resolution electrical model of monolithic thin-film PV modules to evaluate the effect of different electrical configurations on array efficiency.

We begin in Methodology by describing the framework and methodology used for the parametric 3D and electrical simulations. In Solar Insolation Analysis, we analyze module shading and resulting solar insolation as a function of tracking system type, façade orientation, and module distance. In Electrical Performance Analysis, we discuss the electrical system design and performance as a function of relevant parameters for the different shading system configurations analyzed. In Experimental Implementations, we show experimental measurement results of thin-film PV module performance in partial shading conditions and progress leading to building integration. In Discussion and Conclusions, we summarize the results and discuss limitations and possible advancements of this work.

Methodology

General approach

In order to study both the appearance and solar insolation on an array of dynamic PV modules attached to the building envelope, we have developed a framework for the parametric 3D design and calculation of module shading as well as solar irradiance at high resolution. The results are analyzed as a function of different input parameters, such as module arrangement and tracking control. For further analysis of the electrical performance, the shading pattern and solar irradiance are coupled with an electrical model which calculates the characteristic current–voltage (I-V) curves of PV modules as a function of time. On the basis of this, we evaluate electric energy yield for different module string interconnections and other electrical design parameters. The general workflow is illustrated in Figure 1.

The developed modeling framework is versatile and principally any configuration in terms of module shape, arrangement, and motion control can be simulated. For practical relevance we focus in this study on a rectangular PV module design, three types of façade-attached shading system types (Fig. 2A), and one or two-axis solar tracking. Parameters varied include façade orientation, distance between modules,

Input parameters

Site & Building Weather data   **PV system geometry Motion control Electrical design**

Parametric 3D model

Module positioning and shading calculation using Rhinoceros / Grasshopper

Solar radiation analysis

Electric model

Simulation and design of thin-film PV modules and array configuration in Matlab

Electrical performance analysis

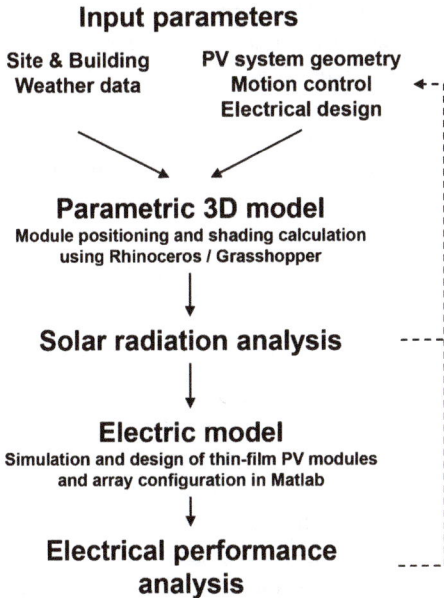

Figure 1. Modeling framework for illustration of the workflow. The dashed line represents ways to optimize geometry, motion control, and electrical design parameters based on system performance.

as well as electrical design parameters. The goal of the simulation is to analyze solar insolation for complex dynamic shading situations and to find the best electrical layout for a given shading system configuration and control strategy. Generally, the integration of dynamic 3D and electrical modeling allows to further optimize module shape, arrangement, and control, minimizing mutual shading and maximizing electric energy yield. In the following, the methods of the parametric 3D and electrical model are explained.

Parametric 3D model

Module positioning

The workflow for the parametric 3D design of shading systems has been implemented in the Rhinoceros 3D software [31] using the Grasshopper plugin [32]. The resulting designs consists of multiple rectangular surfaces that represent PV modules, each of which can be rotated in the solar azimuth and/or the altitude direction, within a specified range of angles. The number of modules and their positions are controlled by indicating coordinates

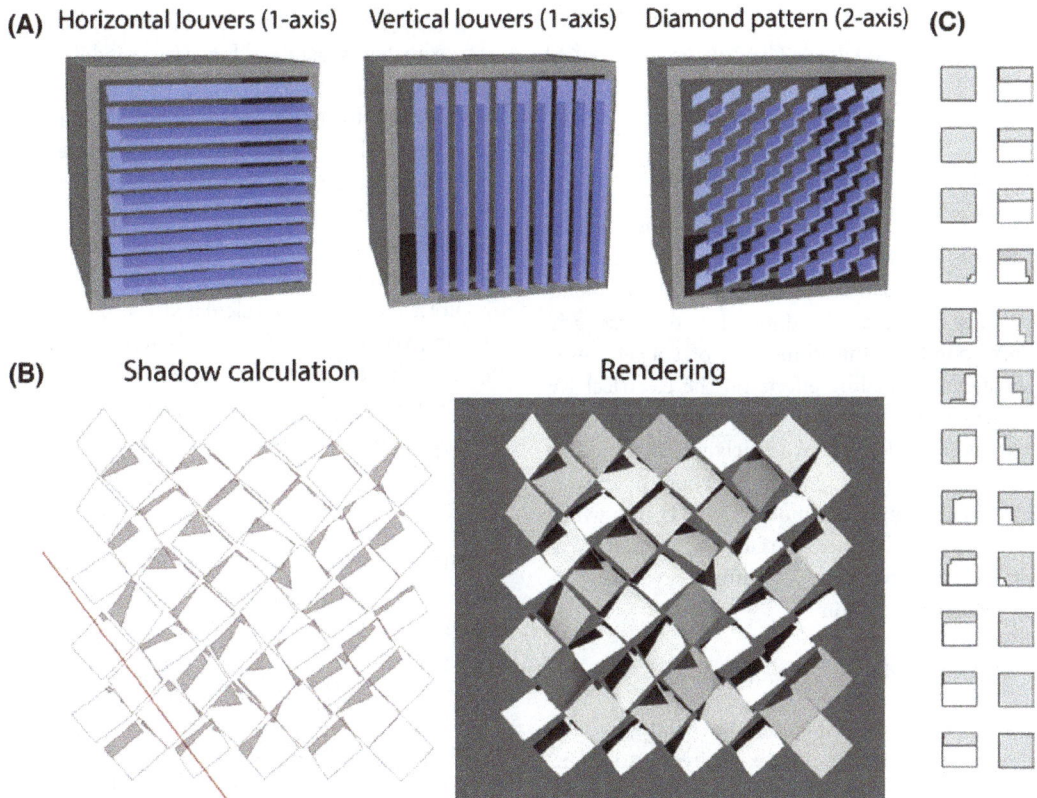

(A) Horizontal louvers (1-axis)   Vertical louvers (1-axis)   Diamond pattern (2-axis)   **(C)**

(B)   Shadow calculation   Rendering

Figure 2. (A) Renderings of analyzed module configurations and mutual shading for a specific sun position. Depending on the system, the modules orient toward the sun using one or two-axis tracking. (B) Screenshot of the shading calculation performed in Rhino-Grasshopper for arbitrary module positions. The red line indicates the direction of the sun. For comparison a rendering of the 3D model is shown on the right. (C) Projection of shadows for a single module over the course of a day; gray represents the shaded area.

in 3D space at which the modules should be placed. This is done by specifying parameters that generate a rectangular or diamond pattern grid.

To exemplify, we generate and simulate different horizontal and vertical louver shading systems with single-axis solar tracking and a diamond configuration with dual-axis tracking in front of a room with a dimension of 3.6 m width and 2.8 m height (see Fig. 2A). The size of individual PV modules is fixed at 40 by 40 cm. In the reference case, the distance between louver blinds and modules in the diamond pattern is fixed at 10% of the module dimension, resulting in seven blinds with nine modules in series for the horizontal and vice versa for the vertical louver system. The diamond configuration is composed of 50 individual modules.

In our framework, one-axis tracking is simulated by rotating the modules in the direction of the sun along a horizontal or vertical axis. The center of rotation lies at an adjustable distance behind the module plane, so that actual mechanical rotation systems can be simulated. The orientation of the modules is defined by aligning the module's normal vector to the projection of the sun's direction on the plane perpendicular to the module's axis of rotation. For two-axis solar tracking, the modules orient exactly perpendicularly to the sun direction. Horizontal (azimuth) and vertical (altitude) tracking angles can be constrained to any angle range, so that tracking systems with physical limits on the range of rotation can be simulated.

Calculation of shading and solar irradiation

For each module, the calculation of solar irradiation is simulated at a resolution of 50 by 50 grid points, corresponding to an irradiance matrix with 2500 elements per module and instance. The dimension of these grid elements corresponds to the dimension of subcells used for the assessment of shading effects on the electrical PV module performance.

Solar irradiation is calculated separately for each module m and grid point p ($G_{tot,m,p}$) as a function of time t based on the three-component model [33, 34], for which the radiation on an inclined surface is calculated as the sum of direct beam ($G_{dir,m,p}$), diffuse sky ($G_{dif,m,p}$), and diffuse reflected radiation ($G_{ref,m,p}$) as

$$G_{tot,m,p}(t) = G_{dir,m,p}(t) + G_{dif,m,p}(t) + G_{ref,m,p}(t). \qquad (1)$$

To assess these components, the module shading fraction or detailed shape (depending on the desired resolution) as well as the amount of radiation coming from the sky versus radiation from the ground needs to be known.

Within the parametric 3D model, the solar position is generated using the DIVA plugin [35], which provides vectors in the direction of the sun for any location on

earth at any moment in time, based on the method described in [36]. The location for our analysis is Zurich, Switzerland (latitude 47.37°N, longitude 8.55°E). The sun direction vectors are used for shadow calculations and solar tracking. The shape of shadows on the modules is calculated based on vector algebra, using the sun light direction, panel geometry, and optionally external geometry as input. First, the module's normal vector is compared to the sun direction to check if sunlight could be received by the module. If so, the input geometry is intersected with the module's plane; the geometry that is located behind the module's plane logically does not generate shadows on the module's front side. The remaining geometry is then projected on the module plane using the method described in [37]. To verify this method, the resulting shadows have been compared with those generated by a physically based rendering engine [38], which is based on [39] (Fig. 2B). By generating sequences of sun positions at fixed time intervals, the geometric behavior of a tracking system can be studied. For any moment in time, shading on the panels (including mutual shading) can be studied both numerically and graphically (Fig. 2C). As the shadow shapes are generated as vector graphics, shading grids at any resolution can be quickly exported.

The sky view factor (or diffuse radiation factor) is approximated separately for each module by creating a hemispherical mesh, connecting the midpoints of the faces of this mesh to the midpoint of the module and checking these connections for intersections with panel geometry and external geometry. The area of the faces belonging to lines that are not intersected with any geometry are then added. Dividing the sum of this addition by the total surface area of the hemisphere results in the approximated view factor.

Direct radiation is calculated as the product of the shading vector ($SV_{m,p}$), which is zero for shaded and one for not shaded areas, the projection of the module area on its normal plane relative to incident solar radiation ($NP_{m,p}$), and direct normal irradiance (G_{dni})

$$G_{dir,m,p}(t) = SV_{m,p}(t) \cdot NP_{m,p}(t) \cdot G_{dni}(t). \qquad (2)$$

Diffuse sky radiation is the product of the sky view factor ($SF_{m,p}$) and diffuse irradiance on the horizontal plane ($G_{dif,hor}$)

$$G_{dif,m,p}(t) = SF_{m,p}(t) \cdot G_{dif,hor}(t). \qquad (3)$$

Diffuse reflected radiation is the product of the fraction of ground light irradiance reflected toward the module, the albedo of reflecting objects (a) and global irradiance on the horizontal plane ($G_{glo,hor}$)

$$G_{ref,m,p}(t) = (1 - SF_{m,p}(t)) \cdot a \cdot G_{glo,hor}(t). \qquad (4)$$

In Eq. 4 it is assumed, that the sum of the fraction of light reflected from the ground to the module and the sky view factor for this module equals one. Note that currently only a single albedo value is used, not distinguishing between different reflecting objects.

The irradiation data used in this work was exported from the Meteonorm software [40] and is based on long-term measurements of a MeteoSwiss weather station in Zurich, Switzerland (Zurich SMA, Fluntern). The total annual global insolation on a horizontal plane at this location is 1120 kWh/m^2. For the analysis of solar insolation presented in Section 3, hourly solar irradiance is averaged per month and only one average day per month is simulated. PV electricity generation is strongly dependent on the specific shading and insolation condition. Therefore, the full hourly resolution was used as input data for the analysis of electric performance in Section 4. The weather data is shown in Figure 3.

While the time resolution of shadow calculations is determined by the time step set in the sun positioning algorithm (which in the DIVA plugin can be set to any value), the time resolution of irradiation on the modules is determined by solar irradiance input (one hour for the used weather data).

Electrical model

One main reason for energy generation losses of BIPV systems is partial shading of PV modules due to surrounding buildings, trees, or other objects [16]. Another reason, particularly for PV integrated in dynamic shading systems, is mutual shading between PV modules. Varying insolation due to partial shading of PV modules leads to different I-V characteristics of PV cells and consequently to electrical mismatch within the module. Furthermore, modules exposed to different shading conditions lead to mismatch and power losses in the whole PV array [8, 41], which can cause hot-spot heating and damage of cell encapsulation materials affecting system reliability [8, 18, 19].

Modeling the effect of shading on PV systems power output has been extensively investigated in the literature [8, 17–28]. The impact of different cell shapes and orientations on the I-V characteristics of monolithic thin-film PV modules can be assessed with a two-dimensional SPICE circuit simulation [18, 41, 42]. We employ a similar model implemented in MATLAB. Figure 4A–D illustrate different levels of this model from array to module to subcell. In this two-dimensional simulation, each module consists of 50 cells in series and each cell is subdivided into 50 parallel connected subcells. The subcell I-V curve is modeled based on the standard equivalent circuit model with a single diode, one series and one shunt resistance [43]. For the parameters of the model we use values of a commercial CIGS (copper indium gallium selenide) module [44, 45] with an exponential increase in the shunt resistance at low irradiance [43, 45]. In case of partial shading, some of the cells are in reverse bias and act as a load. In order to account for this effect, the reverse characteristic of cells is modeled according to [46, 47], similar to the measured characteristic for CIGS solar cells [48]. The module I-V curve is calculated by summing the current of parallel connected subcells and subsequently the voltage of series connected cells. Since the module shading matrix and the network of subcells within a module have the same spatial resolution, both models can be directly coupled to simulate module electric

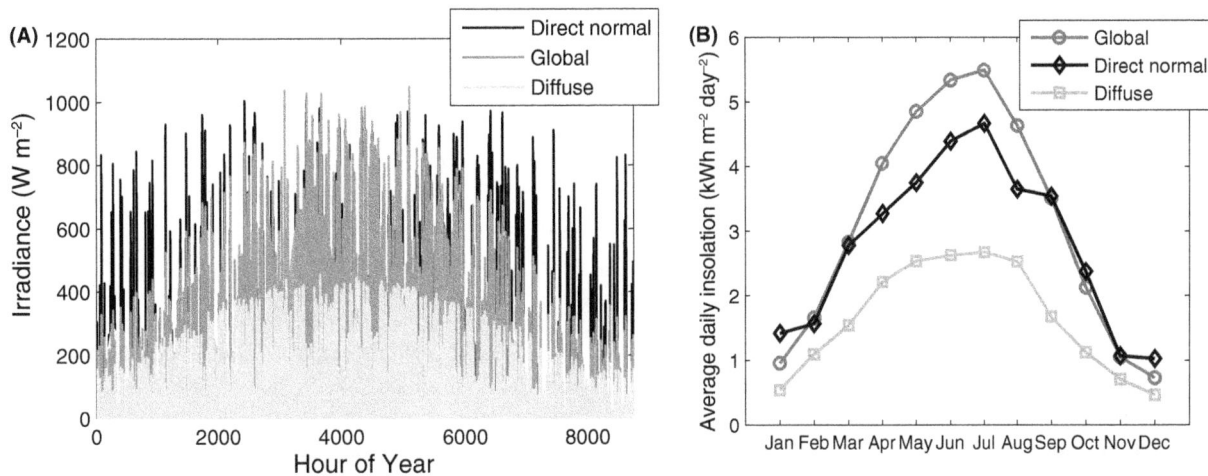

Figure 3. (A) Direct normal, diffuse horizontal, and global horizontal irradiance over 1 year at the reference location with hourly resolution. (B) Average daily insolation per month.

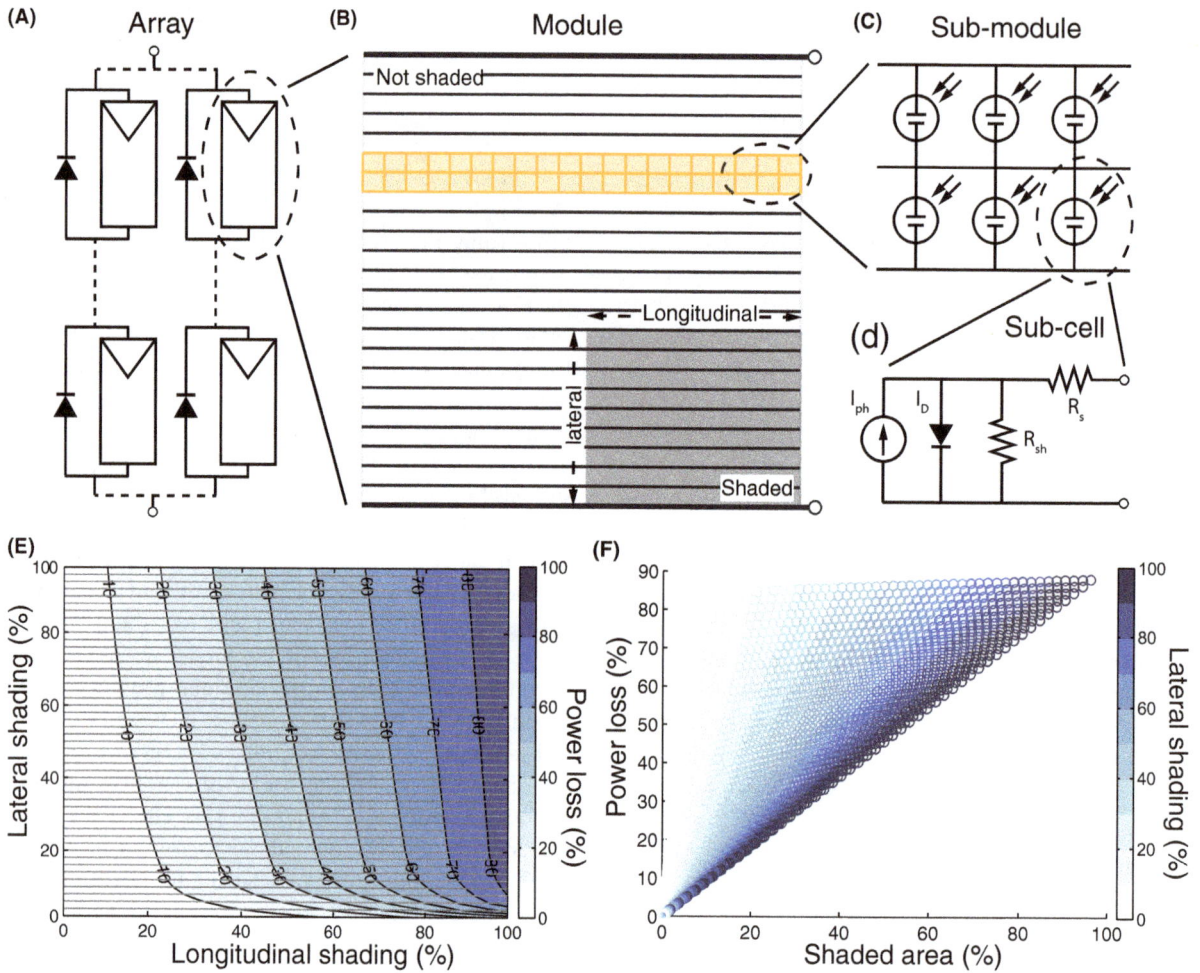

Figure 4. (A) Array of several PV modules with external bypass diodes in series-parallel connection. (B) Typical thin-film PV module with long rectangular cells connected in series. The module is partially shaded. (C) The module is modeled as a 2D grid of parallel-series connected subcells. (D) Each subcell is modeled based on the standard equivalent circuit. (E) Simulation of power loss as a function of lateral and longitudinal shading. In this case, the irradiance in the shaded region is reduced by ca. 80% relative to the full irradiance in the unshaded region. Horizontal lines indicate the orientation of PV cells. (F) Power loss as a function of shaded area. Color scale indicates the amount of lateral shading. Power loss is linear at 100% lateral shading, but deviates significantly at small lateral shading.

performance for a specific shading pattern and solar irradiance. Figure 4E and F show the simulated module power loss at the MPP (maximum power point) as a function of shading parallel (longitudinal) and perpendicular (lateral) to cell elongation.

Note that different from [18, 41, 42] contact sheet resistances between individual subcells have been neglected. Another limitation of this work includes the assumption of homogenous temperature distribution of PV cells in shaded and unshaded regions. Previous studies analyzed the feedback of power dissipation on module temperature distribution and showed that due to electrical mismatch from partial shading, temperature differences between cells in shaded and unshaded regions occur [9, 18]. Considering such electro-thermal effects is important for the prediction

of hot-spots, potentially leading to PV module degradation and permanent damage [9]. Infrared thermal imaging of partially shaded PV modules analyzed in this study has shown, that module temperature distribution is dominated by direct light absorption. The model we developed offers subcell temperature resolution, however, further analysis and experimental validation will be required to accurately account for module temperature distribution based on both effects, direct light absorption and electrical dissipation due to cell mismatch. With the current model setup we observe good agreement between measured and simulated data and consider both simplifications to be of minor influence on the general conclusions of this study.

The array I-V curve is simulated by summing the voltage of modules connected in series strings and then the current of strings connected in parallel. Usually modules have integrated bypass diodes that conduct in case of reverse bias due to shading and limit the voltage drop over a module to the voltage at which the diode starts to conduct (ca. 0.6 V for silicon diodes). In this work, we consider three cases for the placement of bypass diodes, which are detailed in Section 4. Note that due to manufacturing tolerances, varying degradation or dirt, PV modules exhibit slightly different current–voltage characteristics. Mismatch losses induced by such phenomena are neglected in the analysis.

Solar Insolation Analysis

Sun-tracking ability, module shading, and consequently solar insolation on PV modules varies significantly as a function of system configuration parameters, such as motional degree of freedom (one/two-axis tracking), façade orientation, and module spacing. In this section, we first analyze solar tracking performance and mutual shading for different shading system configurations using one and two-axis solar tracking and then present results of the average daily and monthly distribution of solar insolation for various façade orientations and module distances.

Figure 5A depicts the projection of module area on its normal plane relative to incident solar radiation over the course of a day for each month of the year and Figure 5B the corresponding visible (not shaded) module area. These two variables are important as they are proportional to direct irradiation on the module surface (see Eq. 2). The values are averaged over all modules of a south facing façade. The simulation is done for one and two-axis tracking systems with 10% module spacing. For comparison also a planar system without tracking ability is shown. As expected, for a fixed plane the area normal to sun radiation exhibits strong seasonal and daily variations. It approaches zero in the morning and evening and is lower in summer than in winter. Due to solar altitude tracking, horizontal louvers have an improved daily performance and balance out seasonal fluctuations. Vertical louvers with azimuth tracking orient toward the sun in the morning and evening, but have a minimum at solar noon as they cannot track solar altitude. The two-axis tracking system completely cancels out seasonal and daily fluctuations. The improvements of tracking systems orienting the module toward the sun are countered by increased mutual shading (Fig. 5B). While for a static planar system there is no mutual shading at all, the effects of shading increase using single-axis tracking and even further with dual-axis

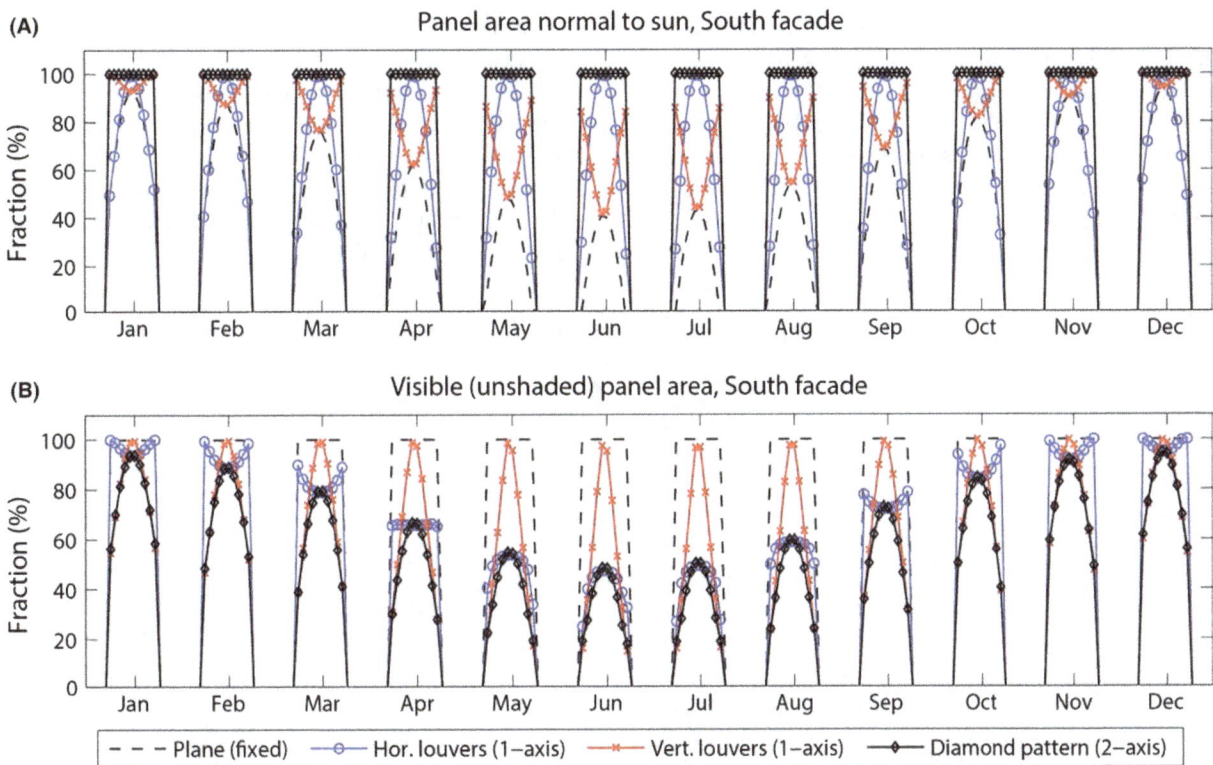

Figure 5. (A) Fraction of the module area oriented normally to the sun over the course of a day for four different solar tracking systems on a south facing façade. (B) Visible, not shaded module area.

tracking. For all systems we observe a trade-off between module mutual shading and solar tracking performance in terms of projected normal area, in particular for narrow spaced module arrangements.

The results of module tracking performance and mutual shading serve as inputs to the calculation of daily and monthly solar insolation shown for a horizontal louver system using solar altitude tracking in Figure 6. The irradiance has a different daily profile depending on façade orientation and season. East, west, or south oriented facades receive most sunlight in the morning, evening or at noon, respectively. As such, daily variations in solar insolation can be partly balanced by using PV modules for different façade orientations. Interestingly for this tracking system on a south facing façade, direct insolation in winter is relatively high but drops during summer months due to mutual shading. This leads to relatively low seasonal variation in solar insolation. The effect is less influential for east or west oriented facades, which exhibit less mutual shading in summer and have strong seasonal variations.

Mutual shading and solar insolation have been evaluated for different façade orientations as a function of the distance between modules, that is, the distance between rows or columns of single-axis or individual modules of dual-axis tracking systems in percent of the module dimension. Figure 7A shows the decrease in mutual shading with increasing module distance for a south and east facing façade. Module shading fraction is calculated as the average amount of mutual shading of all modules within the façade assembly at times, when there is direct light irradiation on the respective façade. For a south facing façade, the mutual shading fraction is high for all systems in summer and low in winter. This trend can be also compared to the results shown Figure 5B. The variation with module distance is smallest for the vertical

and horizontal louver system in summer and winter, respectively. For the east oriented façade, shading is high in winter, particularly for the vertical louver and diamond configuration tracking system. This is because an east facing façade receives direct light in winter only for a short amount of time, during which these two systems generate high mutual shading. The horizontal louver tracking system is highly shaded in summer and less in winter. The dependence is opposite for the vertical louver system.

The amount of shading and its reduction with increasing module distance lead to a variation in insolation per module area as shown in Figure 7B. At small module distance, the absolute amount of insolation is similar for all systems, but at a higher level for the east façade in summer and for the south façade in winter, respectively. In turn, the variation in insolation is higher for the south façade in summer and for the east façade in winter. Overall the increase in insolation with module distance is highest for the two-axis tracking system.

Further analysis not shown in Figure 7 reveals, that even though the insolation per module area increases with increasing module distance, the absolute amount of insolation on all modules of the façade decreases. This is because absolute module area decreases more rapidly than relative insolation increases. Note that an increase in the module spacing corresponds to less reduction in module area for the one-axis than for the two-axis tracking system, due to a variation in the module distance in only one instead of two dimensions. For conventional ground-mounted PV tracking systems [49–51], the ratio of module relative to ground area is known as the ground cover ratio and the trade-off between land occupation and energy yield is discussed in [10, 52]. Comparing the performance of the different tracking systems analyzed

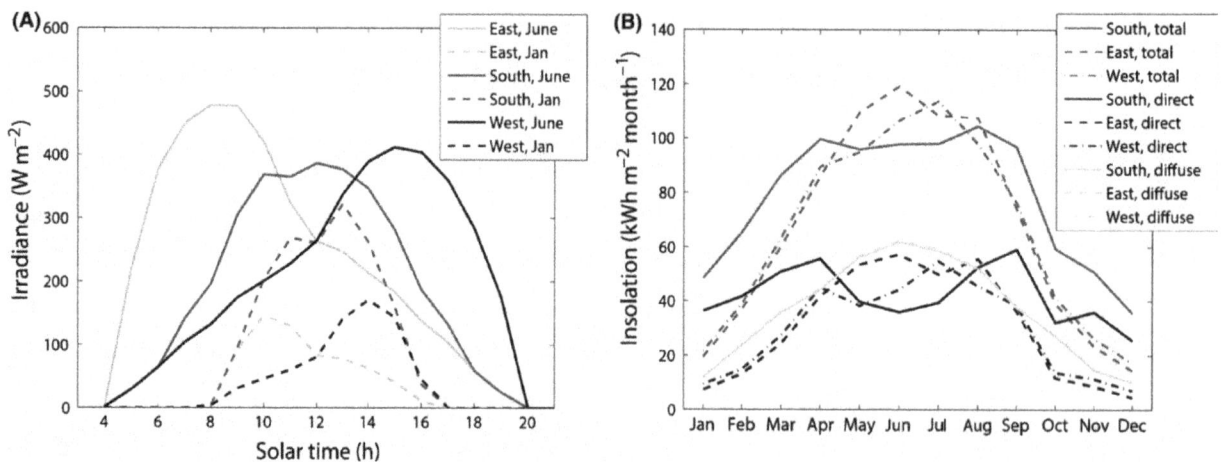

Figure 6. (A) Average daily solar irradiance per module area for a horizontal louver solar tracking system in June and January at the location of Zurich. (B) Average monthly insolation per module area by façade orientation.

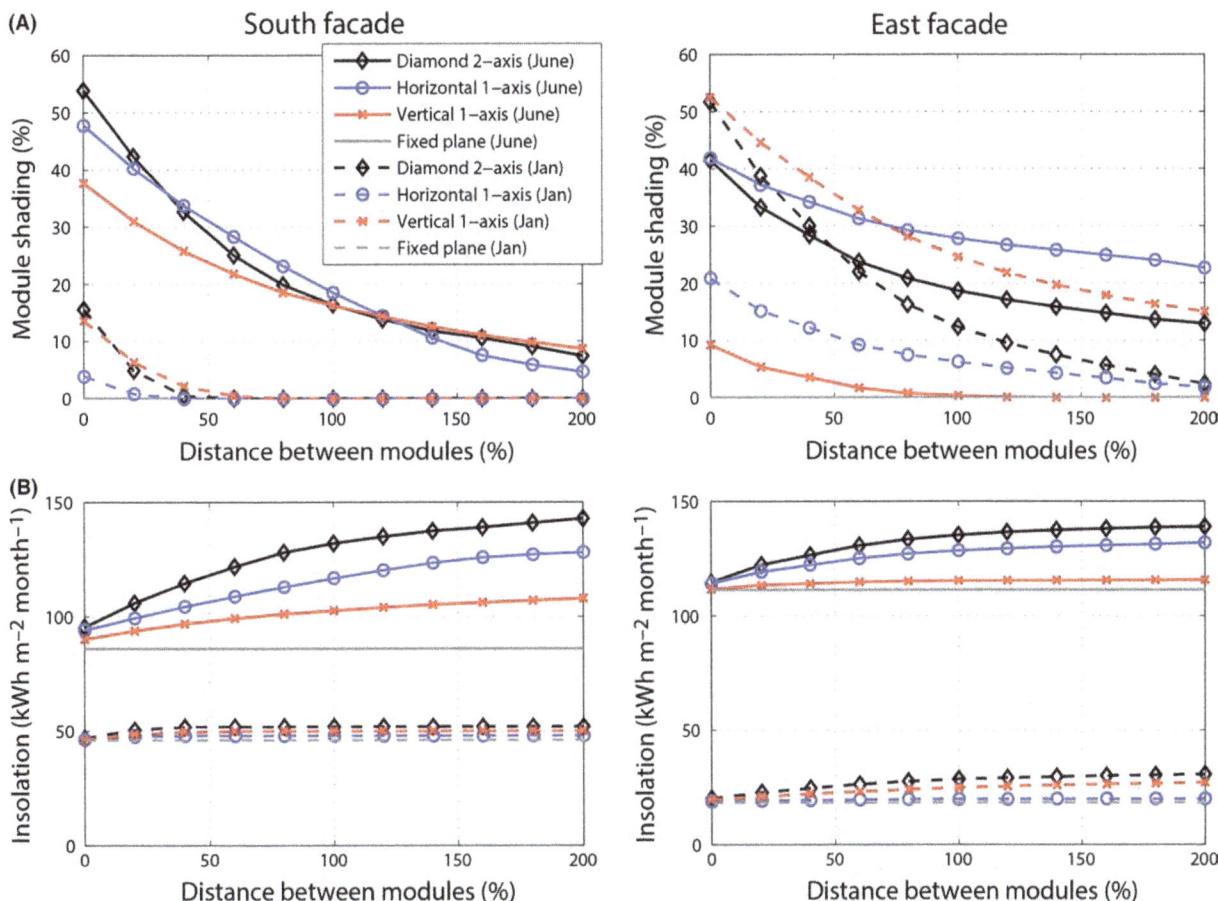

Figure 7. (A) Average module mutual shading and (B) insolation per module area as a function of the distance between modules for south (left) and east (right) oriented façade shading systems in June and January.

in this work relative to the ground cover ratio instead of the module distance, shows that the horizontal louver one-axis and diamond two-axis tracking systems achieve the highest insolation per façade area in summer. In winter, the best performance is reached by vertical louver and two-axis tracking systems. On an annual basis, the two axis-tracking system reaches highest insolation per façade area.

Electrical Performance Analysis

Knowledge of the irradiance per module can be used to design electrical configurations that reduce power losses. Due to mutual shading, the irradiance per module can vary significantly depending on the shading system type and the location of the module within the façade assembly. In this section, we first evaluate the shading patterns by configuration for a south oriented facade shading system. On the basis of this, we analyze suitable electrical designs and present simulation results of power production.

Shading analysis and system design

In shaded conditions the photocurrent produced by a PV module decreases significantly. Because series connected PV modules are forced to operate at the same current, shading of a single module within a string of modules connected in series limits the current of other not shaded modules. On the hand, parallel-connected modules are forced to operate at the same voltage. Since the voltage decrease during partial shading is comparably low, parallel connection of modules leads to lower mismatch and power losses in partially shaded conditions than series connection [8, 16, 19]. There is, however, a limit to the degree of parallelism, because currents in parallel-connected modules, and hence resistive (ohmic) losses, are higher. In addition, wiring is more complex for parallel connected modules and usually for the inverter a certain minimum input voltage is required. For this reason, modules are often connected in series-parallel arrays by first forming series connection of modules in strings and then connecting those strings in parallel. In

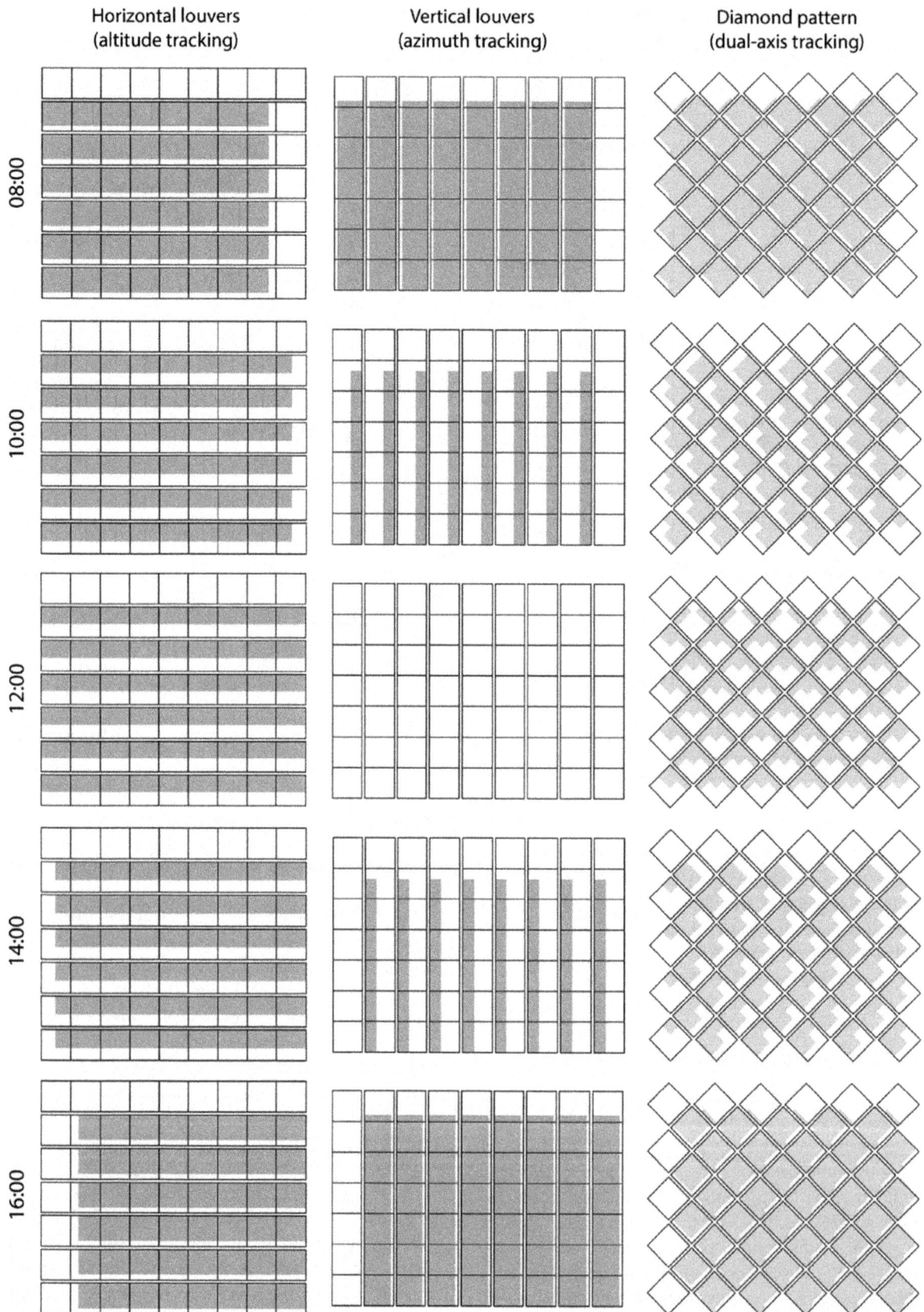

Figure 8. Projected mutual shading patterns over one clear day in June for a south oriented façade.

order for these strings to work most efficiently, the current at the maximum power point should be similar for PV modules connected in a string. For the analyzed PV shading systems, we therefore connect modules with similar shading pattern and solar irradiance over the course of a day in series strings. In particular, we try to avoid the situation that only one or few modules within a string are shaded, while others are fully illuminated. Figure 8 depicts calculated mutual shading among modules of a south oriented façade for the three shading systems analyzed.

The figure shows the 2D projection of mutual shading for five instances from morning to afternoon of a clear day in June. For horizontal and vertical louver systems, shadows follow the direction of the axis of rotation. Therefore, it is straightforward to connect modules within louver blinds in series, which also facilitates wiring. For the diamond configuration, the shading pattern is more complex. Generally, modules located in the top and subsequent row, the left and right column, as well as all central modules receive similar irradiance. For the analyzed arrangement of 50 modules, it is, however, impossible to find string connections of more than two modules exactly matching this shading pattern. For this reason, we constrain the analysis to two cases of either five or ten modules in series as shown in Figure 9. Note that configuration B is unsymmetrical and favors the shading situation in the morning. The string connections shown

in Figure 9 will be analyzed further in the next section.

Electrical performance of different configurations

In this section, we analyze the electrical performance in terms of array I-V characteristic and maximum power as a function of relevant electrical design parameters. In addition to possible string connections shown in Figure 9, these parameters include PV cell orientation and the placement of bypass diodes. For each string configuration, we model three cases for the placement of bypass diodes: (1) none; (2) one bypass diode in parallel to the entire 50 cells of a module; (3) two submodules of 25 cells in series, each with one bypass diode in parallel. Furthermore, we study the effect of a rotation of cell orientation by 90°.

Figure 10 shows the resulting maximum array power per module area for one clear day in June. For the horizontal louver system, vertical cell orientation (indicated as 90°) leads to significantly higher electricity generation than horizontal cell orientation. This can be explained by shading of the upper part of most modules (Fig. 8) and the stronger decrease in module MPP for lateral than longitudinal shading (Fig. 4E and F). In contrast, electricity generation of the vertical louver system, is higher for horizontal cell orientation. This can be explained by the

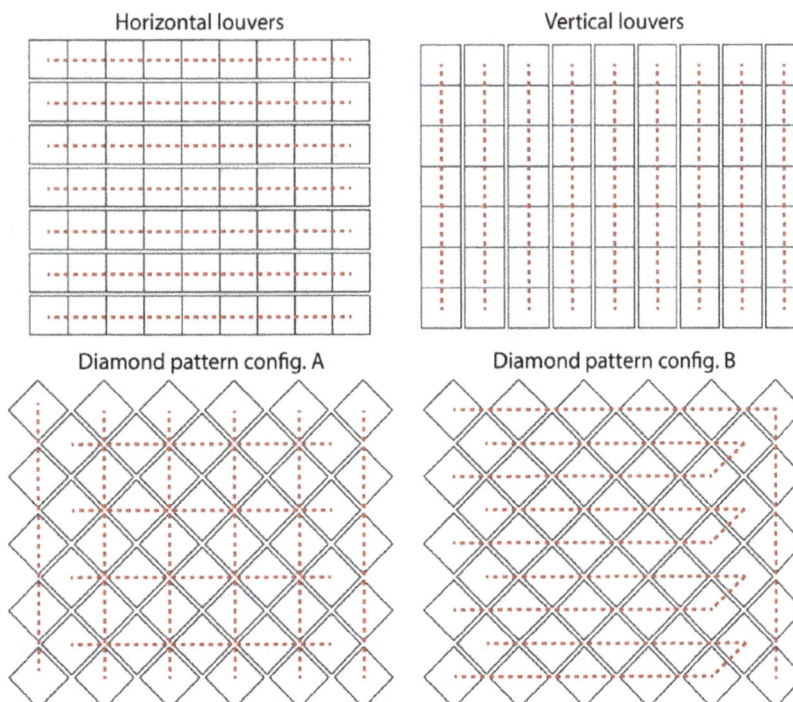

Figure 9. Possible string connection of modules based on shading analysis.

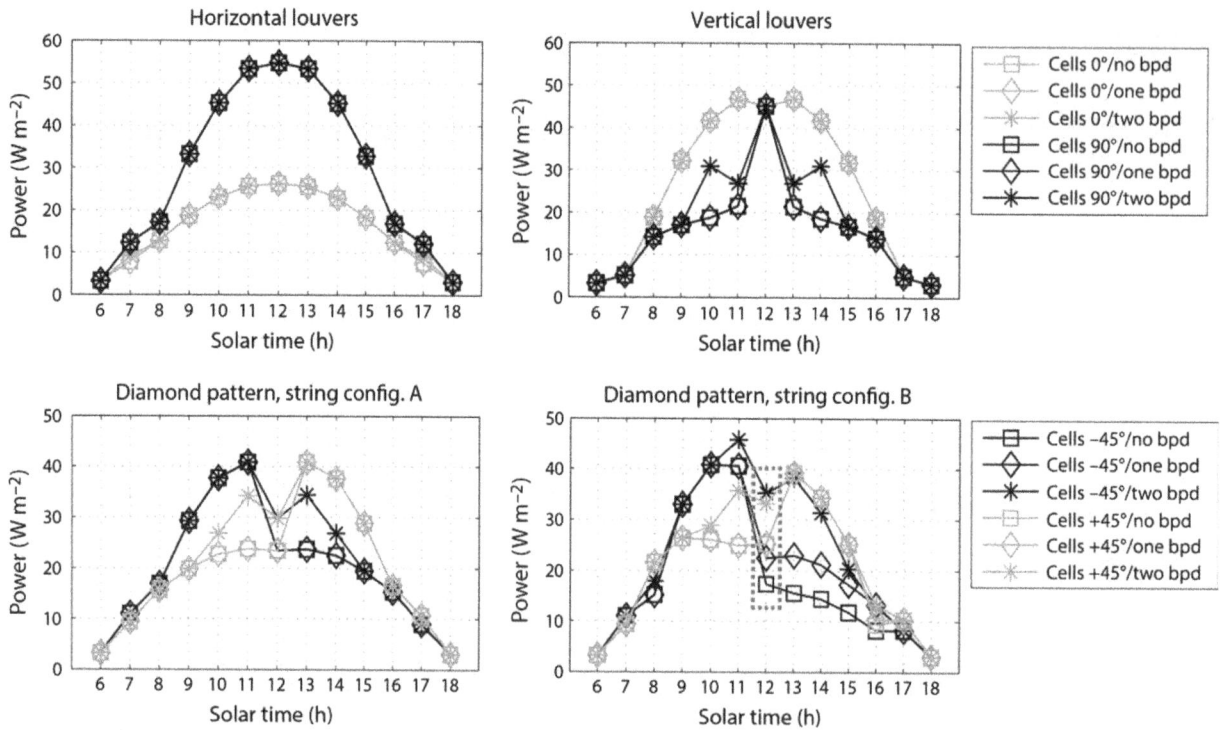

Figure 10. Maximum power per module area during a clear day in June. For each configuration, two different cell orientations and three options for the placement of bypass diodes (bpd) are simulated. The highlighted area is analyzed further in Figure 11.

inverted shading pattern. At solar noon, cell orientation is irrelevant for the vertical louver system, because modules are not shaded (Fig. 8). For the diamond pattern system in both string configurations, −45° cell orientation leads to higher electricity generation in the morning and +45° in the afternoon. Our analysis shows, that power losses caused by differing morning and evening shading conditions, can be reduced by integration of two bypass diodes per module. While string configuration A leads to a symmetrical electricity generation profile for different cell orientations, it is asymmetric for configuration B due to the nonuniform string design (Fig. 9).

The results shown in Figure 10 are on a highly aggregate level. For many applications it is of interest to know the exact I-V characteristic of modules, strings, and array, e.g. for the design of optimal electrical interconnections [14–16] and MPPT (maximum power point tracking) methods [30, 53–55]. Figure 11A shows the I-V curves of all 50 modules for one specific instance in Figure 10 (diamond configuration, string layout B, at 12o'clock). With one integrated bypass diode, reverse bias losses are limited, because the diode conducts below certain a threshold voltage. With two bypass diodes, module I-V curves exhibit a stepwise profile for some shading conditions. The integration of module bypass diodes leads to multiple steps for the I-V curves of strings (Fig. 11B)

and arrays (Fig. 11C), as well as multiple peaks in the P-V (power-voltage) characteristic of arrays.

Varying shading conditions and solar irradiation influence the relative performance of the configurations analyzed in this section. In an analysis of monthly electricity generation of those different configurations, we find that the order from lowest to highest electric yield remains the same as the comparison for a single clear day in Figure 10 suggests. However, the absolute difference varies significantly throughout the year. As an example, Figure 12 shows monthly electricity production of the horizontal louver tracking system for horizontal and vertical PV cell orientation. Results are based on hourly weather data and shading simulations. The difference in electric energy yield between horizontal and vertical PV cell orientation is low from November to January and high from March to October. The main reason for this trend is a decrease in module shading during winter (Fig. 7A).

System comparison

In this section, we compare the performance of the three PV shading system configurations analyzed (cf. Fig. 2A) with regard to solar insolation, system efficiency, and electric energy yield. As can be seen from Figure 7, raising

(A)

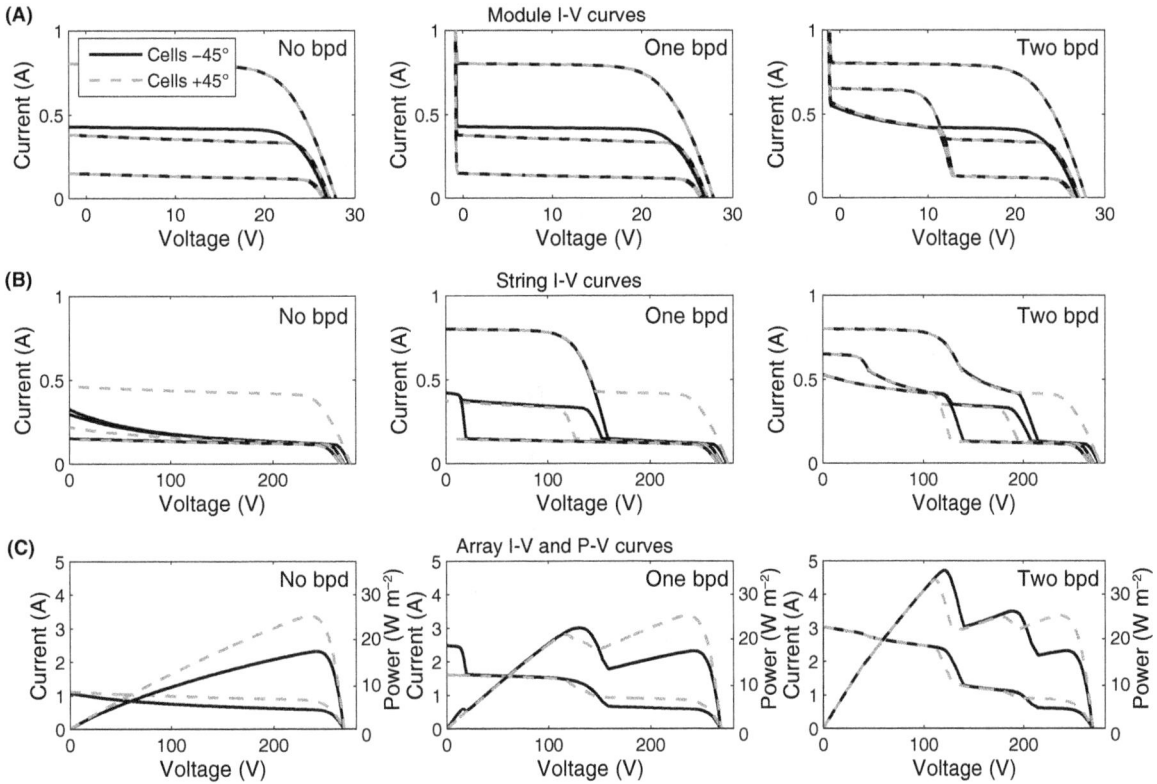

Figure 11. (A) I-V curves of individual modules for one specific instance shown in Figure 10 (diamond configuration, string layout B, at 12 o'clock). The diagram shows the I-V curves of all 50 modules (mostly overlapping) for either −45° or +45° cell orientation. From left to right, the number of bypass diodes per module increases from zero to two. (B) I-V curves of different strings with 10 modules in series. (C) Resulting array I-V (left) and P-V curves (right axis).

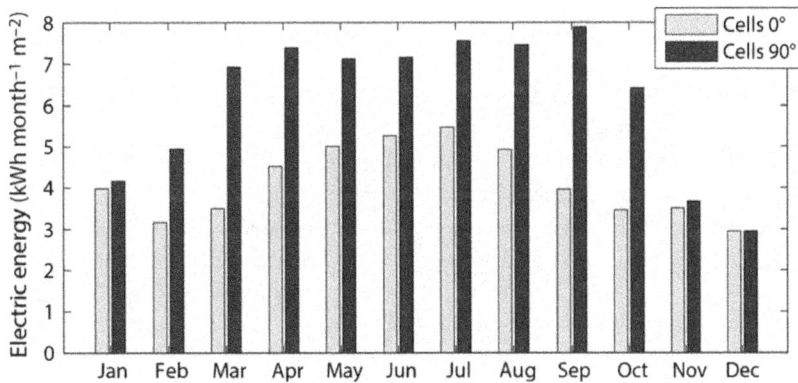

Figure 12. Monthly electricity generation of horizontal louver system for horizontal and vertical PV cell orientations with one integrated bypass diode per module.

the distance between PV modules reduces mutual shading and increases solar irradiance on the module. In addition, reducing the amount of shading leads to less mismatch between cells and improves the electrical conversion efficiency. On the other hand, increasing the distance between modules reduces the FCR (façade cover ratio). The FCR is a measure of the density of PV modules on the façade and defined as the PV module area relative to the total area of the façade. Based on the FCR, the performance of different system configurations can be compared at the same module density. Figure 13A shows the relation between number of modules, FCR and distance between

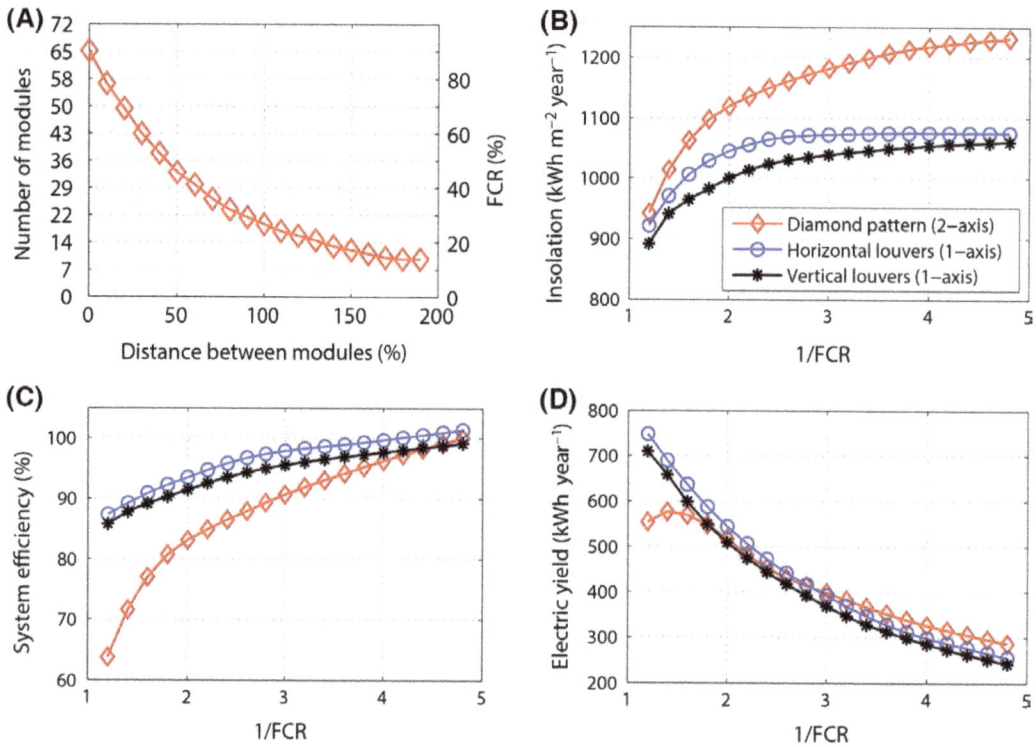

Figure 13. (A) Number of modules in the diamond pattern configuration and façade cover ratio as a function the distance between modules. (B) Annual solar insolation per module area as a function of the inverse façade cover ratio for three different system configurations. (C) Corresponding normalized electrical efficiency and (D) total electric yield of the façade shading systems.

modules for the diamond pattern configuration. In the analysis, the FCR is reduced from ca. 90% to 15% by varying the spacing between individual modules of the diamond pattern or rows and columns of the louver systems, but keeping the total façade area constant. Figure 13B–D illustrate corresponding performance characteristics for a south oriented façade as a function of the inverse FCR for the three different shading system configurations analyzed. Note that each module is fully laminated with thin-film CIGS cells (used in the most efficient orientation) and equipped with one external bypass diode.

Figure 13B shows solar insolation per module area as a function of 1/FCR. Due to mutual shading it decreases with increasing module density. Only at large module spacing, the full potential of the solar tracking system can be exploited, in particular for the two-axis system. Figure 13C shows the corresponding average efficiency for conversion of solar radiation to electric energy. It is normalized to the maximum efficiency of the diamond pattern system. With increased mutual shading at narrow module distance, electrical efficiency decreases. This is in particular the case for the diamond pattern system, due to inefficient alignment of the module shading pattern relative to cell orientation. At wide distance between

modules, the efficiency of all systems is similar. Total electric yield of the façade system is the product of specific solar insolation, PV module area, and electrical efficiency. It is depicted for constant façade area in Figure 13D. Electric yield continuously decreases with increasing module distance for the horizontal and vertical louver systems. In the diamond pattern configuration, it exhibits a maximum at ca. 70% FCR or 30% distance between modules. Up to this value, the increase in electrical efficiency and specific solar insolation offsets the reduction in active module area. Overall, the highest electric yield is achieved with the horizontal louver system up to a FCR of ca. 35% and with the diamond pattern system at lower FCR.

Note that the results shown in this section only apply to a south facing façade at the location of Zurich and modules covered fully by PV. Efficiency loss at narrow module distance can be reduced by PV module design with improved shading tolerance [41], the integration of multiple bypass diodes per module, or placement of PV cells in module areas with little shading. The efficiency degradation in the diamond configuration relative to the louver systems at narrow module distance is also less critical for east and west relative to south oriented facades.

(A)

(B)

(C)

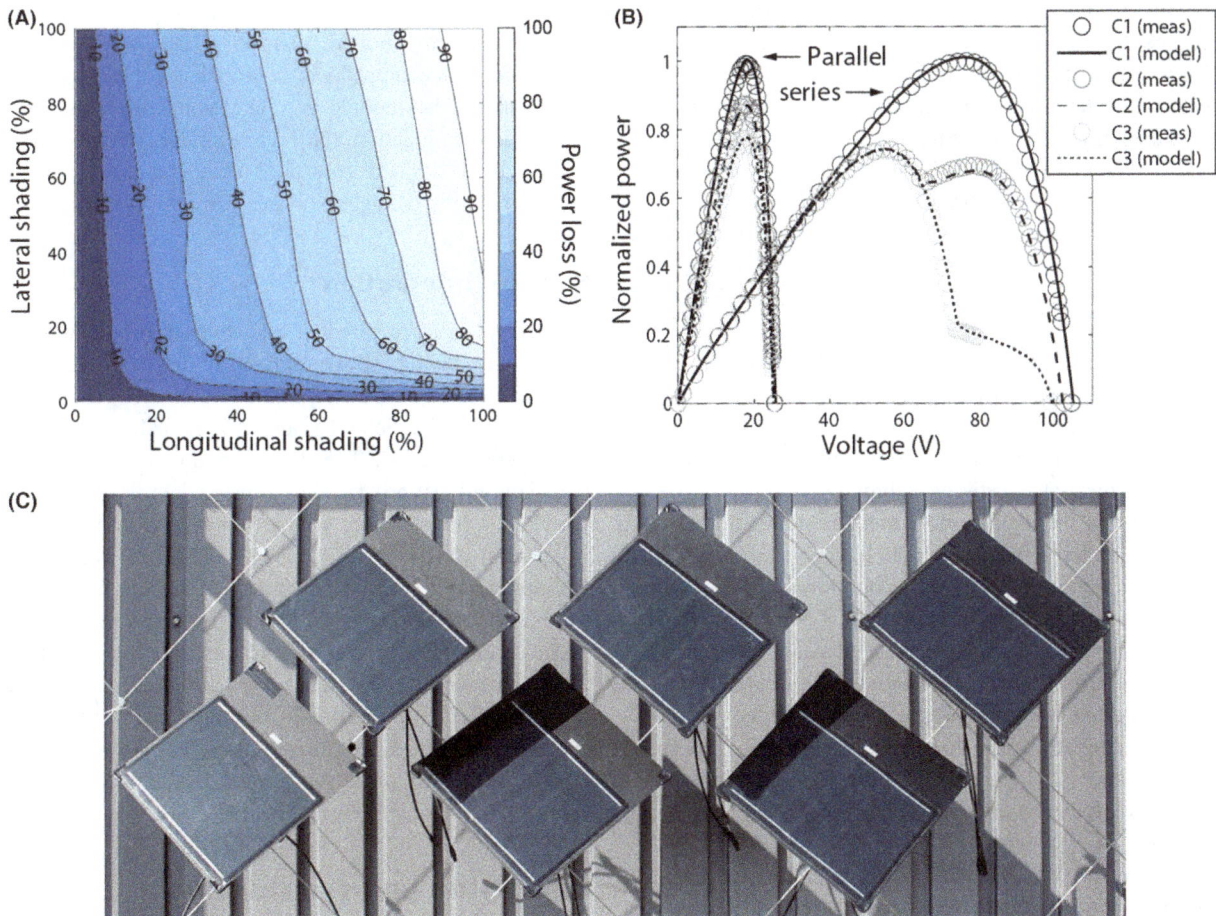

Figure 14. (A) Measured power loss of a single module as a function of longitudinal and lateral shading. (B) Normalized measured and modeled P-V curves of four modules in parallel and series connection for different shading conditions: C1 – no module shaded, C2 – one module shaded 100% in lateral and 50% in longitudinal direction, C3 – one module shaded 50% in lateral and 100% in longitudinal direction. (C) Thin-film copper indium gallium selenide module prototypes installed in diamond pattern on cable-net structure. The spacing between modules is increasing from left to right, leading to asymmetric mutual shading. Each module consists of three submodules connected with a bus bar in parallel.

Experimental Implementations

Measurement of I-V characteristics in partially shaded conditions

To validate the electrical model, outdoor measurements have been performed using six CIGS module prototypes produced by Flisom AG (www.flisom.ch). Each module has a size of 40 by 40 cm with one external bypass diode integrated in the junction box. Currently only approximately two-third of the module area is covered by PV cells and each module consists of three submodules connected in parallel (Fig. 14C). The next generation of modules is planned to be entirely laminated with PV cells that span the full width of the module.

I-V curves have been measured for different shading conditions with a PVPM2540C measurement device from PV Engineering. I-V curves were obtained in controlled shading experiments for single modules at clear sky. In these experiments, a module was mounted horizontally on the ground and an appropriate shading object moved in approximately 30 cm distance from the module to incrementally adjust the lateral and longitudinal shading fraction with a step size of 5–10%, depending on the influence on output power. The irradiance in the shaded and unshaded region was measured using an external reference cell and a pyranometer. For each shading condition, the MPP was inferred from the P-V characteristic curve. The measured power loss relative to the unshaded condition is shown Figure 14A. In order to compare these measurements to simulation results, we applied nonlinear least squares fitting of the model to the measured, unshaded module I-V curve to extract parameters relevant for the employed equivalent circuit [43]. Based on this and the assumed cell reverse characteristic, the subcell I-V curve in the shaded and unshaded region is modeled and the module I-V curve reconstructed. Note that it is not possible to directly relate the measurement results of Figure 14A to the simulation results shown in Figure 4E, because the underlying module characteristics are different. For comparison of measurement and simulation, the specific design of the measured module prototype as a parallel connection of three submodules has been considered in the model. The deviation of measured and simulated MPP is below 5% for most shading conditions and largest for regions with strong variation in the MPP, that is, at low lateral and high longitudinal shading. The percentage root mean square deviation between measured and simulated values is 4.8% and the mean absolute deviation 4.6%. This deviation may be caused by the inaccuracy in positioning the shading object and potential unknown mismatch between submodules.

In addition to single module measurements, I-V curves of arrays under different partial shading conditions and module interconnections were obtained. To compare measurements and model, the array I-V curve was reconstructed based on a fit of each unshaded module I-V curve, the specific shading pattern, and the interconnection of modules. As an example, Figure 14B shows measured and simulated normalized P-V curves of four modules in series and parallel for three different shading conditions. As expected, power loss in shaded conditions is lower for parallel-connected than series connected modules, however, currents and resistive losses are higher. For both module interconnections, the power loss with one module shaded 50% in lateral and 100% in longitudinal direction is higher than in the opposite case of 100% lateral and 50% longitudinal shading. For series connected modules, this has no influence on the MPP though. Multiple extrema of the P-V curve are caused by bypass diodes, which limit the power loss of partially shaded strings.

Figure 14C shows CIGS module prototypes mounted in diamond pattern configuration on a cable-net structure. Each module is equipped with a novel soft robotic actuator, which allows the module to rotate along two axes based on pneumatic inflation [56]. This can be used for solar tracking and daylight control. Measurements of the PV modules in this configuration are currently performed.

Building integration

The framework presented in this study is currently being applied in a living lab experiment at the ETH House of Natural Resources (HoNR, www.honr.ethz.ch), in which an ASF (Adaptive Solar Façade) is designed, constructed, and operated [2, 4]. The design of the ASF is close to the diamond pattern façade discussed in this study. The HoNR has been inaugurated in June 2015 and the ASF installation is shown in Figure 15. The ASF consists of 50 modules on a modular frame placed in front of a south facing, two-person office, in order to evaluate its thermal and electric performance in an office environment. For reference, the adjacent office with a conventional shading system is monitored. Continuous monitoring will allow us to study the interconnection between power generation, solar intake, user preferences, and heating and ventilation systems energy demand.

Discussion and Conclusions

In this work we presented a parametric modeling framework of solar insolation and electric energy yield for building integrated dynamic PV systems. The methodology is applied to evaluate the performance of different façade-mounted PV shading system configurations with one and two-axis solar tracking. We furthermore study the influence of façade orientation and module distance.

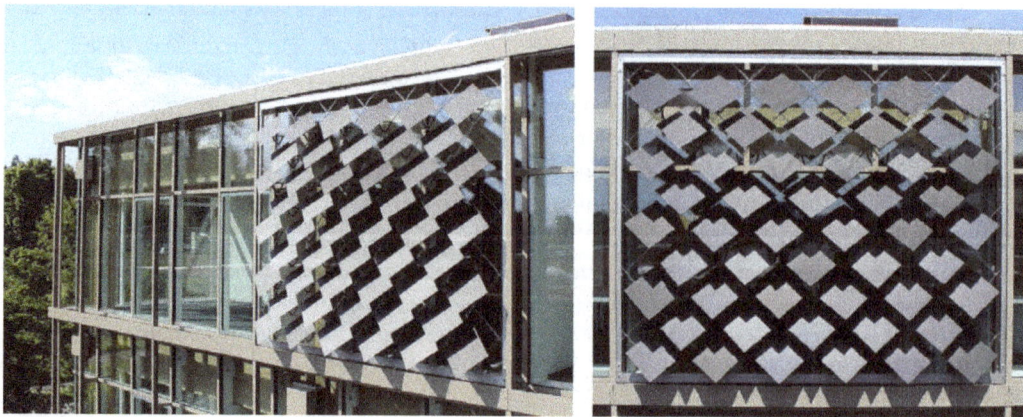

Figure 15. Adaptive Solar Façade installed at ETH House of Natural Resources. The right picture shows the shadow pattern on the modules for highest sun position, which can be compared to the calculated shadow pattern in Figure 8.

The analysis shows, that there is a trade-off between tracking performance, in terms of the amount of incident solar radiation perpendicular to module surface, and mutual shading between modules, in particular for closely positioned modules. The orientation of the tracking axis strongly influences timing and spatial distribution of shading patterns on the modules, which has immediate consequences on the optimal electrical configuration. The distance between modules is a critical parameter influencing the amount of mutual shading and hence limiting solar irradiation on shading modules using solar tracking. For a certain module spacing, highest solar insolation per module area is reached for two-axis tracking, followed by one-axis tracking, and finally fixed systems. The difference between the systems is negligible at very small distances, but increases for larger module distances. The relative performance between horizontal and vertical louver façade tracking systems depends on the season, façade orientation, and geographical location.

With regard to PV electricity generation, mutual shading between PV modules reduces the conversion efficiency from incident solar radiation to electrical energy. At narrow module spacing, the highest electric yield on a south facing façade is reached by horizontal louver systems. Only at wider module spacing with less shading, the full potential of the two-axis tracking system can be exploited. The methodology we have developed offers high spatial resolution, which helps to minimize electrical mismatch losses in case of strongly shaded conditions. We show that careful planning of module string configuration, PV cell orientation, and location of bypass diodes reduces electrical mismatch losses induced by partial shading and can result in more than 50% higher energy yield compared to uninformed design strategies. Knowledge of the precise shading pattern and electrical performance is furthermore useful to evaluate potential benefits of power electronic converters using advanced MPPT [30, 53–55] and for the development of thin-film PV cell geometries with improved shading tolerance [41].

The analysis presented in this study focuses on solar insolation and electric energy yield of different BIPV shading system configurations. However, these systems will be used as adaptive shading elements in exchange with users and the building climate and energy system. Analysis taking into account building lighting, heating, and cooling energy demand in addition to PV electricity production will be performed for a complete assessment of the different systems. In this regard, it is possible to couple dynamic shading system simulations in Rhinoceros/ Grasshopper to building energy modeling in EnergyPlus [57]. Previous work of dynamic shading systems has shown that specific module control sequences minimize building energy demand for heating, cooling, and lighting [2].

While solar tracking achieves good glare protection and high insolation on PV modules, it does not take into account energy required for heating, cooling, and lighting. In future work, solutions balancing PV electricity production and building energy demand will be investigated.

Single and dual-axis tracking systems have the potential to significantly enhance power generation of PV modules [49–51]. While solar tracking is adequate to determine the potential of systems with large module distances, there may be other control strategies that are more suitable for panel configurations with smaller module distances and high mutual shading. The methodology presented in this study will help us to develop such advanced control methods minimizing power losses. One control strategy used for tracking PV plants is the so-called back-tracking method, for which shading is completely avoided and losses due to small angle of incidence are minimized. Solutions of back-tracking for one and two-axis tracking systems exist [10, 11], however, the parametric 3D modeling framework developed in this work may help to find solutions for complex system geometries. In addition to back-tracking, strategies with motion control of individual modules (or module clusters) and their impact on electricity generation can be evaluated. Further integration of 3D parametric, PV electric, and building system simulation will enhance integrated design and open new optimization possibilities.

The work presented in this study is currently applied in the context of the ASF project. The developed methodology is used to investigate the performance of different geometrical arrangements, advanced module control, and PV system electrical design.

Acknowledgments

This research has been financially supported by CTI within the SCCER FEEB&D (CTI.2014.0119) and by the Building Technologies Accelerator program of Climate-KIC. We gratefully acknowledge Flisom AG for provision of high-efficiency CIGS PV modules.

Conflict of Interest

None declared.

References

1. Perez, M. J., V. Fthenakis, H. C. Kim, and A. O. Pereira. 2012. Façade–integrated photovoltaics: a life cycle and performance assessment case study. Prog. Photovoltaics Res. Appl. 20:975–990.

2. Jayathissa, P., Z. Nagy, N. Offedu, and A. Schlueter. 2015. Numerical Simulation of Energy Performance, and

Construction of the Adaptive Solar Façade. *Proceedings of Advanced Building Skin Conference.*

3. Velasco, R., A. P. Brakke, and D. Chavarro. 2015. Computer-aided architectural design futures. The next city-new technologies and the future of the built environment. Springer, Berlin Heidelberg, 172–191.

4. Rossi, D., Z. Nagy, and A. Schlueter. 2012. Adaptive distributed robotics for environmental performance, occupant comfort and architectural expression. Int. J. Arch. Comp. 10:341–360.

5. Yoo, S. H., and E. T. Lee. 2002. Efficiency characteristic of building integrated photovoltaics as a shading device. Build. Environ. 37:615–623.

6. Yoo, S. H., and H. Manz. 2011. Available remodeling simulation for a BIPV as a shading device. Sol. Energy Mater. Sol. Cells 95:394–397.

7. Mandalaki, M., K. Zervas, T. Tsoutsos, and A. Vazakas. 2012. Assessment of fixed shading devices with integrated PV for efficient energy use. Sol. Energy 86:2561–2575.

8. Roche, D., H. Outhred, and R. J. Kaye. 1995. Analysis and control of mismatch power loss in photovoltaic arrays. Prog. Photovoltaics Res. Appl. 3:115–127.

9. D'Alessandro, V., A. Magnani, L. Codecasa, F. Di Napoli, P. Guerriero, and S. Daliento. 2015. Dynamic electrothermal simulation of photovoltaic plants. *International Conference on Clean Electrical Power (ICCEP)*, 682–688.

10. Narvarte, L., and E. Lorenzo. 2008. Tracking and ground cover ratio. Prog. Photovoltaics Res. Appl. 16:703–714.

11. Lorenzo, E., L. Narvarte, and J. Munoz. 2011. Tracking and back-tracking. Prog. Photovoltaics Res. Appl. 19:747–753.

12. Castellano, N. N., J. A. G. Parra, J. Valls-Guirado, and F. Manzano-Agugliaro. 2015. Optimal displacement of photovoltaic array's rows using a novel shading model. Appl. Energy 144:1–9.

13. Diaz-Dorado, E., A. Suarez-Garcia, C. J. Carrillo, and J. Cidras. 2011. Optimal distribution for photovoltaic solar trackers to minimize power losses caused by shadows. Renewable Energy 36:1826–1835.

14. Mermoud, A. 2012. Optimization of Row-Arrangement in PV Systems, Shading Loss Evaluations According to Module Positioning and Connexions. *Proceedings of the 27th European Photovoltaic Solar Energy Conference.*

15. Tian, H., F. Mancilla-David, K. Ellis, E. Muljadi, and P. Jenkins. 2013. Determination of the optimal configuration for a photovoltaic array depending on the shading condition. Sol. Energy 95:1–12.

16. La Manna, D., V. Li Vigni, E. R. Sanseverino, V. Di Dio, and P. Romano. 2014. Reconfigurable electrical interconnection strategies for photovoltaic arrays: a review. Renew. Sustain. Energy Rev. 33:412–426.

17. Alonso-Garcia, M. C., J. M. Ruiz, and W. Herrmann. 2006. Computer simulation of shading effects in photovoltaic arrays. Renew. Energy 31:1986–1993.

18. Dongaonkar, S., C. Deline, and M. A. Alam. 2013. Performance and reliability implications of two-dimensional shading in monolithic thin-film photovoltaic modules. IEEE J. Photovol. 3:1367–1375.

19. Bishop, J. W. 1988. Computer simulation of the effects of electrical mismatches in photovoltaic cell interconnection circuits. Solar Cells 25:73–89.

20. Deline, C., A. Dobos, S. Janzou, J. Meydbray, and M. Donovan. 2013. A simplified model of uniform shading in large photovoltaic arrays. Sol. Energy 96:274–282.

21. Patel, H., and V. Agarwal. 2008. MATLAB-based modeling to study the effects of partial shading on PV array characteristics. IEEE Trans. Energy Convers. 23:302–310.

22. Karatepe, E., M. Boztepe, and M. Colak. 2007. Development of a suitable model for characterizing photovoltaic arrays with shaded solar cells. Sol. Energy 81:977–992.

23. Maeki, A., S. Valkealahti, and J. Leppaaho. 2012. Operation of series-connected silicon-based photovoltaic modules under partial shading conditions. Prog. Photovol. 20:298–309.

24. Celik, B., E. Karatepe, N. Gokmen, and S. Silvestre. 2013. A virtual reality study of surrounding obstacles on BIPV systems for estimation of long-term performance of partially shaded PV arrays. Renewable Energy 60:402–414.

25. Tsai, H.-L. 2010. Insolation-oriented model of photovoltaic module using matlab/simulink. Sol. Energy 84:1318–1326.

26. Ishaque, K., Z. Salam, and Syafaruddin. 2011. A comprehensive MATLAB simulink PV system simulator with partial shading capability based on two-diode model. Sol. Energy 85:2217–2227.

27. Quaschning, V., and R. Hanitsch. 1996. Numerical simulation of current-voltage characteristics of photovoltaic systems with shaded solar cells. Sol. Energy 56:513–520.

28. d'Alessandro, V., F., Di Napoli, P., Guerriero, and S. Daliento, et al. 2015. An automated high-granularity tool for a fast evaluation of the yield of PV plants accounting for shading effects. Renew. Energy 83:294–304.

29. Capdevila, H., A. Marola, and M. Herrerias. 2013. High resolution shading modeling and performance simulation of sun-tracking photovoltaic systems. *Proceedings of the 9th Int. Conf. on Concentrator Photovoltaic Systems.*

30. Poshtkouhi, S., V. Palaniappan, M. Fard, and O. Trescases. 2012. A General approach for quantifying the benefit of distributed power electronics for fine grained MPPT in photovoltaic applications using 3-D modeling. IEEE Trans. Power Electron. 27:4656–4666.

31. www.rhino3d.com (accessed 30 June 2015).

32. www.grasshopper3d.com (accessed 30 June 2015).

33. Häberlin, H. 2012. Photovoltaics system design and practice. John Wiley & Sons, Hoboken, New Jersey, USA.

34. Goswami, D. Y., F. Kreith, and J. F. Kreider. 2015. Principles of solar engineering, 3rd ed. Taylor and Francis, CRC Press, Boca Raton, Florida, USA.

35. Jakubiec, J. A., and C. F. Reinhart. 2011. DIVA 2.0: Integrating daylight and thermal simulations using Rhinoceros 3D, Daysim and EnergyPlus. *Proceedings of the 12th Conference of International Building Performance Simulation Association.*

36. Duffie, J. A., and W. A. Beckman. 2006. Solar engineering of thermal processes (Vol. 3). Wiley, New York, NY.

37. Salomon, D.. 2011. The computer graphics manual: oblique projections. Springer Science & Business Media, London, England.

38. www.luxrender.net (accessed 30 June 2015).

39. Pharr, M., and G. Humphreys. 2010. Physically based rendering: from theory to implementation. Morgan Kaufmann, Burlington, Massachusetts, USA.

40. www.meteonorm.com (accessed 30 June 2015).

41. Dongaonkar, S., and M. A. Alam. 2015. Geometrical design of thin film photovoltaic modules for improved shade tolerance and performance. Prog. Photovoltaics Res. Appl. 23:170–181. doi:10.1002/pip.2410.

42. Koishiyev, G. T., and J. R. Sites. 2009. Impact of sheet resistance on 2-D modeling of thin-film solar cells. Sol. Energy Mater. Sol. Cells 93:350–354.

43. Mermoud, A., and T. Lejeune. 2010. Performance assessment of a simulation model for PV modules of any available technology. *Proc. of the 25th European Photovoltaic Solar Energy Conference.*

44. Product datasheet Shell Solar ST40. Available at http://www.gehrlicher.com/fileadmin/content/pdfs/de/produktarchiv/Shell_ST40.pdf (accessed 30 June 2015).

45. www.pvsyst.com (accessed 30 June 2015).

46. Spirito, P., and V. Abergamo. 1982. Reverse bias power dissipation of shadowed or faulty cells in different array configurations, *Proc. of the 25th European Photovoltaic Solar Energy Conference*, 296–300.

47. Alonso-Garcia, M. C., and J. M. Ruiz. 2006. Analysis and modelling the reverse characteristic of photovoltaic cells. Sol. Energy Mater. Sol. Cells 90:1105–1120.

48. Mack, P., T. Walter, R. Kniese, D. Hariskos, and R. Schäffler. 2008. Reverse Bias and Reverse Currents in CIGS Thin Film Solar Cells and Modules, *Proc. of the 23rd European Photovoltaic Solar Energy Conference.*

49. Huld, T., M. Suri, and E. D. Dunlop. 2008. Comparison of potential solar electricity output from fixed-inclined and two-axis tracking photovoltaic modules in Europe. Prog. Photovoltaics Res. Appl. 16:47–59.

50. Koussa, M., A. Cheknane, S. Hadji, M. Haddadi, and S. Noureddine. 2011. Measured and modelled improvement in solar energy yield from flat plate photovoltaic systems utilizing different tracking systems and under a range of environmental conditions. Appl. Energy 88:1756–1771.

51. Drury, E., A. Lopez, P. Denholm, and R. Margolis. 2014. Relative performance of tracking versus fixed tilt photovoltaic systems in the USA. Prog. Photovoltaics Res. Appl. 22:1302–1315.

52. Perpinan, O. 2012. Cost of energy and mutual shadows in a two-axis tracking PV system. Renewable Energy 43:331–342.

53. Garcia, M., J. M. Maruri, L. Marroyo, E. Lorenzo, and M. Perez. 2008. Partial shadowing, MPPT performance and inverter configurations: observations at tracking PV plants. Prog. Photovoltaics Res. Appl. 16:529–536.

54. Sarvi, M., S. Ahmadi, and S. Abdi. 2015. A PSO-based maximum power point tracking for photovoltaic systems under environmental and partially shaded conditions. Prog. Photovoltaics Res. Appl. 23:201–214.

55. Ishaque, K., and Z. Salam. 2013. A review of maximum power point tracking techniques of PV system for uniform insolation and partial shading condition. Renew. Sustain. Energy Rev. 19:475–488.

56. Svetozarevic, B., Z. Nagy, D. Rossi, and A. Schlueter. 2014. Experimental Characterization of a 2-DOF Soft Robotic Platform for Architectural Applications. *Proceedings of the Robotics Science and Systems Conference*, UC Berkeley, California.

57. Roudsari, M., M. Pak, and A. Smith. 2014. Ladybug: A Parametric Environmental Plugin for Grasshopper to Help Designers Create an Environmentally-Conscious Design. *Proc. of Int. Conf. of IBPSA*, Chambery, France.

5

Al_2O_3/TiO_2 double layer anti-reflection coating film for crystalline silicon solar cells formed by spray pyrolysis

Hiroyuki Kanda[1], Abdullah Uzum[1], Norihisa Harano[1], Seiya Yoshinaga[2], Yasuaki Ishikawa[2], Yukiharu Uraoka[2], Hidehito Fukui[3], Tomitaro Harada[3] & Seigo Ito[1]

[1]Department of Materials and Synchrotron Radiation Engineering, Graduate School of Engineering, University of Hyogo, 2167 Shosha, Himeji, Hyogo 671-2280, Japan
[2]Graduate School of Materials Science, Nara-Institute of Science and Technology, 8916-5 Takayama, Ikoma, Nara 630-0192, Japan
[3]Daiwa Sangyo Co. Ltd., 3-4-11, Nakayasui, Sakai, Osaka, Japan

Keywords
Al_2O_3, anti-reflection coating, Cz-Si, spray pyrolysis, TiO_2

Correspondence
Seigo Ito, Department of Materials and Synchrotron Radiation Engineering, Graduate School of Engineering, University of Hyogo, 2167 Shosha, Himeji, Hyogo 671-2280, Japan.
E-mail: itou@eng.u-hyogo.ac.jp

Funding Information
No funding information provided.

Abstract

An Al_2O_3/TiO_2 double layer anti-reflection coating (ARC) film formed by spray pyrolysis was introduced for monocrystalline silicon solar cells as the nonvacuum processing method. The thickness of the Al_2O_3 layer and TiO_2 compact layer was controlled by the volume of deposited precursor solution and confirmed by ellipsometry and scanning electron microscopy. The average photovoltaic properties of photocurrent density (J_{sc}), open-circuit photovoltage (V_{oc}), fill factor (FF), and photo energy conversion efficiency (η) were 37.0 mA/cm^2, 590 mV, 0.712, and 15.5%, respectively. A significant improvement on J_{sc} and η could be confirmed owing to the Al_2O_3/TiO_2 ARC. The results of Fourier transform infrared (FTIR) spectroscopy and optical simulation with modeling for the reflectance properties confirmed that C-H-based organics remained after the deposition of thin films.

Introduction

Reducing optical losses in solar cells is a major factor for achieving high-efficiency silicon solar cells, and improving the absorption properties is one of the key features counteracting such loss. Various materials and thin films, including SiO_2, SiN_x, TiO_2, Al_2O_3, etc., have been used for anti-reflection coating (ARC) purposes depending on the type of solar cells.

In earlier decades, TiO_2 was introduced as an anti-reflection layer for crystalline silicon solar cells [1, 2], due to its good optical properties that enhance light absorption capability, and because of the high refractive index for silicon solar cells. Today, SiN_x has dominated the silicon solar cell industry, due to even better optical characteristics and a passivation effect which leads to high bulk lifetimes [3, 4]. SiN_x or the stack layers (e.g. SiN_x/SiO_x) are deposited by a well-known plasma-enhanced chemical vapor deposition technique [5, 6]. However, the needs for vacuum processing of toxic and hazardous gases such as SiH_4 and NH_3 in CVD operation are major drawbacks, due to the considerable costs involved.

In order to meet industrial requirements with simple low-cost technologies, high-throughput, cost-effective methods need to be investigated and adapted to the solar cell manufacturing process. A considerable amount of literature has been published on sol–gel processes including TiO_2, TiO_2-SiO_2 [1, 7], and some Al_2O_3-based Ti-doped mixed sol–gel sources as well [8, 9]. However, further investigations are needed to apply these kinds of low-cost materials to the solar cell fabrication processes.

In this work, as a low temperature and nonvacuum method, spray-pyrolysis deposition was performed for TiO_2

layer and Al_2O_3 layers, and the Al_2O_3/TiO_2 double layer was introduced as an alternative anti-reflection coating film. Film structures were modeled and confirmed by experimental studies. Applying appropriate film thicknesses for each layer and the suitable order of the layers according to their refractive indexes are the key points [10, 11]. From that point of view, after confirming the structure of TiO_2 and Al_2O_3 thin films separately, <air/Al_2O_3/TiO_2/Si> double layer structure was built considering the gradual increasing order of refractive indexes – 1.7, 2.3, 4.3 – for Al_2O_3, TiO_2, and silicon, respectively, where the refractive index of air is equal to 1. Owing to the advantage of this gradual increase in refractive index on the light path, much lower reflectance can be expected [10, 11]. Finally, silicon solar cells were fabricated with and without using Al_2O_3/TiO_2 double coating film.

Experimental

At first, TiO_2 and Al_2O_3 films were formed on flat crystalline silicon substrates and primary evaluation of the coating films was carried out. After optical optimizations, Al_2O_3/TiO_2 double layer anti-reflection coating films were applied to the silicon solar cells fabricated on textured wafers.

For the optical measurements, square-shaped p-type Si wafers (25×25 mm^2) were cut out from 6-inch Cz-Si p-type wafers. All wafers were processed by HF dip (20 v/v %) for 1 min and DI (deionized) water rinse. UV/O_3 surface treatment was carried out for a complete cleaning process. TiO_2 and Al_2O_3 films were formed on the silicon wafers by spray pyrolysis deposition using precursor solutions which were sprayed on to the surface of silicon wafers using a glass atomizer. The Si wafers were set on a hot plate heated at a deposition temperature of 450°C. The TiO_2 precursor solution was prepared by 10-fold dilution of titanium bis-isopropoxidebis-acethylacetone (TAA) (prepared by mixing titanium (VI) isopropoxide and acetylacetone in a 1:2 mole ratio) to ethanol. The Al_2O_3 precursor solution was 0.03 mole L^{-1} of aluminum (III) acetylacetonate ($Al(acac)_3$) in ethanol solution. Deposited film thicknesses were mainly controlled by the amount of sprayed precursor solutions. Each layer was analyzed by an ellipsometer (Uvisel ER Agms-nsd, HORIBA, Ltd., Kyoto, Japan).

Fittings of ellipsometer spectrum were performed by DeltaPsi2 software, using Horiba Jobin Yvon, Uvisel ErAgmsnds, considering the rough structures and the mixtures with organic residues. Scanning electron microscope (SEM) measurements (JSM-6510, JEOL), reflection analysis by ultraviolet–visible spectroscopy (Lambda 750 UV/VIS Spectrometer; Perkin Elmer, Waltham, Massachusetts, USA), and Fourier transform infrared spectroscopy (FTIR) (Frontier FTIR, Perkin-Elmer) were performed for the analysis.

Silicon solar cells were fabricated in four different sets with a variety of surface conditions: with/without ARC on the silicon surface, with/without texture. For the fabrication of silicon solar cells, 25 ×25 mm p-type Cz-Si wafers were used, which were also cut out from 6-inch wafers. For surface-textured Si solar cells, alkaline texturing was performed in KOH (5.19 g) solution in H_2O (100 mL) with Alka-Tex (GP Solar [Hainbuchenring, Neuried, Germany]; 0.28 mL) at 80°C (set: 100°C) for 30 min. In order to prevent texturing the the back side of the Si wafer was coated with polysilazane by a spin coater twice at 1500 rpm for 20 sec, followed by annealing at 600°C for 60 min in O_2 gas flow. The Si wafers were processed by HF dip (20 v/v %) for 10 min and DI water rinse. RCA cleaning [12] was then carried out to remove contaminant particles on the surface of the wafers using a $NH_4OH/H_2O_2/H_2O$ (1:1:5 in volume) solution, for 10 min at 80°C. After removal of the natural oxide films by HF dip (20 v/v %), a mixed solution of $HCl/H_2O_2/H_2O$ (1:1:5 in volume) was used to remove metallic contaminations and mobile ions on the surface by dipping the wafers for 10 min at 80°C. In order to prevent phosphorous diffusion on the back side of the Si wafer, silicon dioxide layer was formed on the back side using precursor solution (polysilazane) which was coated by a spin coater twice at 1500 rpm for 20 sec, followed by annealing at 600°C for 60 min in O_2 gas flow. For n^+-emitter formation, $POCl_3$ diffusion was performed at 880°C for 40 min with 0.2 $Lmin^{-1}$ N_2 flow. Afterward, the wafers were treated with an HF dip (10 v/v %) and a DI water rinse, successively. Front and back contacts were formed by screen-printing Ag and Al, respectively. Screen-printed pastes were dried at 125°C for 5 min (the dried Al paste was ca. 40 μm thick) and co-firing was carried out at 800°C for 1 min in an oven. After optimization by optical analysis, a double layer ARC film was formed with a 90 nm Al_2O_3 film over a 40 nm TiO_2 film using 100 and 4 mL precursor solutions, respectively.

Results and Discussion

Evaluation of thin films

In order to analyze and optimize TiO_2 and Al_2O_3 spray on deposited layers for the processing of Al_2O_3/TiO_2 double layer film, single-layered TiO_2 and Al_2O_3 films were analyzed. The fitting models were considered for the simulation. First, a fitting model with a "flat layer (main body of the layer)" and a "roughness layer (surface residue)" was defined according to the SEM images depicted in Figure 1A, named as "Fitting Model #1." It was then improved by considering "organic elements" in the

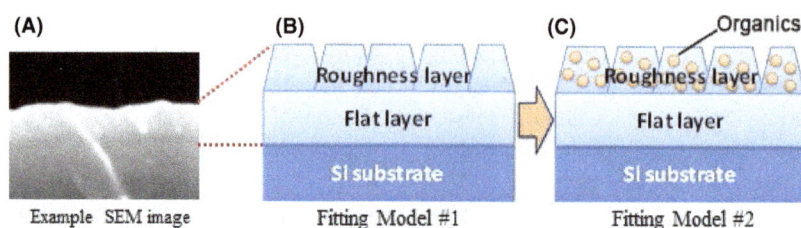

Figure 1. (A) Example of a scanning electron microscope (SEM) image; (B) Fitting Model #1 consists of a "flat layer" and a "roughness layer"; (C) Fitting Model #2 consists of a "flat layer" and a "roughness layer containing organic elements."

Table 1. Ellipsometry measurement of 6 mL deposited TiO_2 films by use of Fitting Model #1 and Fitting Model #2.

	Fitting Model #1		Fitting Model #2	
	Thickness/nm	Material ratio	Thickness/nm	Material ratio
Flat layer	60.1	TiO_2 (100%)	54.4	TiO_2 (100%)
Roughness layer	2.16	TiO_2 (31.7%)/void (68.3%)	10.2	TiO_2 (33.4%)/void (25.5%)/ organic (36.1%)
Matching factor	6.88		0.251	

Table 2. Ellipsometry measurement of 100 mL deposited Al_2O_3 films by use of Fitting Model #1 and Fitting Model #2.

	Fitting Model #1		Fitting Model #2	
	Thickness/nm	Material ratio	Thickness/nm	Material ratio
Flat layer	11.3	Al_2O_3 (100%)	5.28	Al_2O_3 (100%)
Roughness layer	66.6	Al_2O_3 (81.3%)/ void (18.7%)	73.3	Al_2O_3 (71.54%)/void (11.45%) /organic (17.03%)
Matching factor	2.79		0.346	

roughness layer, shown in Figure 1B, named as "Fitting model #2." The "organic elements" was a parameter in the fitting software of the ellipsometer (Horiba Jobin Yvon, Uvisel ErAgms-nds).

Tables 1 and 2 show the fitting results of 60 mL TiO_2 and 100 mL Al_2O_3 precursor deposited on Si wafer, respectively, using the models in Figure 1. "Matching factor" stands for the perfectness of the fitting between the experimental data and the fitting data, which is expected to be below "1" for a good matching [13]. "Void" and "organic" in the material ratio column and the rough-layer line show the percentage of air and organic material in the rough layer, respectively. Hydrocarbon-based organics can remain inside the formed layer due to organics contained in the precursor solution. The difference between Fitting Models #1 and #2 is the presence and absence of "organic" elements in the rough layer. Owing to the organics, the matching factor of the TiO_2 film and Al_2O_3 film decreased significantly from 6.88 to 0.251 and from 2.79 to 0.346, respectively, resulting in a significantly more reliable fitting for the estimation of film thickness of sprayed TiO_2 and Al_2O_3 in this work. TiO_2 film was

occupied by a flat layer of 84.2%. On the other hand, Al_2O_3 film was occupied by a rough layer of 93.3%.

For further investigation, FTIR spectra of deposited thin films were measured in a wide spectrum from 400 to 4000/cm to analyze optical film properties. Figures 2A and B show the FTIR spectra of single-layer TiO_2/Si and Al_2O_3/Si structures after spray pyrolysis deposition, respectively. Hydrocarbon-based organic peaks were observed at 2850, 2866.3, 2929.9 and 2967/cm in the FTIR spectra of TiO_2 films formed on silicon wafers. These multiple peaks from 2850 to 2970/cm can be assigned to C–H absorptions including CH_2 and CH_3-based compounds [14] which show the presence of organics. Similar peaks at 2851.3, 2865.8, 2928.3 and 2964/cm were also observed in Al_2O_3 films formed on silicon substrates.

Furthermore, dependences of the TiO_2 and Al_2O_3 film thicknesses including flat and rough layers were estimated by changing the deposition volume of the precursor solution based on ellipsometry measurements. Figures 3A and B present the estimated thickness data of TiO_2 and Al_2O_3 films, respectively. Total thickness was calculated by the sum of rough and flat layers. According to these results,

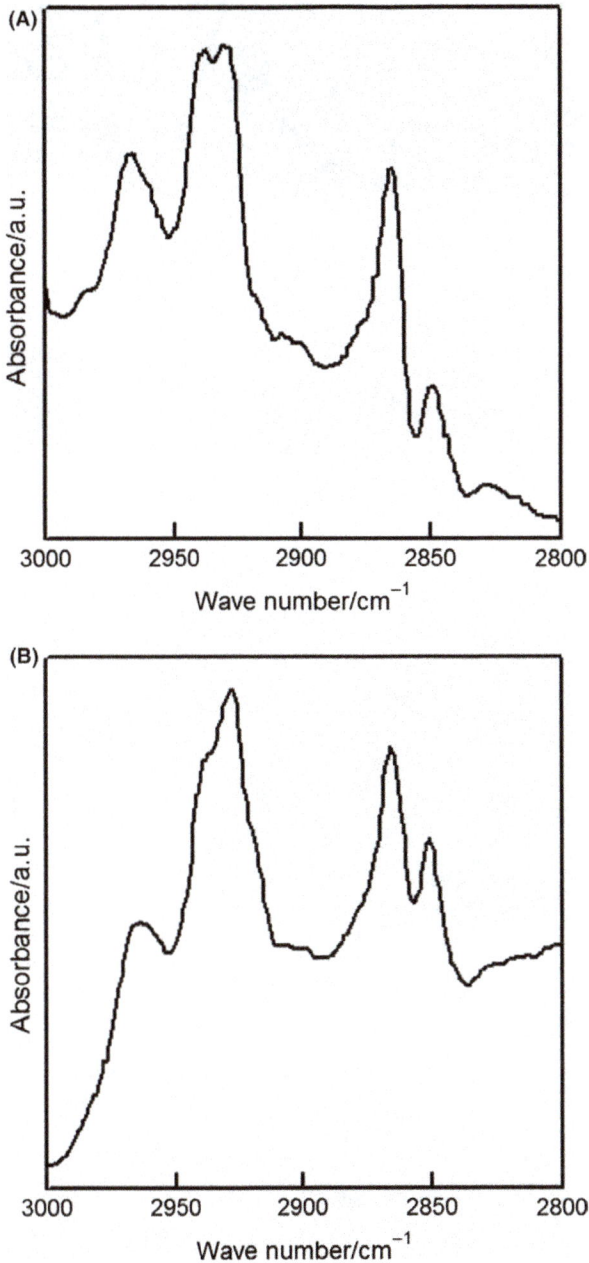

Figure 2. TiO_2 (A) and Al_2O_3 (B) films formed on Si substrate by the spray pyrolysis deposition method.

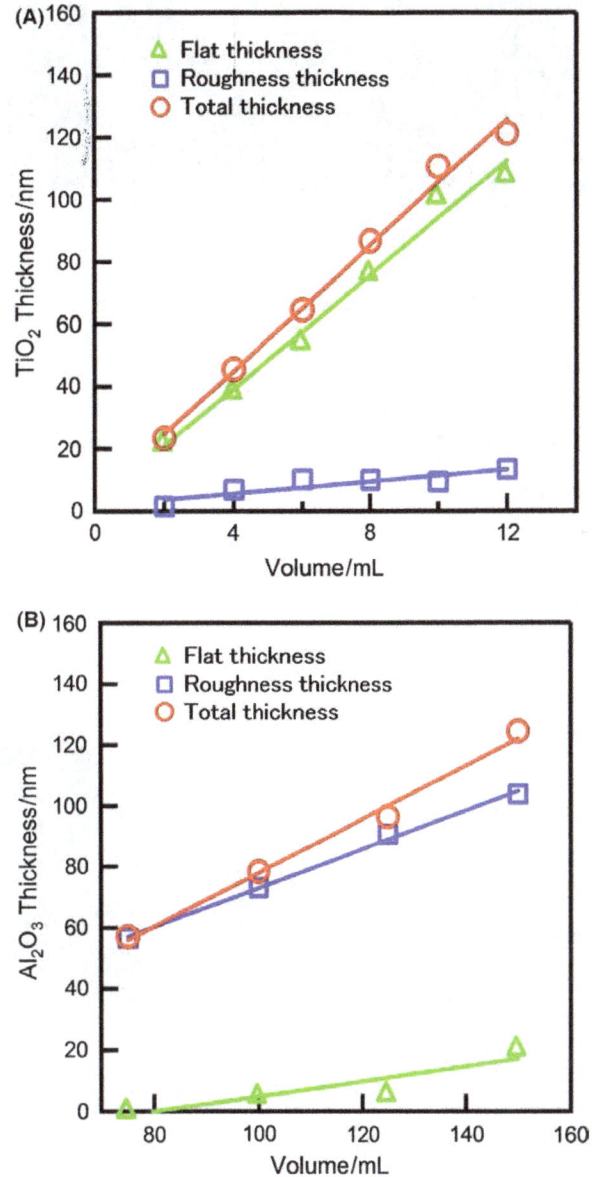

Figure 3. TiO_2 (A) and Al_2O_3 (B) films formed by changing the amount of precursor solutions.

deposition rates of TiO_2 and Al_2O_3 films can be calculated as 10.1 nm mL^{-1} and 0.88 nm mL^{-1}, respectively.

In order to confirm the deposition speed, the cross-sectional SEM images were also observed as shown in Figure 4 using thick TiO_2 and Al_2O_3 films. TiO_2/Si (Fig. 4A), and Al_2O_3/Si (Fig. 4B) structures formed by spray pyrolysis depositions using 30 mL TiO_2 and 400 mL Al_2O_3 precursor solutions had approximately film thicknesses of 260 and 380 nm, respectively, according to the images. Based on these values, the deposition rates can

be calculated as 8.7 nm mL^{-1} for TiO_2 and 0.95 nm mL^{-1} for Al_2O_3 films. However, according to the ellipsometer measurements, deposition rates were found as 10.1 nm mL^{-1} for TiO_2 deposition and 0.88 nm mL^{-1} for Al_2O_3 deposition. This difference may be attributed to the effect of surface residue and roughness after deposition when the deposition is a considerably thicker layer.

For the optimization process of <Al_2O_3/TiO_2> double layer, the thickness of the TiO_2 film was varied between 20 and 120 nm (precursor solution 2–12 mL) while the thickness of the Al_2O_3 film was changed from 68 to 135 nm (precursor solution 75–150 mL).

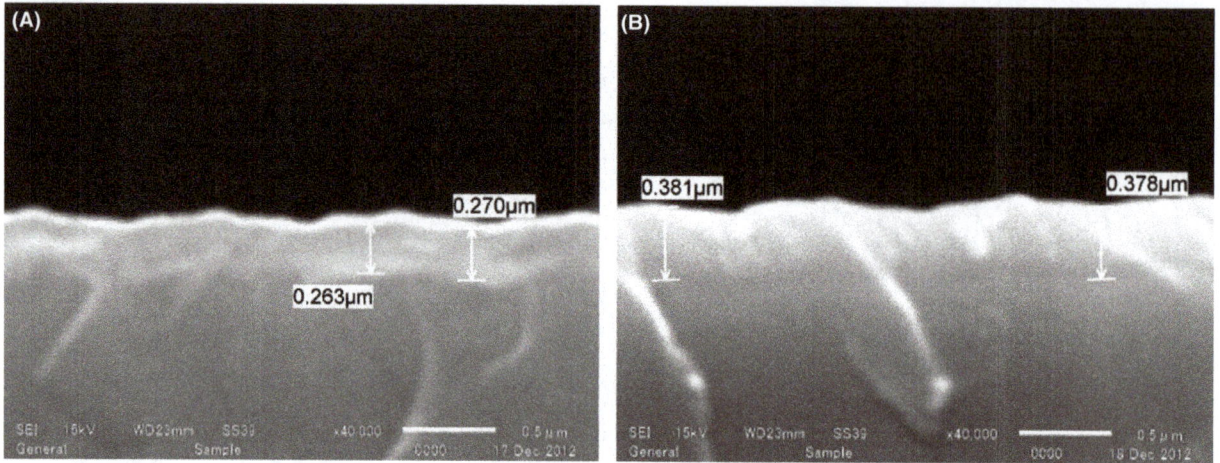

Figure 4. (A) 30 mL TiO₂ precursor solution layer on silicon substrate; (B) 400 mL Al₂O₃ precursor solution layer. Each layer was deposited by the spray pyrolysis deposition method.

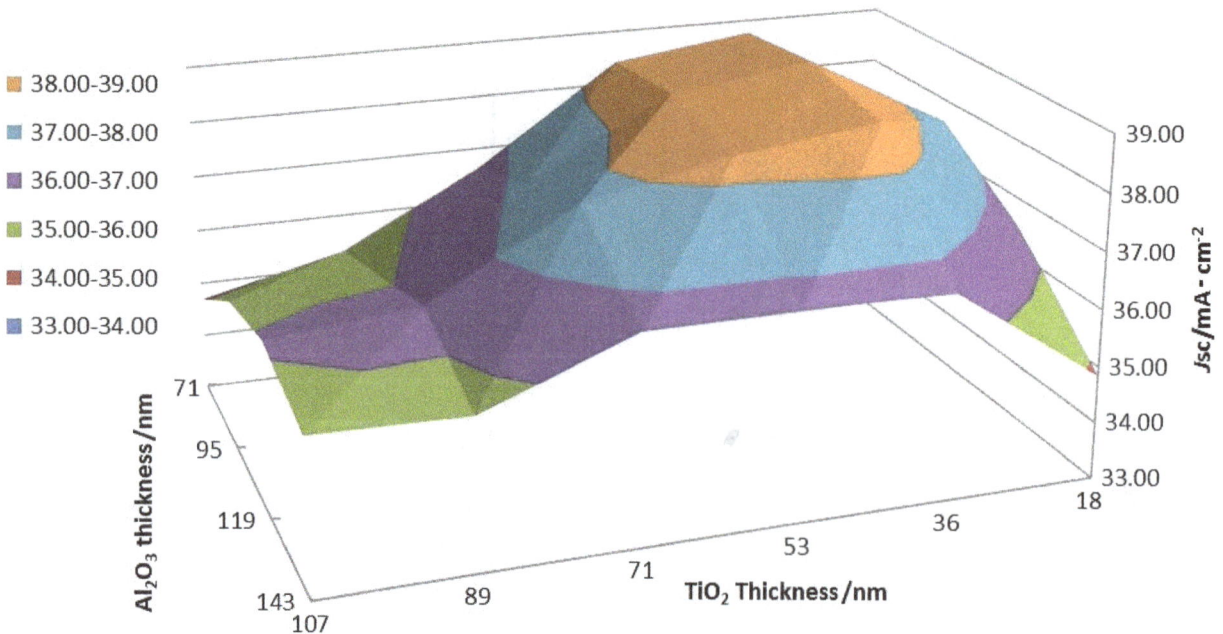

Figure 5. Estimated J_{sc} calculated by the experimental reflectivity data using <Al₂O₃/TiO₂/Si> layers with various thicknesses of Al₂O₃ and TiO₂ films.

After the reflectance measurements of the cells with ARC, maximal short circuit current densities from the reflectance results were calculated using equation (1) in order to observe the optimum Al₂O₃ and TiO₂ film thicknesses to obtain efficient Al₂O₃/TiO₂ double layer ARC film.

$$J_{sc,max} = \frac{eS}{hc} \int_{300}^{1200} P_{in} \times (1 - \text{reflectance}(\lambda))\, d\lambda \qquad (1)$$

where "$J_{sc,max}$" is the maximum short circuit current density from the reflectance results in mA cm⁻², "e" stands

for the charge of an electron (1.602×10^{-19} C), "h" is Planck's constant (6.626×10^{-34} J s⁻¹), "c" is the speed of light (2.998×10^8 m s⁻¹), "P_{in}" is the input power of solar irradiation (AM1.5, W (m² nm)⁻¹), and "λ" is the wavelength in nm.

Figure 5 presents the short circuit current density ($J_{sc,max}$) calculated by equation (1) using experimental reflectance spectra of silicon substrate with Al₂O₃/TiO₂ ARC film for various thicknesses of Al₂O₃ and TiO₂ films. According to the calculations using experimental absorption spectra, the maximum value of $J_{sc,max}$ of

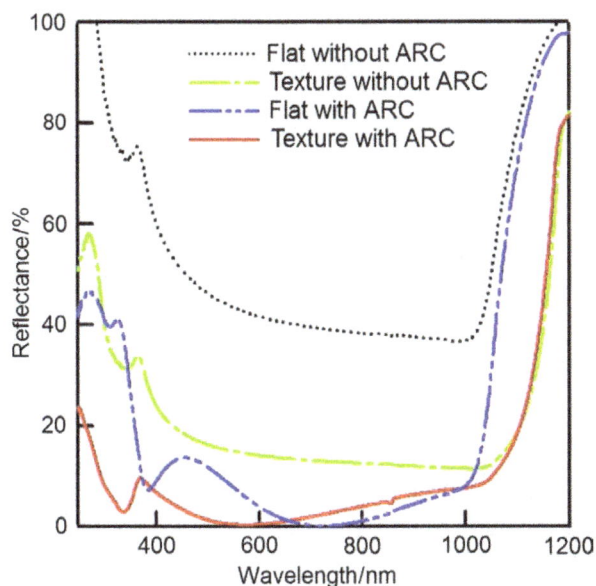

Figure 6. Reflectance spectra of silicon wafers with various surface structures.

38.9 mA cm^{-2} was achieved with a 90 nm Al$_2$O$_3$/40 nm TiO$_2$ double layer ARC film (data not shown). Based on the experimental finding, the 90 nm Al$_2$O$_3$/40 nm TiO$_2$ double layer ARC film was applied when fabricating silicon solar cells.

Solar cells with Al$_2$O$_3$/TiO$_2$ ARC film

Reflectivity measurements of four types of surfaces (with/without ARC, with/without texture) are given in Figure 6. The average reflectivity of the flat silicon surface without ARC was around 40% and decreased to around 20% after texturing. By applying the double layer ARC film on the flat and textured silicon surfaces, the reflectivity was decreased significantly to a minimum of 0.2% at around 700 nm and to 0.4% at around 600 nm, respectively. As a result, the average reflectance values with the ARC on flat and textured Si surfaces were 20.1% and 9.71%, respectively. In reference also, additional increase in J_{SC} can be expected by texture [15].

The electrical characteristics of fabricated solar cells shown in Figure 7 are summarized in Table 3. Owing to the double layer ARC film, J_{SC} of the solar cells improved remarkably. Without ARC, J_{SC} of solar cells with flat surface was around 27.6 mA cm^{-2} where it improved up to 33.5 mA cm^{-2} by texturing. The average J_{SC} of the flat cells with Al$_2$O$_3$/TiO$_2$ double layer ARC was 34.3 mA cm^{-2}, which improved to 37.0 mA cm^{-2} by the textured silicon cell with Al$_2$O$_3$/TiO$_2$ double layer ARC. Consequently, the Cz-Si p-type

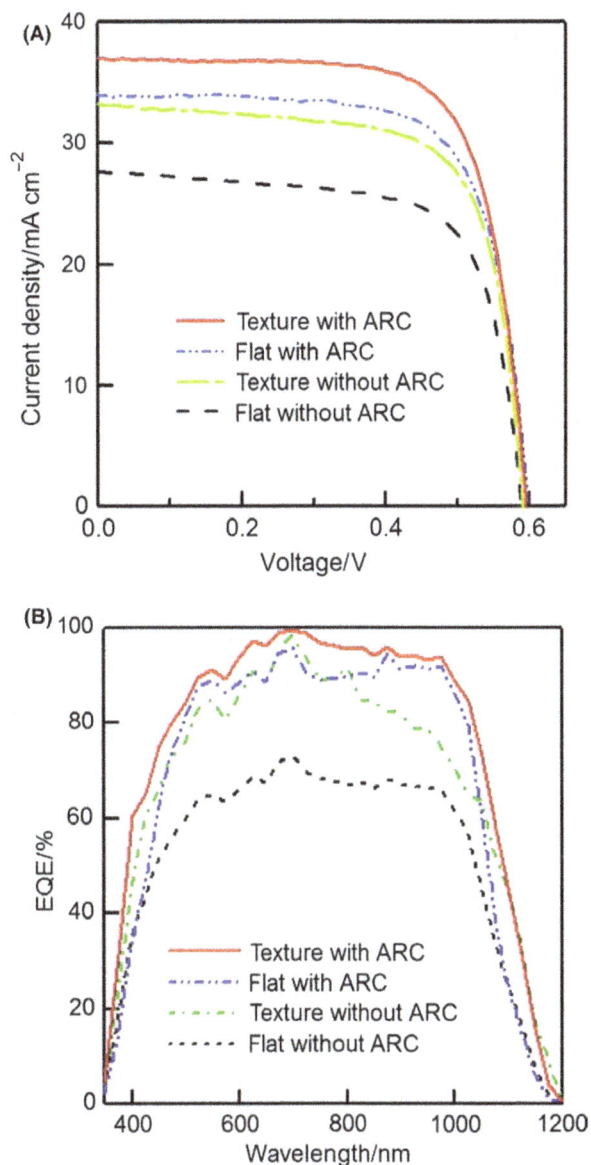

Figure 7. Current–voltage curves (A) and external quantum efficiency (EQE) spectra (B) of four fabricated sets of solar cells with/without texture and anti-reflection coatings (ARC).

solar cells with Al$_2$O$_3$/TiO$_2$ double layer ARC on textured surface reached a η of 15.5% with a V_{OC} of 590 mV and FF of 71.2. A significant improvement on J_{SC} and η could be confirmed owing to the Al$_2$O$_3$/TiO$_2$ ARC when compared to the textured cells without ARC. The increase in J_{SC} was related to the suppression of reflectance (Fig. 6) and the improvement of external quantum efficiency (EQE) (Fig. 7B), clearly. At short wavelength in Figure 7B, EQE was suppressed especially under 400 nm may be due to the absorption of the TiO$_2$ layer which has a high extinction coefficient at

Table 3. Average data of photovoltaic characteristics of fabricated Si solar cells with various surface structures with/without surface texturing and ARC.

Surface structure	ARC	J_{sc} (mA cm^{-2})	V_{oc} (V)	FF	η (%)
Flat	Without	27.6 ± 0.1	0.588 ± 0.02	0.708 ± 0.012	11.5 ± 0.2
Flat	With	34.3 ± 0.8	0.596 ± 0.03	0.706 ± 0.014	14.4 ± 0.5
Texturing	Without	33.5 ± 0.6	0.588 ± 0.05	0.702 ± 0.004	13.8 ± 0.2
Texturing	With	37.0 ± 0.2	0.590 ± 0.09	0.712 ± 0.026	15.5 ± 0.8

FF, fill factor; ARC, anti-reflection coating.
The average data and the errors have been calculated using three different samples.

short wavelength. The small fluctuation in the range of 500–1000 nm could be due to measurement error. It is well known that the passivation effect of TiO$_2$ would be degraded by annealing over 350°C, which affects the decrease in carrier lifetime and open circuit voltage [16, 17]. Therefore, another passivation layer like a silicon dioxide should be inserted between silicon and TiO$_2$ to achieve higher open circuit voltage, which is a future subject of study in our group. We have also confirmed the calculated J_{SC} of single layer ARC with TiO$_2$ and SiN$_x$ to be 39.50 and 40.67 mA cm^{-2}, respectively, which were obtained from theoretical reflectance and equation (1) on flat silicon substrates. However, the J_{SC} of double layer ARC with TiO$_2$/Al$_2$O$_3$ was calculated as 43.46 mA cm^{-2}. Previous research also showed the advantage of double layer ARC with low reflectance less than single layer ARC [18–21]. Moreover, another advantage of TiO$_2$/Al$_2$O$_3$ is non-vacuum spray pyrolysis deposition, which can realize the lower cost production system than batch vacuum process like a SiN$_x$ in terms of installation and running costs [22].

Conclusion

Al$_2$O$_3$/TiO$_2$ thin films were prepared by a low-cost spray pyrolysis deposition method as an alternative ARC film for crystalline silicon solar cells. High-quality Al$_2$O$_3$/TiO$_2$ thin films formed at 450°C and a reflectivity lower than 0.4% was achieved at 600 nm. After the evaluation of Al$_2$O$_3$/TiO$_2$ thin films, the Cz-Si crystalline silicon solar cells were fabricated with/without Al$_2$O$_3$/TiO$_2$ ARC. The average conversion efficiency was 15.5% with a J_{SC} of 37.0 mA cm^{-2}, V_{OC} of 590 mV and FF of 71.2 at the optimized condition. Improvements on J_{SC} of 3.5 mA cm^{-2} and η of 1.7% could be confirmed owing to the Al$_2$O$_3$/TiO$_2$ ARC when compared to the textured cells without ARC. Considering the nonvacuum process and simplicity, Al$_2$O$_3$/TiO$_2$ double layer ARC films formed by the spray pyrolysis deposition method were found to be very attractive and promising for crystalline silicon solar cells with good optical performance.

References

1. Brinker, C. J., and Harrington, M. S. 1981. Sol-gel derived antireflective coatings for silicon. Solar Energy Materials 5:159–172.
2. Richards, B. S. 2003. Single-material TiO$_2$ double-layer antireflection coatings. Solar Energy Materials and Solar Cells 79:369–390.
3. Richards, B. S. 2004. Comparison of TiO$_2$ and other dielectric coatings for buried-contact solar cells: a review. Progress in Photovoltaics: Research and Applications 12:253–281.
4. Dekkers, H. F. W., G. Beaucarne, M. Hiller, H. Charifi, and A. Slaoui. 2006. Molecular hydrogen formation in hydrogenated silicon nitride. Applied Physics Letters 89:211914.
5. Aberle, A. G., and R. Hezel. 1997. Progress in low-temperature surface passivation of silicon solar cells using remote-plasma silicon nitride. Progress in Photovoltaics: Research and Applications 5:29–50.
6. Wan, Y., K. R. McIntosh, and A. F. Thomson. 2013. Characterisation and optimisation of PECVD SiN$_x$ as an antireflection coating and passivation layer for silicon solar cells. AIP Advances 3:032113.
7. Pettit, R. B., C. J. Brinker, and C. S. Ashley. 1985. Sol-gel double-layer antireflection coatings for silicon solar cells. Solar Cells 15:267–278.
8. Vitanov, P, Loozen, X, Harizanova A, Ivanova T, and Beaucarne G. 2008. A study of sol-gel deposited Al$_2$O$_3$ films as passivating coatings for solar cells application. 23rd European Photovoltaic Solar Energy Conference and Exhibition, 1–5 September, Valencia, Spain 1596–1599.
9. Liang, Z., D. Chen, C. Feng, J. Cai, and H. Shen. 2011. Crystalline silicon surface passivation by the negative charge dielectric film. Physics Procedia 18:51–55.
10. Kobiyama, M. 2002. Basic theory of thin film optics (in Japanese), Optronics Co. Ltd., Shinjuku, Tokyo, Japan. (ISBN-10: 4-900-47496-7), (ISBN-13: 978-4900474963).
11. Born, M., and Wolf, E. 1999. Principles of optics, 7th ed., Cambridge Univ. Press, Cambridge, U.K. p. 67. (ISBN-10: 0-521-64222-1).

12. Kern, W. 1990. The evolution of silicon wafer cleaning technology. Journal of the Electrochemical Society 137:1887–1892.

13. Information from Horiba Ltd. (Japan) http://www.horiba.com/us/en/scientific/products/ellipsometers/

14. Pavia, D. L., G. M. Lampman, G. S. Kriz, and J. R. Vyvyan. 2009. Introduction to spectroscopy, 4th ed. Brooks/Cole Cengage Learning, Boston, MA, USA (ISBN-10: 0-495-11478-2).

15. Meiners, B. M., S. Holinski, P. Schafer, S. Hohage, and D. Borchert. 2014. Investigation of anti-reflection-coating stacks for silicon heterojunction solar cells. 29th European PV Solar Energy Conference and Exhibition Amsterdam, Netherlands, 2AV.3.22.

16. Yu, I. S., Y. W. Wang, H. E. Cheng, Z. P. Yang, and C. T. Lin. 2013. Surface passivation and antireflection behavior of ALD on n-type silicon for solar cells. International Journal of Photoenergy 2013(431614):p7.

17. Jhaveri, J., S. Avasthi, G. Man, W. E. McClain, K. Nagamatsu, A. Kahn, J. Schwartz, et al. 2013. Hole-blocking crystalline-silicon/titanium-oxide heterojunction with very low interface recombination velocity. IEEE 39th Photovoltaic Specialists Conference (PVSC), June, 3292–3296.

18. Ali, K., S. A. Khan, and M. Z. M. Jafri. 2014. Effect of double layer (SiO_2/TiO_2) anti-reflective coating on silicon solar cells. International Journal of Electrochemical Science 9:7865–7874.

19. Moradi, M., and Z. Rajabi. 2013. Efficiency enhancement of Si solar cells by using nanostructured single and double layer anti-reflective coatings. Journal of NanoStructures 3:365–369.

20. Wright, D. N., E. S. Marstein, and A. Holt. 2005. Double layer anti-reflective coatings for silicon solar cells. Conference record of the thirty-first IEEE Photovoltaic Specialists Conference, January, 1237–1240.

21. Wang, W. C., M. C. Tsai, J. Yang, C. Hsu, and M. J. Chen. 2015. Efficiency enhancement of nanotextured black silicon solar cells using Al_2O_3/TiO_2 dual-layer passivation stack prepared by atomic layer deposition. ACS Applied Materials & Interfaces 7:10228–10237.

22. Yang, C. H., S. Y. Lien, C. H. Chu, C. Y. Kung, T. F. Cheng, and P. T. Chen. 2013. Effectively improved SiO_2-TiO_2 composite films applied in commercial multicrystalline silicon solar cells. International Journal of Photoenergy 2013: 823254, p8.

6

Multi angle laser light scattering evaluation of field exposed thermoplastic photovoltaic encapsulant materials

Michael D. Kempe[1], David C. Miller[1], John H. Wohlgemuth[1], Sarah R. Kurtz[1], John M. Moseley[1], Dylan L. Nobles[1], Katherine M. Stika[2], Yefim Brun[2], Sam L. Samuels[2], Qurat (Annie) Shah[3], Govindasamy Tamizhmani[3], Keiichiro Sakurai[4], Masanao Inoue[4], Takuya Doi[4], Atsushi Masuda[4] & Crystal E. Vanderpan[5]

[1]National Renewable Energy Laboratory, 1617 Cole Boulevard, Golden, Colorado 80401
[2]DuPont Company, 200 Powder Mill Road, Wilmington, Delaware 19803
[3]Polytechnic Campus, Arizona State University, 7349 East Unity Avenue, Mesa, Arizona
[4]National Institute of Advanced Industrial Science and Technology, 1-1-1 Umezono, Tsukuba, Ibaraki 305-8568, Japan
[5]Underwriters Laboratories, 455 East Trimble Road, San Jose, California

Keywords
Adhesives, creep, EVA encapsulant, polymer, qualification standards, thermoplastic

Correspondence
Michael D. Kempe, National Renewable Energy Laboratory, 1617 Cole Boulevard, Golden, CO 80401.
Tel: 303-384-6325;
E-mail: michael.kempe@nrel.gov

Funding Information
This work was supported by the U.S. Department of Energy under Contract No. DE-AC36-08-GO28308 with the National Renewable Energy Laboratory.

Abstract

As creep of polymeric materials is potentially a safety concern for photovoltaic modules, the potential for module creep has become a significant topic of discussion in the development of IEC 61730 and IEC 61215. To investigate the possibility of creep, modules were constructed, using several thermoplastic encapsulant materials, into thin-film mock modules and deployed in Mesa, Arizona. The materials examined included poly(ethylene)-co-vinyl acetate (EVA, including formulations both cross-linked and with no curing agent), polyethylene/poly-octene copolymer (PO), poly(dimethylsiloxane) (PDMS), polyvinyl butyral (PVB), and thermoplastic polyurethane (TPU). The absence of creep in this experiment is attributable to several factors of which the most notable one was the unexpected cross-linking of an EVA formulation without a cross-linking agent. It was also found that some materials experienced both chain scission and cross-linking reactions, sometimes with a significant dependence on location within a module. The TPU and EVA samples were found to degrade with cross-linking reactions dominating over chain scission. In contrast, the PO materials degraded with chain scission dominating over cross-linking reactions. Although we found no significant indications that viscous creep is likely to occur in fielded modules capable of passing the qualification tests, we note that one should consider how a polymer degrades, chain scission or cross-linking, in assessing the suitability of a thermoplastic polymer in terrestrial photovoltaic applications.

Introduction

In the manufacturing of the photovoltaic modules there is a desire to use new thermoplastic encapsulant materials. This is motivated by the desire to reduce lamination time or temperature, reduce moisture permeation with new materials [1], use less corrosive materials [2, 3], improve electrical resistance [4, 5], or facilitate the reworking of a module after lamination. However, the use of any thermoplastic material in a high-temperature outdoor environment, where creep may occur, could raise safety and performance concerns. Therefore, there has been an increased concern in the photovoltaic (PV) community regarding the possibility of viscoelastic creep and has brought about concerns for the testing of modules according to IEC 61730 and IEC 61215 [6–8]. Small areas of a module may reach much higher temperatures (>150°C) during the "hot-spot" test or during partial shading of a module without bypass-diode protection [9, 10]; but the localized nature of this occurrence is different from the situation of prolonged operation in the hottest environments and mounting configurations. In very hot environments, modules are known to reach temperatures in excess of 100°C [11, 12]. One could envision an encapsulant with a melting point near 85°C with a highly thermally activated drop in viscosity, resulting in significant creep at 100°C. Creep is distinguished here from delamination, which may also occur when the degree of cure is less than intended, where a primer may not activate as intended [13].

Some early work with poly(ethylene)-co-vinyl acetate EVA encapsulation performed at the Jet Propulsion Laboratory (JPL) did consider the issue of creep during operation at high temperature [6, 14]. PV technology developers at that time were concerned that there was the possibility of displacement of the components within a heated module operating in the field, but did not formally investigate to verify creep using a variety of modules deployed in a hot location. To specifically prevent creep, EVA, that was cross-linked via a peroxide-initiated reaction, was advocated at that time. The use of 65% gel was found to facilitate passing the sales qualification tests (which included the "melt/freeze" test at that time) and was therefore suggested by JPL [6]. The use of EVA with at least 65% gel content was reaffirmed by Springborn Laboratories (later known as Specialized Technology Resources, Inc., or STR) [15], and currently, the use of EVA with 60–90% gel content after lamination is commonly used in the industry. However, if one wants to switch to a thermoplastic material, gel content is not a useful metric to evaluate creep. Therefore, one must find other ways to verify proper processing of a new encapsulant material.

Examination of EVA in fielded modules reveals gel contents above 90% [16, 17]. Often when EVA degrades, it leaves a clear border around the perimeter of the cell and a yellow/brown inner area. It has been observed that in the clear portion, the amount of oxygen incorporation and UV absorber (Cyasorb 531) was higher, but the gel content was slightly lower. Here, oxygen ingress is slow enough such that it is consumed on the cell perimeter and the oxygen serves to increase the rate of chain scission.

At elevated temperatures, EVA degrades by thermal deacetylation with the formation of acetic acid [16, 17]. However, at ambient or use conditions, this reaction pathway is unimportant and long-term fielded EVA samples have been shown to have <1% change in the amount of residual acetate groups [2, 18]. However, even this small amount of acid formation can lower the pH sufficiently to increase degradation rates of PV cells and interconnections.

Prior work studied the possible hazards associated with creep in modules, including how to screen for creep and how to test for the results of creep [19, 20]. Modules were fabricated with a variety of encapsulant materials, including polyethylene vinyl acetate (EVA, both cross-linked and with no curing agent), polyethylene/polyoctene copolymer (PO), poly(dimethylsiloxane) (PDMS), polyvinyl butyral (PVB), and thermoplastic polyurethane (TPU). These modules were subjected to high temperatures using outdoor aging (in Mesa, AZ or Golden, CO) or indoor aging (in a temperature step-stress test performed in a chamber), and the resulting creep was documented. The greatest creep was observed for EVA with no curing system present (3 mm in Mesa, AZ and 0.3 mm in Golden, CO), while the next greatest displacement was on the order of 30 μm. Curiously, the rate of creep for the fielded EVA specimens was seen to decrease, even though the enabling ambient temperature remained warm through the summer months of the exposure period. The observed creep was correlated with material-level tests to identify the best way to characterize phase transitions that could be predictive of creep in the field. The creep measured outdoors and indoors was found to correlate with each other, where the nonuniform temperature distribution within a module as well as the packaging components (frame or mounting clips) was found to greatly limit creep. Creep was also particularly correlated with the melting point, the phase transitions and rheological properties measured for the encapsulant materials. These data suggested that some noncuring thermoplastics were cross-linking, but did not present conclusive evidence to that effect.

The effort described here adds to the aforementioned study [19–21]. We demonstrate that, in some cases, most notably noncuring EVA (NC-EVA), a significant factor in the reduction of encapsulant creep is the formation of cross-links in the absence of cross-linking agents. The ability to cross-link polymers in the absence of specific chemistry to do so very significantly reduces the likelihood of damage resulting from creep in a field-deployed module. We explore the importance of proximity to the edge of the module to assess the role of diffusion of reactants and/or reaction products in the dominance of chain scission over cross-linking reactions.

Experimental Procedure

The specifics of the materials and construction are more thoroughly covered in M. D. Kempe et al. [21], but we will present a brief description of the sample preparation here.

Module construction

Encapsulant materials being used, or under investigation for use, in PV modules were obtained from industrial manufacturers. The nomenclature for as well as the phase

Table 1. Differential scanning calorimetry (DSC) and dynamic mechanical analysis (DMA) determined phase transitions.

Encapsulant material type and designation		Phase transitions					
		DSC			DMA at 0.1 rad/sec		
		T_g (°C)	T_m (°C)	T_f (°C)	T_g (°C)	T_m (°C)	T_c (°C)
Cured commercial PV EVA resin	EVA	−31	55	45	−30	47	
Commercial PV EVA resin with all components but the peroxide	NC-EVA	−31	65	45	−28	69	
Polyvinyl butyral	PVB	15			17		121
Aliphatic thermoplastic polyurethane	TPU	2			3		84
Pt catalyzed, addition cure polydimethyl siloxane gel (mock modules)	PDMS-M	−158	−40	−80			
Pt catalyzed, addition cure polydimethyl siloxane gel (Si modules)	PDMS-Si	−150	−40	−80			
Thermoplastic polyolefin #1	TPO-1	−43	93	81	−35	105	
Thermoplastic polyolefin #3	TPO-3	−44	61	55	−41	79	
Thermoplastic polyolefin #4	TPO-4	−34	106	99	−21	115	

DMA glass transitions (T_g) were determined as the peak in the phase angle, and the DMA melting and freezing transitions (T_m and T_f, respectively) were determined when the phase angle was 45° except for the cross-linked PDMS [poly(dimethylsiloxane)] and poly.EVA where an inflection point in the modulus was used. Because PVB and thermoplastic polyurethane (TPU) did not have melt transitions, the critical temperature for significant flow (T_c) was determined as the point where the phase angle was 45°. Data are taken from M. D. Kempe,et al. [21].

characteristics of these materials are summarized in Table 1. A noncuring poly(ethylene-vinyl acetate) (NC-EVA) was formulated identically to a standard EVA formulation but without the inclusion of a peroxide to promote curing during lamination.

In previous experiments [19, 21], both crystalline silicon and thin-film mock samples were tested. Because we wanted to redeploy the silicon modules, only the thin-film mock modules were destructively evaluated in this work. The thin-film mock modules were constructed using two pieces of 3.18-mm thick, 61 × 122 cm glass obtained from a thin-film PV manufacturer. The rear surface of the back plate was painted black to simulate the optical absorption of a thin-film module. Thin-film mock modules were mounted by adhesively attaching fiberglass channel on the back, allowing the front piece of glass to move freely.

For the thin-film mock modules, the creep (displacement of the front glass relative to the back glass) was measured using a high-precision electronic depth gauge that was incremented ±1 μm. This gauge was mounted to a flat plate to ensure that it was positioned perpendicular to the side of the module, and in the plane of one of the glass plates. With this setup, the repeatability of creep measurement was better than ±20 μm.

Modules were deployed in Mesa, Arizona at the Arizona State University from May until September 2011 on a rack tilted at 33°latitude tilt and a 255 azimuth to more directly face the sun at the hottest part of the day. Additionally, a single NC-EVA thin-film mock module was exposed in Golden, Colorado at a 180 azimuth and 40°latitude tilt [8]. Insulation was placed on the backside

to simulate a close-roof installation (resulting in maximum measured temperatures between 102°C and 104°C in Mesa, Arizona, AZ or Golden, CO)).

The modules were also examined indoors in environmental chambers in a step-stress experiment using a replicate set of modules. The temperatures applied during the test ranged from 65 to 110°C, with a 5°C increment, then from 110 to 140°C in a 10°C increment using a 200-h exposure time at each step.

Size exclusion chromatography and multiangle laser light scattering

Following field deployment, the formation of polymer chain cross-links on the mock modules was evaluated using size exclusion chromatography (SEC) in conjunction with multi-angle laser light-scattering (MALLS, Waters Corporation GPCV 2000 instrument 34 Maple StreetMilford, MA 01757 USA) and viscometric detection (using a capillary viscometer detector CV from Waters). For the tetrahydrofuran (THF) solvent, four SEC styrene-divinyl benzene 8 × 300 mm columns from Shodex (4-1, Shiba Kohen 2-chome, Minato-ku, Tokyo, 105-8432, Japan) were used for separation. These were two linear GPC KF806M, one GPC KF802, and one GPC KF-801 columns. Columns were run at 40°C, with a flow rate of 1.0 mL/min, an injection volume of 0.219 mL, and a run time of 90 min.

Samples were cut out of the mock modules using a ceramic saw blade (Wale Apparatus Co., Inc.400 Front StreetHellertown, PA 18055) enabling samples to be taken at different distances from the edge. For the TPU and EVA, polymer was removed from the glass by extraction for 72 h at 40°C, utilizing mild agitation, in a solution of THF; however, this typically left behind some EVA as an insoluble fraction. For EVA, the residual glass-polymer specimens were then additionally soaked in trichlorobenzene (TCB) overnight at 140°C, with slight agitation, to solubilize some of the remaining polymer. The TCB soluble measurements of EVA represent material not soluble in THF, but soluble in TCB. For the thermoplastic polyolefins (TPO), the SEC measurements were done using only TCB solvent.

The TCB-based system utilized two Styragel HT 6E and one Styragel HT 2 styrene-divinyl benzene columns from Waters for SEC separation. The TCB columns used a temperature of 140°C, a flow rate of 1.01 mL/min, an injection volume of 0.2095 mL, and a run time of 80 min.

Samples of "virgin" (unaged) polymer film samples were dissolved in THF and TCB, as appropriate, at room temperature and 150°C, respectively, to serve as controls. All solutions were made with approximately 1 mg/mL polymer concentration.

Gel content measurement

Gel content measurement was accomplished using a Soxhlet extractor with toluene, THF, mixed isomeric xylenes, or methanol as the solvent. Test material was obtained in approximately 1 g samples by cutting out small samples using a ceramic wet saw, and subsequently separating the polymer from the glass manually using a blade. Polymer samples were taken at various distances from the edge of the glass.

Results and Discussion

NC-EVA

Three NC-EVA samples were removed from the mock modules deployed in Golden, Colorado and Mesa, Arizona. One cut from the edge of the module, one about 2 cm from the edge, and another about 4 cm. The unexposed sheets were completely soluble in both THF and TCB. For the exposed samples, a qualitative assessment during the preparation for SEC experimentation indicated that only ~40% of the NC-EVA was soluble in THF. Then, of this THF-insoluble fraction another ~40% (relative to initial amount) was soluble in TCB, leaving ~20%, which was not soluble in either THF or TCB. Qualitatively, this is in agreement with rigorous gel content measurements using toluene in a Soxhlet extractor (Fig. 1), where 5.5–35.4% gel was measured at various locations. This NC-EVA was formulated with EVA pellets containing trace amounts of butylated hydroxyl toluene (BHT) to which was added an hindered amine light stabilizer (HALS), methacryloxypropyl trimethoxysilane (Z6030), a phenylphosphonite, a benzophenone-based UV absorber, but no peroxide [5, 18].

The cross-linking of the NC-EVA in the fielded module was further investigated using SEC in conjunction with MALLS and viscometry detection (Fig. 2). For the THF soluble fraction, very little change in the molecular weight distribution was seen for either location within the specimens. The sample taken 4 cm from the edge for the Colorado module has a high-molecular-weight fraction, but otherwise, the THF soluble fractions are similar to the unexposed material. The EVA specimens contain

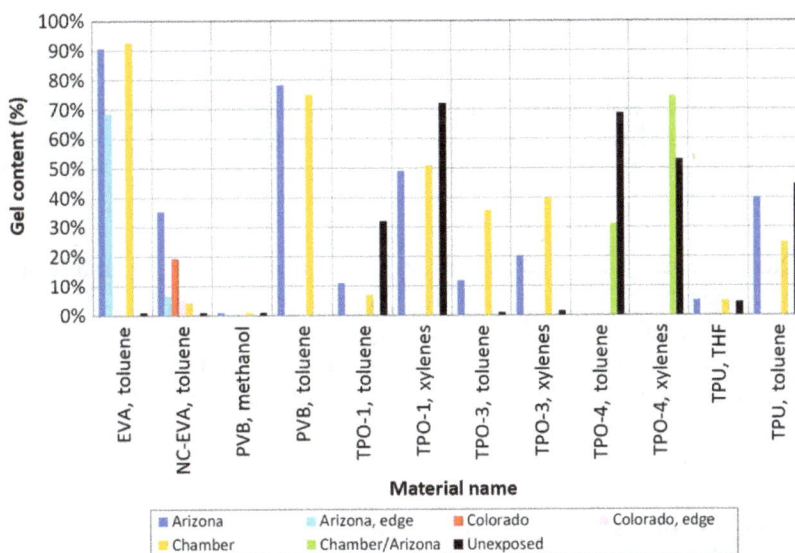

Figure 1. Results of gel content measurements using a Soxhlet extractor. "Edge" samples were taken <1 cm from the side of the module, all other samples were taken about 4 cm from the side.

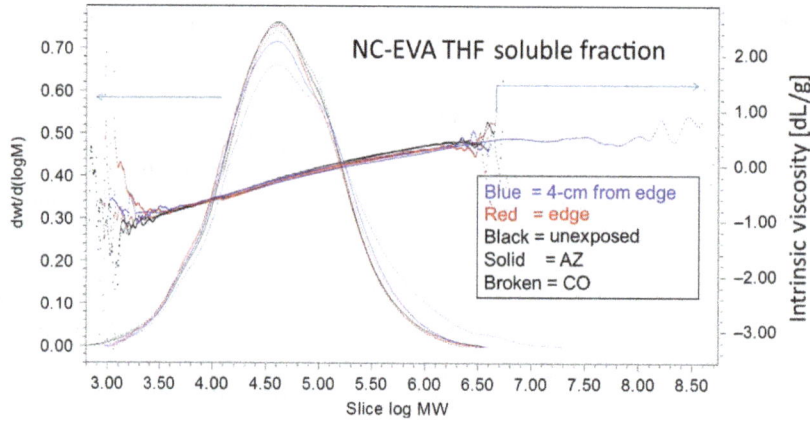

Figure 2. Weight fraction determined by SEC-MALLS (SEC, size exclusion chromatography) and viscometry for the NC-EVA (poly[ethylene]-co-vinyl acetate) thin-film mock module after exposure in Golden, Colorado (dotted lines) and Mesa, Arizona (solid lines). Samples dissolved in tetrahydrofuran (THF) by extraction at 40°C overnight. Average molecular weight (MW) of each slice of polymer distribution is in units of g/mol as determined by SEC-MALLS. Right axis, intrinsic viscosity (dL/g) of each slice. Lines are normalized to the area under the curve.

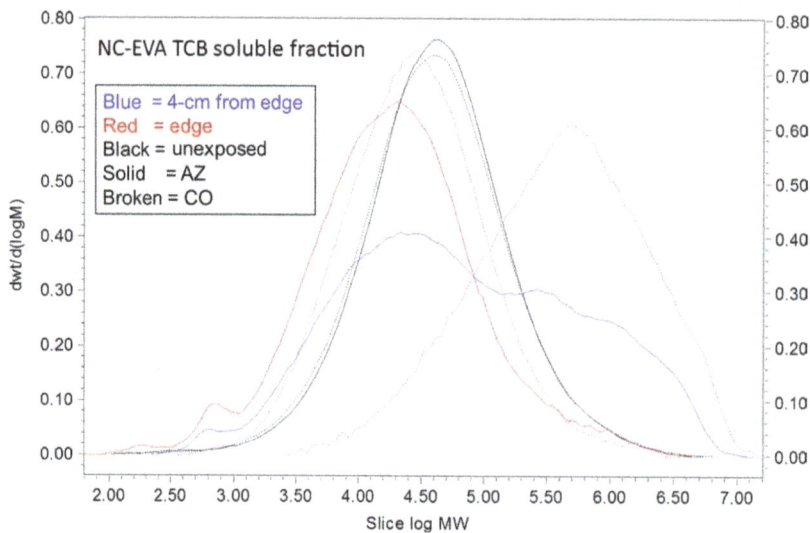

Figure 3. Weight fraction determined by SEC-MALLS for the NC-EVA poly(ethylene-co-vinyl acetate) thin-film mock module after exposure in Golden, Colorado (dotted lines) and Mesa, Arizona (solid lines). Samples obtained from tetrahydrofuran (THF) insoluble fraction by extraction in TCB at 150°C overnight. Molecular weight (MW) is in units of g/mol determined by (SEC, size exclusion chromatography) SEC-MALLS. Lines are normalized to the area under the curve.

33 wt% vinyl acetate. Qualitative experiments indicate that above about 30 wt% the solubility of EVA in THF is good, but below 30 wt% the solubility in THF is poor [22]. This indicates that even very slight modification/degradation of EVA molecules is sufficient to render them significantly less soluble in THF consistent with our data.

In contrast, the SEC profiles of the TCB-solubilized samples varied greatly depending on the exposure site, and on the distance from the module edge (Fig. 3). Samples taken at distances of 2 and 4 cm from the edge have a higher molecular weight fraction than the control indicating that significant cross-linking is occurring (Fig. 4).

The molecular weight distributions for Colorado and Arizona are very different. Of note in Table 2 are the apparently lower values for M_n in TCB versus THF. It is believed that this is an experimental artifact resulting from the silane component being isorefractive in THF but not in TCB. Because of the significantly lower temperature experienced in Colorado [19, 21], one would expect to see less degradation, that is cross-linking, than for the Arizona sample. Also of significance is the low-molecular-weight fractions, which if unmodified should have been previously extracted in THF. The absence of low-molecular-weight fractions at a distance of 4 cm from

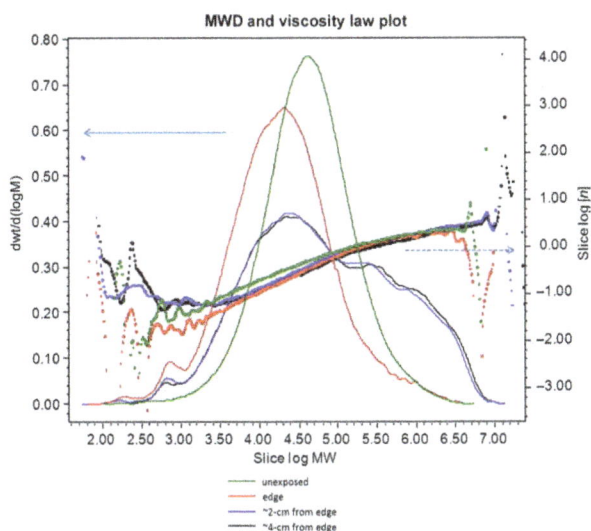

Figure 4. Weight fraction determined by (SEC, size exclusion chromatography) SEC-MALLS and viscometry for the NC-EVA thin-film mock module after exposure in Arizona. Samples dissolved in TCB by extraction at 150°C overnight. Molecular weight (MW) is in units of g/mol as determined by SEC-MALLS. Right axis, intrinsic viscosity (dL/g) of each slice. Lines are normalized to the area under the curve.

Table 2. NC-EVA poly (ethylene-co-vinyl acetate) size exclusion chromatography, multiangle laser light scattering

Sample	M_n (g/mol)	M_w (g/mol)	M_z (g/mol)	PDI
Tetrahydrofuran (THF)	20,000	98,000	450,000	4.8
Unexposed	20,000	98,000	450,000	4.8
Trichlorobenzene (TCB)	15,000	100,000	620,000	6.5
Unexposed	16,000	98,000	510,000	6.3
Mesa, Arizona				
TCB soluble fraction				
Edge	20,000	89,000	380,000	4.5
~2 cm from the edge	20,000	85,000	330,000	4.2
~4 cm from the edge	21,000	100,000	440,000	4.9
THF soluble fraction				
Edge	5,500	67,000	630,000	12.1
~2 cm from the edge	8,700	390,000	2,200,000	45.2
~4 cm from the edge	11,000	380,000	2,100,000	35.9
Golden, Colorado				
TCB soluble fraction				
Edge	20,000	88,000	340,000	4.4
~4 cm from the edge	21,000	190,000	2,100,000	8.9
THF soluble fraction				
Edge	13,000	66,000	390,000	5.2
~4 cm from the edge	100,000	980,000	3,100,000	8.7

M_n is the number average molecular weight, M_w is the weight average molecular weight, M_z is the Z average molar mass, and PDI is the polydispersity index which is equal to M_w/M_n.

the edge in the Colorado sample indicates that the process of dissolution in TCB at 140°C is not significant in forming these fractions. Noting that low-molecular-weight fractions are more prevalent for the edge samples suggest that O_2 may play an important role in the formation of low molecular weight THF-insoluble fractions.

Gel content measurements of both the Colorado and the Arizona test modules (Fig. 1) demonstrate a significantly lower gel content near the edge and a higher gel content away from the edge for NC-EVA. The same trend was also seen in the EVA with peroxide where the gel content of the edge sample was 68.5%, and 4 cm from the edge it was 90.7%, consistent with other observations [16, 17]. This is presumably because of an interaction with O_2 at the periphery.

Numerous observations of fielded modules have observed discoloration of EVA in the center area of a cell; however, greater oxygen incorporation is observed around the cell perimeter [18]. Similarly, it has been observed that the gel content in the more discolored EVA at the center of the cell is higher than EVA at the perimeter in aged samples [16]. Thus, yellowing and cross-linking are correlated with lower oxygen incorporation, and lower cross-link densities are correlated with higher oxygen incorporation and oxidative bleaching. We hypothesize that oxygen can act to increase the rate of chain scission, but that its absence allows cross-linking chemistries to dominate more. Of particular note, is the observation that samples at a distance of 2 and 4 cm from the edge have nearly identical molecular weight profiles (Fig. 4). This implies that oxygen availability is the same at 2 and 4 cm, therefore the polymer environment is very likely anaerobic at distances >2 cm from the edge of the cells, consistent with the observed size of brown/yellow areas in some field deployed EVAs.

Marais and Hirata [23] measured the solubility of O_2 in 33 wt% EVA at 25°C at 0.0023 cm^3 (STC)/cm^3/cm Hg (or 5.6×10^{-5} g/cm^3) and a diffusivity (D) of about 4.0×10^{-7} cm^2/sec. For the 140-day exposure, the characteristic distance for oxygen ingress ($\sqrt{DT}$) at a temperature of 25°C is 4.8 cm. For comparison, at 25°C the diffusivity of water in EVA is comparable at 4.8×10^{-7} cm^2/sec, but its solubility at 100% RH is 100 times higher at 0.0021 g/cm^3 [1, 24–26]. Considering that the perimeter of 2 cm is where all of this diffusant is consumed, a linear gradient at 25°C would result in 0.07 g/cm^3 of water consumed in the perimeter of 2 cm over a 10-year period. Because EVA is not observed to be this highly degraded when fielded for long periods, water is not completely consumed so quickly upon diffusion into EVA. At the higher module temperatures in field deployment, oxygen would diffuse further than this estimate at 25°C. Thus, it is highly likely that oxygen ingress is enhancing chain scission in EVA, and is limited by its low solubility in EVA.

If the degree of branching is constant as a function of molecular weight, then the lines for intrinsic viscosity would be straight lines when plotted against Log(MW). The slight concave down curvature in Figures 2 and 4 indicates increased amounts of branching at higher molecular weights. For the THF soluble fractions, the difference in intrinsic viscosity between the aged samples is undetectable. But for the TCB soluble fraction there is a relative reduction in the intrinsic viscosity for low-molecular-weight fractions, indicating that they are becoming more compact and branched.

For the NC-EVA samples deployed in both Mesa, Arizona and Golden, Colorado, cross-linking was observed resulting from combined thermal/UV aging. Both SEC and gel content measurements indicate an increase in the frequency with which degradation results in chain scission for locations where, presumably, oxygen is able to diffuse. Very small amounts of degradation are able to significantly reduce the solubility of EVA in THF. Regardless of the initial degree of cure, or even the total absence of cure agents, typical PV EVA formulations are therefore expected to cure as they age.

Thermoplastic polyolefins

In this work three different thermoplastic polyolefins were evaluated. Thermoplastic polyolefin #1 (TPO-1) was designed to be a material that might be susceptible to creep, but still pass the qualification test because its melting transition was above 85°C. However, when fielded, the movement of the glass was barely detectable (~30 μm) [19, 20]. Thermoplastic polyolefin #4 (TPO-4) is similar to TPO-1, but with a higher melting point and viscosity, which resulted in no detectable movement. Finally, thermoplastic polyolefin #3 (TPO-3) was made by a different manufacturer and has been used in the construction of commercial modules. TPO-3 has a melting point between 61°C and 79°C (depending on the measurement method) which is just a bit higher than EVA. Because it had a much higher viscosity than NC-EVA [20], the 0.090 mm of creep was not sufficient to see the effects of chain scission/cross-linking in the creep response curve for the outdoor exposed modules. However, a significant increase in viscosity was seen in the indoor stressed samples at temperatures >90°C [21]. All three of these materials were analyzed in SEC to evaluate cross-linking and chain scission reactions.

All three TPO samples were completely soluble in TCB at 150°C and were analyzed in this solvent using the multidetector SEC system (Fig. 5). Similar to NC-EVA, for all three materials the sample taken from the edge showed the greatest reduction in molecular weight indicating that a similar O$_2$-induced chain scission mechanism may be active at the periphery. By far, the greatest shift was for TPO-4, followed by TPO-1, then TPO-3. Similarly,

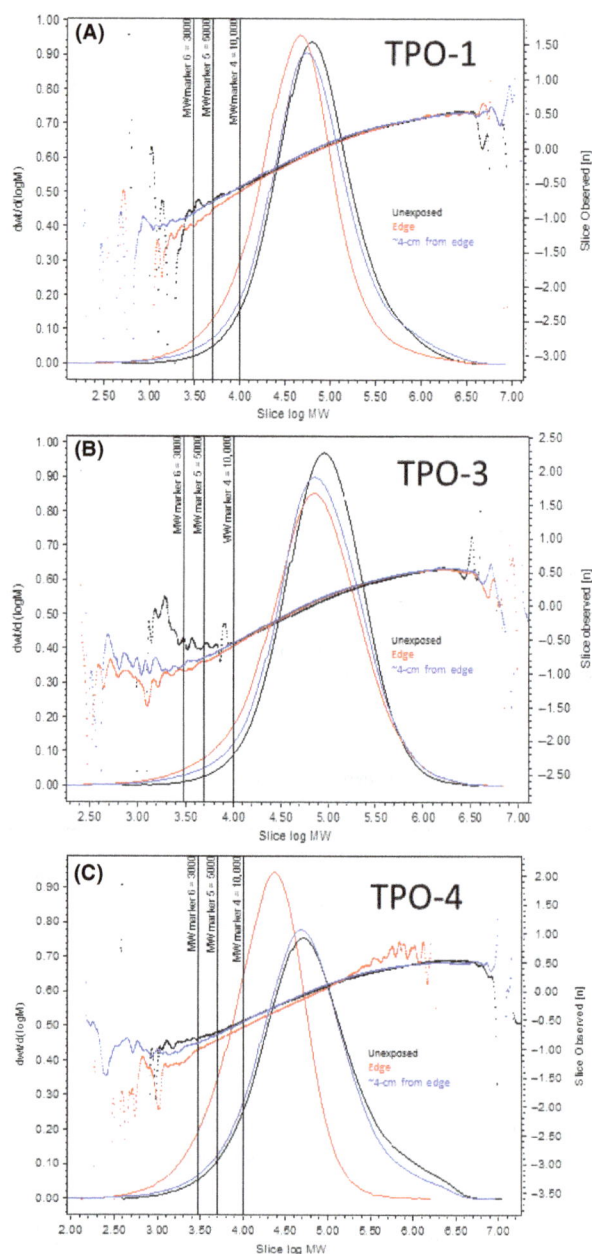

Figure 5. Weight fraction determined by SEC-MALLS (SEC, size exclusion chromatography) and viscometry for the TPO-1 thin-film mock module after exposure in Arizona. Samples dissolved in TCB by extraction at 150°C overnight. Molecular weight (MW) is in units of g/mol. Right axis, intrinsic viscosity (dL/g) in slice, horizontal lines. Blue: 4 cm from the edge; red: edge sample; black: unexposed. (A) TPO-1, (B) TPO-3, and (C) TPO-4. Lines are normalized to the area under the curve.

the samples taken at a distance of 4 cm from the edge showed a small drop in molecular weight. This is in stark contrast to the NC-EVA samples which experienced significant cross-linking in the interior anaerobic regions. This difference in cross-linking behavior between EVA and these TPOs could be a result of the stabilization

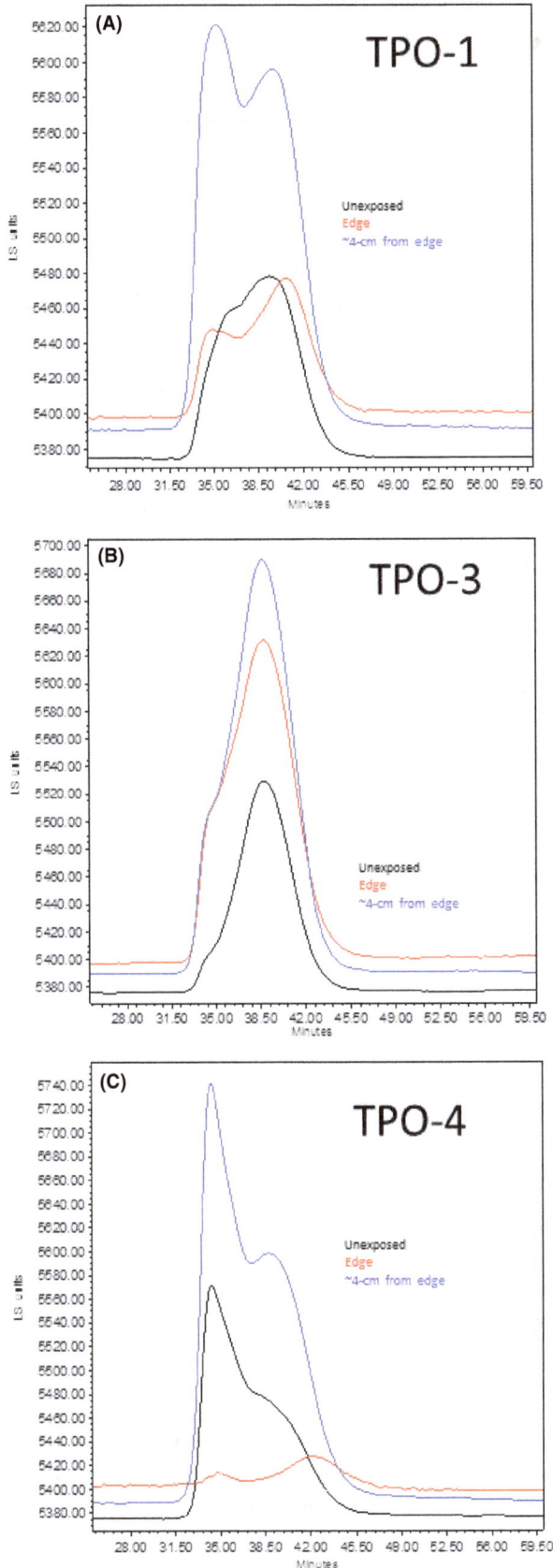

chemistry used, or more likely there could be a greater ability of EVA to form cross-links through a mechanism involving the acetate monomers as seen elsewhere [27]. Tertiary carbons in the backbone are prone to experiencing chain scission. Therefore, it is likely that the chain scission reactions dominate throughout the polyolefin copolymers. In combination, these effects can explain the difference in the cross-linking/chain scission behavior of EVA as compared to TPOs.

Similar to EVA, the slight concave down curvature in Figure 5 indicates the presence of increased amounts of branching at higher molecular weights. For TPO-1 and TPO-4, and to a lesser extent TPO-3, the aged samples showed a relative reduction in intrinsic viscosity near the edge but not at 4 cm distance from the perimeter. Thus, O_2 is degrading the polymers, making them more compact and branched.

The light-scattering measurements at an angle of 90° also provide insight into the cross-linking behavior of the materials (Fig. 6). Here, the strength of the signal is related to the product of concentration and molecular weight. Thus, when two elution peaks are seen in light-scattering detectors, the first peak consists of a small amount of very large molecular weight material, and the second peak is of a much lower molecular weight material where the signal strength is attributable to significantly larger quantities of polymer. Looking at the changes in the relative ratios of these peaks gives a sensitive indication of changes in the large molecular weight fraction.

For TPO-1, the unaged material has a small amount of large molecular weight material. Upon exposure, the position of this large molecular weight fraction moves to larger molecular weights and becomes more pronounced (Fig. 6). Comparison with Figure 5 indicates that this larger molecular weight fraction is only a very small part of the total polymer. Also of note is the fact that the sample taken from a distance of 4 cm from the edge demonstrates more cross-linking than the sample on the edge of the module. Although chain scission dominates the degradation for TPO-1, there is still some evidence cross-linking reactions have occurred. Overall, these changes produced decreases in the mass average molecular weight (M_w) near the edge, and slight increases further in the module, while increasing the polydispersity of the sample as it ages (Table 3).

For TPO-4, there is initially a significant amount of large molecular weight material (Fig. 5) such that the light-scattering chromatograph is dominated by the large molecular weight fraction (Fig. 6). Upon exposure, the

Figure 6. Light-scattering chromatographs of TPOs measured at an angle of 90°. Blue: 4 cm from the edge; red: edge sample; black: unexposed. (A) TPO-1, (B) TPO-3, and (C) TPO-4.

Table 3. TPO size exclusion chromatography, multiangle laser light scattering in TCB.

Sample	M_n (g/mol)	M_w (g/mol)	M_z (g/mol)	PDI
TPO-1				
Unexposed	36,200	124,000	472,000	3.4
Edge	20,200	81,200	454,000	4.0
~4 cm from the edge	26,100	134,000	738,000	5.1
TPO-3				
Unexposed	50,400	135,000	337,000	2.7
Edge	27,300	130,000	461,000	4.8
~4 cm from the edge	35,000	137,000	439,000	3.9
TPO-4				
Unexposed	24,400	172,000	1,040,000	7.0
Edge	8,980	30,100	96,200	3.4
~4 cm from the edge	18,300	135,000	868,000	7.4

M_n is the number average molecular weight, M_w is the weight average molecular weight, M_z is the Z average molar mass, and PDI is the polydispersity index which is equal to M_w/M_n.

small molecular weight fraction becomes more distinct. In the edge sample, M_w is reduced much more than the number average molecular weight (M_n), indicating a higher probability that large chains are reduced, resulting in a reduction in the PDI (Table 3). Neither the light-scattering data nor the molecular weight distribution can clearly demonstrate the formation of cross-links. However, considering that TPO-4 and TPO-1 were made by the same manufacturer and intended to differ primarily in their melting temperatures, it is likely that cross-linking is occurring in TPO-4 but that it is masked by a much stronger chain scission reaction.

For TPO-3, there is initially a weak signal for a large molecular weight fraction in the light-scattering data. After exposure, the large molecular weight fraction becomes more pronounced. Here, M_w remains relatively unchanged after exposure, but M_n is reduced (Table 3). For M_w to be constant, chain scission and cross-linking reactions must be occur at a similar rate. In Figure 5B, both the edge and 4-cm distance from the edge samples demonstrate the formation of a large molecular weight tail which is more pronounced for the edge sample. Thus, oxygen may be enhancing both cross-linking and chain scission reactions slightly in this material.

While TCB at 150°C was a good solvent for all the TPOs, the xylenes and toluene solvent did not produce complete solubility in the Soxhlet extractor (Fig. 1). For TPO-1, toluene was a better solvent than xylenes; but for TPO-4, xylenes were a marginally better solvent than toluene. Similarly, toluene was a poor solvent for PVB and TPU producing gel contents between 28.8% and 78.2%. But when PVB and TPU were tested using methanol and THF in the Soxhlet extractor, gel contents of 0% and ~4.9% were obtained, respectively. This illustrates

how solvent quality, if it is marginal or poor, can give misleading results. Additionally, one must be cautious with TCB at 150°C because the thermal history of the extraction could alter the results, though we did not see evidence for this in our experiments. In the Soxhlet extractor, even with xylenes boiling around 140°C, the liquid is condensed and significantly cooled when in contact with the polymer. One would therefore not expect to see significant degradation of the polymers, even when xylenes are used.

For TPO-1 and TPO-4, a higher gel content was seen for the unexposed sample (Fig. 1). A reduction in the molecular weight is consistent with the SEC experiments, but the poor quality of the solvent leaves open the question of the amount of gel in the materials both before and after the experiment.

Toluene was a better solvent than xylenes for TPO-3 (Fig. 2), but both solvents indicate nearly zero gel content initially. Even though the aged samples indicate significant increases in gel content, the SEC data indicated there was only a small change in the molecular weight distribution (Table 3). Toluene is probably just barely able to dissolve this material such that the small amount of cross-linking occurring is able to significantly affect the measurement of the gel content without really changing the molecular weight distribution significantly.

For all three TPOs, chain scission seems to dominate over cross-linking reactions and is enhanced by the presence of oxygen in the perimeter regions. For TPO-1 and TPO-3 (and possibly TPO-4) there are also some cross-linking chemistry occurring but with a less noticeable dependence on the presence of oxygen. Dominance of chain scission over cross-linking was unexpected because indoor experiments [19, 21] with TPO-1 and TPO-3 mock modules clearly demonstrate a reduction in creep rates above 110°C and 90°C, respectively. This could be because UV light is largely responsible for chain scission, or that the small amount of cross-linked fractions here has a large effect on the rheology of the polymers.

Thermoplastic polyurethane

Even after exposure in Arizona region, the thermoplastic polyurethane (TPU) was highly soluble in THF and could be run in the SEC system using THF. Gel content measurements in THF (Fig. 1) indicate that 5.1, 4.9, and 4.5% of the polymer were not soluble for samples at a distance of 4 cm from the edge in the Arizona region, environmental chamber, and unexposed conditions, respectively. Upon exposure, a large molecular weight tail was seen on the GPC curve due to cross-linking of polymer chains, and there is significant indication of cross-linking in the light-scattering chromatographs (Fig. 7B). However, no signs of chain scission were seen in the mass

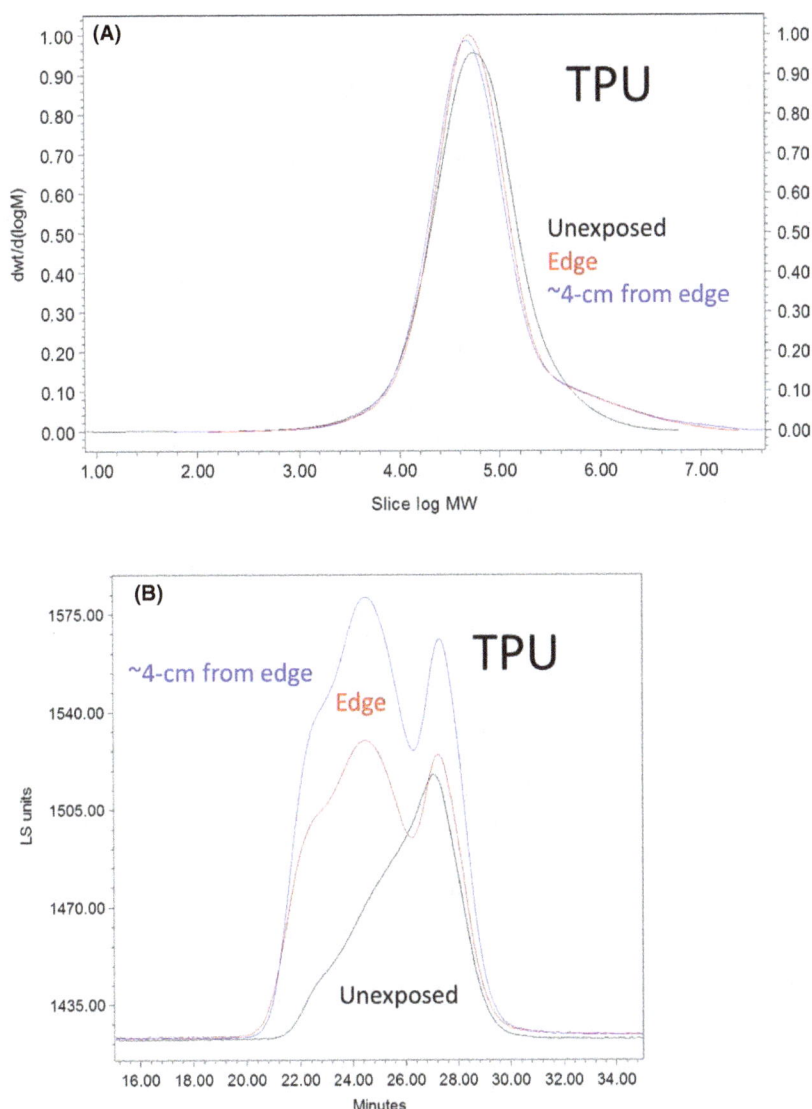

Figure 7. Weight fraction determined by SEC-MALLS (SEC, size exclusion chromatography) and viscometry for the thermoplastic polyurethane thin-film mock module after exposure in Arizona. Samples dissolved in tetrahydrofuran (THF) by extraction at 40°C overnight. Molecular weight (MW) is in units of g/mol. Blue: 4 cm from the edge; red: edge sample; black: unexposed. Lines are normalized to the area under the curve.

chromatograms (Fig. 7A). The absence of small scale for cross-linking is further supported by the fact that M_n actually increased upon exposure (Table 4). Thus, a significant amount of polymer is becoming highly cross-linked.

Initially, the TPU seems to have three bumps in the light-scattering plot (Fig. 7). After exposure, the two larger molecular weight fractions become more pronounced and shifted to larger molecular weights. The shift is slightly larger for the sample near the edge indicating that oxygen has a small additional effect, increasing the rate of cross-link formation. This could be indicative of the use of a block copolymer or a polymer blend where each different polymer types have differing ability to form cross-links.

Table 4. Thermoplastic polyurethane (TPU) size exclusion chromatography, multiangle laser light scattering in a THF (tetrahydrofuran) solvent.

Sample	M_n (g/mol)	M_w (g/mol)	M_z (g/mol)	PDI
TPU				
Unexposed	20,700	89,900	370,000	4.3
Edge	24,700	176,000	2,540,000	7.1
~4 cm from the edge	23,900	229,000	5,160,000	9.6

Conclusions

Modules were deployed in a very hot environment with insulation on their backside in an effort to reproduce one of the worst case situations where creep of the encapsulant might occur. In previous experiments [19, 21], even when noncuring polymer encapsulant formulations were used, only insignificant creeping of the cells or other module components was observed. This was attributed to several factors such as some nonuniformity in the temperature distribution creating cold spots limiting creep. Measurable creep, occurring asymptotically with time, was observed for the more homogeneously constructed thin-film mock modules, particularly for NC-EVA.

For the NC-EVA and the TPOs, the rate of cross-linking and chain scission reactions vary with position in the sample indicating that a chemical species must be entering or leaving the module package affecting the kinetics. It is plausible that oxygen ingress is increasing the rate of chain scission relative to cross-linking and increasing the overall degradation rate.

The EVAs and the TPU tested here increased in their polymer molecular weight attributable to the formation of cross-links. For TPU, no evidence for chain scission was observed, but for EVA chain scission was observed and was enhanced presumably by the presence of oxygen in the perimeter areas. In TPU, these changes did not result in an increase in gel content, but in EVA an increase in gel content was seen even in the chain scission-prone perimeter regions with the NC-EVA sample.

In contrast, the TPOs experienced an overall decrease in molecular weight with aging. SEC data for TPO-1 and TPO-3 indicate some cross-linking reactions are occurring, but this is much less prominent than chain scission reactions. The overall reduction in molecular weight is a small effect and not likely to result in module creep as demonstrated elsewhere [19–21]. When using any thermoplastic material in a PV module, one should determine if cross-linking or chain scission dominates degradation reactions sufficiently to cause a concern for a reduction in molecular weight and subsequent creep.

The materials and "module" designs examined here help to identify the requirements for a module that could be subject to creep during a 25-year field lifetime. It would be difficult to construct an encapsulant that would experience creep when deployed. The encapsulation would have to have a low melting transition (≤70°C as in NC-EVA) providing a low viscosity at operating conditions, the module would have to use a mounting scheme that did not fix the relative positions of the front and back sheets, the module would have to be installed in a very hot climate in a hot mounting configuration, and the encapsulation would have to age predominantly by chain scission rather than cross-linking. Such a module construction might still pass 1000 h of 85°C exposure without creep but fail in the field. Examination at 90°C or 105°C for 200 h (MST 56) has

been recently added to IEC 61730–2 to more rigorously screen for modules that might be prone to creep over prolonged time.

Acknowledgments

This work was part of a large collaborative effort of a number of people working on standards developed at many institutions. The authors gratefully acknowledge the support of the following individuals: Adam Stokes, Alain Blosse, Ann Norris, Bernd Koll, Bret Adams, Casimir Kotarba (Chad), David Trudell, Ed Gelak, Greg Perrin, Hirofumi Zenkoh, James Galica, Jayesh Bokria, John Pern, Jose Cano, Kartheek Koka, Keith Emery, Kent Terwilliger, Kolapo Olakonu, Masaaki Yamamichi, Mowafak Al-Jassim, Nick Powell, Niki Nickel, Pedro Gonzales, Peter Hacke, Ryan Smith, Ryan Tucker, Steve Glick, Steve Rummel, and Tsuyoshi Shioda. This work was supported by the U.S. Department of Energy under Contract No. DE-AC36-08-GO28308 with the National Renewable Energy Laboratory.

Conflict of Interest

None declared.

References

1. Kapur, J., K. Proost, and C. A. Smith. 2009. Determination of moisture ingress through various encapsulants in glass/glass laminates. Pp. 001210–001214 in Photovoltaic Specialists Conference (PVSC), 2009 34th IEEE.
2. Kempe, M. D., G. J. Jorgensen, K. M. Terwilliger, T. J. McMahon, C. E. Kennedy, and T. T. Borek. 2007. Acetic acid production and glass transition concerns with ethylene-vinyl acetate used in photovoltaic devices. Sol. Energy Mater. Sol. Cells 91:315–329.
3. Meyer, S., S. Timmel, S. Richter, M. Werner, M. Gläser, S. Swatek, et al. 2014. Silver nanoparticles cause snail trails in photovoltaic modules. Sol. Energy Mater. Sol. Cells 121:171–175.
4. Reid, C. G., S. ferrigan, I. Fidalgo, and J. T. Woods. 2013. Contribution of PV encapsulant composition to reduction of potential induced degradation (PID) of crystalline silicon PV cells. 28th European Photovoltaic Solar Energy Conference and Exhibition.
5. Reid, C. G., J. G. Bokria, and J. T. Woods. 2013. UV Aging and Outdoor exposure correlation for EVA PV encapsulants. Pp. 882508-882508-11 in 2013 SPIE, San Diego, California.
6. Cuddihy, E. F., A. Gupta, C. D. Coulbert, R. H. Liang, A. Gupta, and P. Willis, et al. 1983. Applications of ethylene vinyl acetate as an encapsulation material for terrestrial photovoltaic modules. DOE/JPL/1012-87 (DE83013509).

7. Dietrich, S., M. Pander, M. Ebert, and J. Bagdahn. 2008. Mechanical assessment of large photovoltaic modules by test and finite element analysis.

8. Wohlgemuth, J., M. Kempe, D. Miller, and S. Kurtz. 2011. Developing standards for PV packaging materials. *SPIE*. Proceedings of SPIE Conference

9. Wohlgemuth, J., and W. Herrmann. 2005. Hot spot tests for crystalline silicon modules. Pp. 1062–1063*31st IEEE Photovoltaic Specialists Conference, 3-7 Jan. 2005*.

10. Wohlgemuth, J.. 2013. Hot Spot Testing of PV Modules. *SPIE, San Diego, California*.

11. Kurtz, S., K. Whitfield, D. Miller, J. Joyce, J. H. Wohlgemuth, and M. D. Kempe, et al. 2009. Evaluation of high-temperature exposure of rack-mounted pohotovoltaic modules. *34th IEEE PVSC*.

12. D. C. Miller, M. D. Kempe, S. H. Glick, and S. R. Kurtz. 2010. Creep in photovoltaic modules: Examining the stability of polymeric materials and components. *35th IEEE Photovoltaic Specialists Conference (PVSC)*, 20–25 June 2010.

13. Miller, D. C. 2013. Examination of a standardized test for evaluating the degree of cure of EVA encapsulation (#1066/0731). National Renewable Energy Laboratory, Golden, Colorado.

14. Cuddihy, E. F., and P. B. Willis. 1984. Antisoiling technology: theories of surface soiling and performance of antisoiling surface coatings. *DOE/JPL/1012-102 (DE85006658)*.

15. Holley, W. W., and S. C. Agro. 1998. Advanced EVA-based encapsulants. Final Report January 1993-June 1997 *NREL/SR-520-25296*.

16. Pern, F. J., and A. W. Czanderna. 1992. Characterization of ethylene vinyl acetate (EVA) encapsulant: effects of thermal processing and weathering degradation on its discoloration. Sol. Energy Mater. Sol. Cells 25:3–23.

17. Patel, M., S. Pitts, P. Beavis, M. Robinson, P. Morrell, N. Khan, et al. 2013. Thermal stability of poly(ethylene-co-vinyl acetate) based materials. Polym. Testing 32:785–793.

18. Klemchuk, P., M. Ezrin, G. Lavigne, W. Holley, J. Galica, and S. Agro. 1997. Investigation of the degradation and stabilization of EVA-based encapsulant in field-aged solar energy modules. Polym. Degrad. Stab. 55:347–365.

19. Kempe, M. D., D. C. Miller, J. H. Wohlgemuth, S. R. Kurtz, J. M. Moseley, and Q.-U.-A. S. J. Shah, et al. 2012. A field evaluation of the potential for creep in thermoplastic encapsulant materials. *Proceedings of the 38th IEEE Photovoltaic Specialists Conferences*.

20. Moseley, J. M., D. C. Miller, Q.-U.-A. S. J. Shah, K. Sakurai, M. D. Kempe, and G. TamizhMani, et al.,2011. The melt flow rate test in a reliability study of thermoplastic encapsulation materials in photovoltaic modules. Pp. 1–20 *NREL/TP-5200-52586*.

21. Kempe, M. D., D. C. Miller, J. H. Wohlgemuth, S. R. Kurtz, J. M. Moseley, and Q.-U.-A. S. J. Shah, et al., 2015. Field Testing of Thermoplastic Encapsulants in High-Temperature Installations. *Submitted to Energy Science & Engineering*.

22. Lux, C., U. Blieske, E. Malguth, and N. Bogdanski. 2013. "Variations in Cross-Link Properties of EVA of Un-Aged and Aged PV-Modules. *29th European Photovoltaic Solar Energy Conference and Exhibition*.

23. Marais, S., Y. Hirata, D. Langevin, C. Chappey, T. Nguyen, and M. Metayer. 2002. Permeation and Sorption of Water and Gases Through EVA Copolymers Films. Mater. Res. Innovations 6:79–88.

24. Kempe, M. D. 2006. Modeling of rates of moisture ingress into photovoltaic modules. Sol. Energy Mater. Sol. Cells 90:2720–2738.

25. Hulsmann, P., D. Philipp, and M. Kohl. 2009. Measuring temperature-dependent water vapor and gas permeation through high barrier films. Rev. Sci. Instrum. 80:113901.

26. McIntosh, K. R., N. E. Powell, A. W. Norris, J. N. Cotsell, and B. M. Ketola. 2011. The effect of damp-heat and UV aging tests on the optical properties of silicone and EVA encapsulants. Prog. Photovoltaics Res. Appl. 19:294–300.

27. Rätzsch, M., and U. Hofmann. 1991. The Crosslinking of Ethylene-Vinyl Acetate Copolymers with Sodium Alcoholates. J. Macromolecular Sci.: Part A - Chemistry 28:145–157.

7

Hot carrier solar cell as thermoelectric device

Igor Konovalov[1] ⓘ & Vitali Emelianov[1,2]

[1]Ernst Abbe University of Applied Science in Jena, Carl Zeiss Promenade 2, 07745 Jena, Germany
[2]Thuringian Postgraduate School of Photovoltaics "Photograd", Institute of Physics, Technical University Ilmenau, 98684 Ilmenau, Germany

Keywords
Double heterostructure, kinetic transport theory, prototype hot carrier solar cell, Seebeck effect

Correspondence
Igor Konovalov, Ernst Abbe University of Applied Science, Carl Zeiss Promenade 2, 07745 Jena, Germany. E-mail: igor.konovalov@fh-jena.de

Funding Information
Carl Zeiss Foundation, Land Thüringen (Grant/Award Number: 'Photograd', 'RFAproPV'), European Commission (Grant/Award Number: 'EFRE RFAproPV').

Abstract

Improvement of solar cell efficiency beyond the Shockley–Queisser limit requires introduction of new physical concepts. One such concept is hot carrier solar cell, proposed more than three decades ago and still not impressively demonstrated in experiment. Here we show that hot carrier solar cell may be considered as thermoelectric device based on Seebeck effect. This enables one to describe the operation of hot carrier solar cell in a simple way. We fabricated a prototype of the hot carrier solar cell showing open circuit voltage at room temperature larger than the band gap in the absorber material. Extrapolation of open circuit voltage to absolute zero temperature results in barrier height depending on light intensity, interpreted by splitting of quasi-Fermi levels between the regions of different carrier temperature. Properties of the prototype solar cell may be described by kinetic transport theory as well as from the point of view of the thermoelectric theory.

Introduction

The idea of hot carrier solar cell was proposed by Ross and Nozik in attempt to overcome Shockley-Queisser efficiency limit of 31% at 1 Sun (or 41% at full concentration) in a simple photovoltaic device [1]. Conceptual implementations of Würfel and other researchers are known [2, 3] and are presented in the following in some detail. Figure 1 shows the classical hot carrier cell concept consisting of light absorber with hot electron gas and two semiconductors with energy selective contacts (ESC) at the interfaces. The ESC filter out carriers having certain energies. The carrier cooling is suppressed, but Auger and radiative recombination mechanisms are considered. The doping concentration in the outer semiconductors can be chosen such that the carrier density in the bands at the energies of selective contacts is sufficient to maintain the same flux in both directions through the energy selective contacts. At this condition, there is no energy flow through the energy selective contacts, so that the electron gas in the absorber is only heated by the Sun but can be cooled only by photonic reemission. The heating will stop at the thermodynamic equilibrium with the Sun, when the radiative recombination reverts the radiation to the Sun (through the concentrator optics not shown in the Figure) in each arbitrary spectral region in accordance with detailed balance principle. The Auger recombination does not change the energy of the gas in the absorber, because all the particles involved remain in the absorber. Similarly, the outer semiconductors maintain the same temperature as the ambient. In the Carnot process, the gas is at thermodynamic equilibrium in turn with the heater and with the cooler during the isothermal processes. Similarly, the process in Figure 1 is reversible process and has the Carnot efficiency. Therefore, Figure 1 represents a perfect solar cell with the thermodynamic efficiency never to be topped because the Carnot efficiency cannot be exceeded. The photovoltaic efficiency is achieved when a reversible heat engine is optimally driven off equilibrium, resulting in the photovoltaic efficiency limit

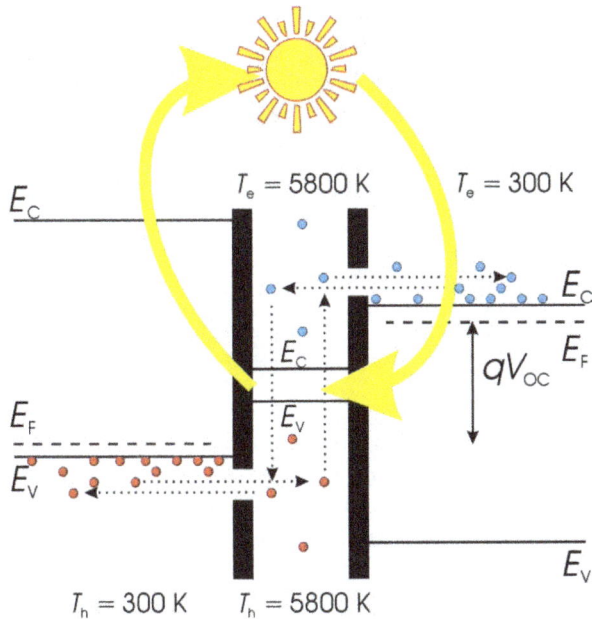

Figure 1. Hot carrier solar cell structure at simultaneous equilibrium with the heater (Sun) and the cooler (ambient) at open circuit condition. $T_{e,h}$ is the temperature of the electron and hole gas.

of at least 85% (Fig. 2A, [4]. This fact is the reason for a continued work at hot carrier solar cell concept throughout the last three decades. Unfortunately, no experimental proof of this hot carrier concept operating at room temperature was attained so far. We modified this concept by replacing the ideal energy selective contacts with the energy barriers being the band offsets at the interfaces (Fig. 2B, [5, 6]. The modified process is, however, thermodynamically nonreversible. A narrow band gap absorber harvests solar radiation in a wide spectral range, and the output voltage is larger than the band gap energy of the absorber semiconductor.

The cell operation is usually described from the point of view of the kinetic transport theory by consideration of quasi-equilibrium thermal emission of the hot carriers from the absorber material into the energy filtering structure. This emission may be described in simple cases by the thermionic emission theory [3, 6–10], being a kinetic transport theory. Kinetic theories are in principle well-suited to describe charge transport in the solar cell; however, the description is rather complex in various specific cases.

Seebeck Theory of Hot Carrier Solar Cell

From the point of view of the thermodynamics, a hot carrier solar cell is a typical thermoelectric device. One can imagine a solar cell being a black body absorber

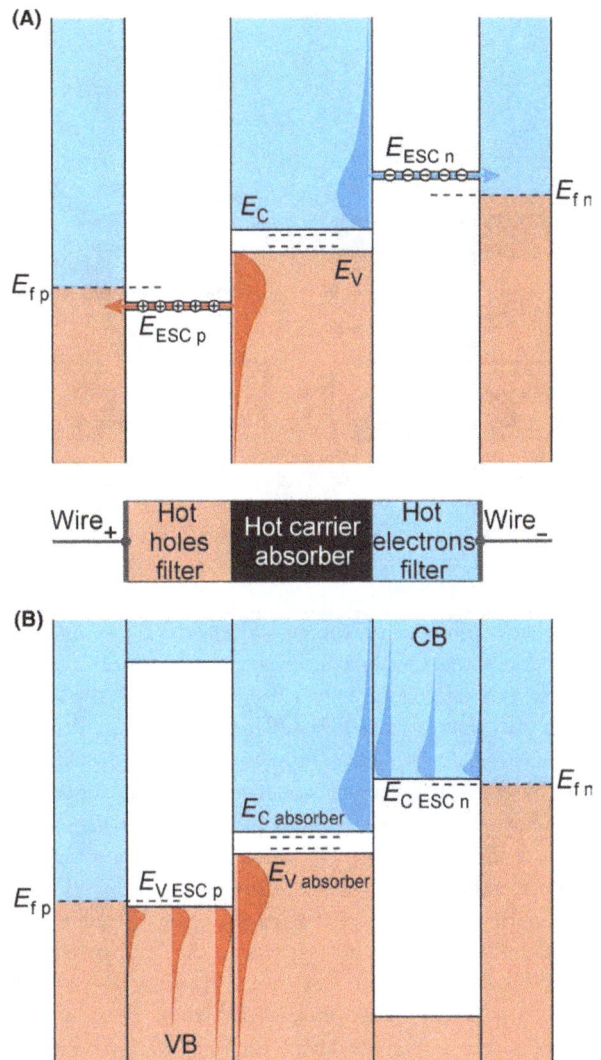

Figure 2. Conceptual implementations of hot carrier solar cells: (A) classical hot carrier cell concept [2]; (B) novel concept [5, 6], discussed here. Energy distribution of hot electrons and holes are shown in the absorber layer: E_V and E_C are energy of the valence and the conduction band edges in the absorber. The energy filters provide energy selective transfer of the hot carriers from the absorber to the metal contacts. VB and CB are the valence and conduction bands in the energy selective layers around the absorber; E_{ESC} are the energy of band edges in the energy selective contacts. Three stages of carrier thermalization in the energy selective contacts are shown: (1) truncated carrier distribution from the absorber being far away from thermal equilibrium; no temperature or Fermi level can be ascribed here; (2) thermalized carrier distribution at a lower temperature than in the absorber; (3) carrier distribution after cooling down to the ambient temperature.

attached to a thermoelectric converter (Fig. 3A). Typical efficiencies of the thermoelectric converters are below 10%, limiting the efficiency of this solar thermal cell. In the solar thermal cell, the lattice of the absorber material is heated by the Sun, and the electron gas of a thermocouple gets heated from the lattice. This circumstance has two

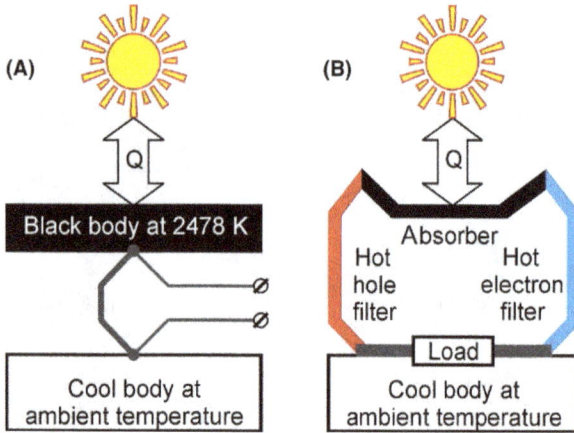

Figure 3. Hot carrier solar cell as thermoelectric device. Model (A) utilizes a conventional thermoelectric converter, attached to a black body absorber. Model (B) is hot carrier solar cell utilizing energy selective filtering of hot carriers from absorber.

drawbacks for efficient energy conversion: (1) in attempt to increase the Carnot efficiency, the temperature of the black body should be high (ideally 2478K, as shown in [11], but most of materials will melt or at least degrade at this high temperature; and (2) Solar heat is partially lost by the lattice thermal conductivity via the phonons. In a hot carrier solar cell both problems may be solved, because the electron gas is heated by the solar radiation directly, rather than via the lattice, which remains cold (Fig. 3B). In both cases the open circuit voltage results from diffusion of the hot carriers in the bulk materials.

Thus, we discuss the hot carrier solar cell as a thermoelectric device, based on Seebeck effect. Being a bulk rather than an interface effect, Seebeck effect determines the open circuit voltage of the hot carrier solar cell independently of the band offsets. Remarkably, Seebeck coefficient shows transitivity, whereas the band offsets do not. The open circuit voltage of a multilayer structure, as it is proposed for the hot carrier solar cells (Fig. 2B), is simply determined by the difference of Seebeck coefficients in the materials where the thermal gradient is large, being typically the energy selective contacts. Seebeck coefficient of the absorber material, staying at constant high electron temperature, makes no influence here. As a result, energy filtering at interfaces between the absorber and the energy selective contacts does not define the open circuit voltage, since it is determined by the bulk Seebeck effect. This finding does not contradict the thermionic theory, because an appropriate band bending may substitute for the band offset during the formation of the energy barrier. Although a specific narrow band energy selective contact material may show large Seebeck coefficient, any other material with as large Seebeck coefficient would also result in as large open circuit voltage.

A theoretical question arises whether it is possible to describe Seebeck effect from the point of view of thermionic emission at a heterojunction. We consider thermionic emission in a simple heterojunction without band bending at open circuit condition by calculation of flux Φ of carriers in both directions through the interface and over an energy barrier (Fig. 2C):

$$|\Phi_{1-2}| = |\Phi_{2-1}|,\qquad(1)$$

$$\frac{n_1 \langle v_1 \rangle}{4}\sqrt{\frac{m_2}{m_1}} = \frac{n_2 \langle v_2 \rangle}{4}\sqrt{\frac{m_1}{m_2}}\cdot\exp\left(-\frac{\Delta E_c}{kT_2}\right).\qquad(2)$$

Here $n_{1,2}$ are electron concentrations, $m_{1,2}$ are effective masses in each of the materials, $\langle v_1 \rangle$ is mean thermal velocity of cold electrons at a temperature T_1, $\langle v_2 \rangle$ is mean thermal velocity in the hot electron gas at a temperature $T_2 > T_1$, ΔE_C is energy barrier for electrons. The flux density $\frac{n \langle v \rangle}{4}$ is known from kinetic molecular theory of ideal gas, applicable to electrons within effective mass approximation. The exponent term describes the probability for carriers to surpass a barrier ΔE_C in approximation of Boltzmann distribution. One readily arrives at the known Richardson formula for thermionic emission by substituting the mean thermal velocity and a classical expression for the electron concentration in the nondegenerated semiconductor within the effective mass approximation [12] into equation (2)

$$\langle v \rangle = \sqrt{\frac{8kT}{\pi m}},\qquad(3)$$

$$n = \frac{2(2\pi mkT)^{3/2}}{h^3}\exp\left(-\frac{E_C - E_F}{kT}\right).\qquad(4)$$

Here h is Planck's constant, E_C and E_F are energy of the conduction band edge and the Fermi level. An additional correction term (square root of the effective mass ratio) is necessary in equation (2) in order to account for a change in the carrier mass after penetration through the interface, an effect not observed for molecules of the ideal gas. The form of the correction term results from consideration of (1) at equilibrium.

If $T_1 \neq T_2$, the electron Fermi level splits at the interface, whereas the split is equal to qV_{OCE}:

$$E_{F_1} = E_{F_2} + qV_{OCE}.\qquad(5)$$

Here V_{OCE} is part of the open circuit voltage due to hot electrons. As it follows from the band diagram in Figure 6:

$$E_{C_1} = E_{C_2} + \Delta E_C.\qquad(6)$$

By substitution equations (3)–(6) into (2) for $m_1 \neq m_2$, $T_1 \neq T_2$ and resolving the equation for V_{OCE} we obtain:

$$V_{OCE} = -\frac{\Delta T}{T}\frac{E_{C_1}-E_{F_1}}{q} - \frac{2kT}{q}\left(1+\frac{\Delta T}{T}\right)\ln\left(1+\frac{\Delta T}{T}\right). \quad (7)$$

Here $\Delta T = T_2 - T_1$. If $\Delta T \ll T_{1,2}$, which is rather common for contemporary hot carrier solar cells, Taylor's series expansion of the logarithmic function up to the linear term results in

$$V_{OCE} = -\frac{k}{q}\left[-\frac{E_{F_1}-E_{C_1}}{kT}+2\right]\Delta T = S\Delta T. \quad (8)$$

Equation in square brackets in (8) is a specific case of Pisarenko relation for Seebeck coefficient of the nonde-generated electron gas [13, 14]:

$$S = -\frac{k}{q}\left\{-\frac{E_{F_1}-E_{C_1}}{kT}+\left(r+\frac{5}{2}\right)\right\}. \quad (9)$$

Here r is the exponent of the energy in the expression for the carrier relaxation time from approximation of the collision term in the Boltzmann kinetic equation:

$$\tau(E) = CE^r, \quad (10)$$

and C is a constant. The value of $r = -1/2$ results from a model of free carriers of a constant scattering cross section, for which the thermal velocity is proportional to the square root of the kinetic energy. In conclusion, the open circuit voltage through thermionic emission of the hot electrons in a heterostructure is equivalent to Seebeck effect in a narrow temperature range, valid at least for the nondegenerated semiconductors.

Experimental

Prototype fabrication

Double side polished epi-ready (100) ZnTe $10 \times 10 \times 0.4$ mm³ wafers doped by phosphorus to a hole concentration of $1 - 1.2 \times 10^{17}$ cm⁻³ were purchased from JX Nippon Mining & Metals. The wafers were split in $5 \times 5 \times 0.4$ mm³ chips. Each chip substrate was etched for 20 sec in 0.4 %v bromine (Alfa Aesar, 99.8 %, CAS: 7726-95-6) solution in methanol (Carl Roth, ≥99.95 %, CAS: 67-56-1) prior to evaporation of a silver (Carl Roth, ≥99.9 %, CAS: 7440-22-4) back contact. Silver was evaporated in a vacuum deposition system B30.2 (VEB Hochvakuum, Jevatec GmbH, Jena, Germany) at P = 3×10^{-5} mbar. The top surface of ZnTe substrates was sequentially polished with 0.1 μm diamond paste and 20 nm SiO$_2$ slurry. Then the

ZnTe chips mounted on a copper tape conductor were temporary sealed by polyimide tape with Ø 3 mm aperture. Open ZnTe surface was etched for 20 sec in the bromine-methanol solution, rinsed in the double deionized ($\sigma < 0.1$ μS) water and then placed into an electrochemical cell under an anodic potential. The aqueous electrolyte contained 0.05 M Pb(NO$_3$)$_2$ + 0.001 M Se + 0.5 M Cd(NO$_3$)$_2$ at pH = 2 corrected by HNO$_3$. It was deaerated with nitrogen in a three-electrode electrochemical cell prior to the deposition. The chemicals were at least 99.9% grade purchased from Carl Roth and Alfa Aesar. The potential of the working electrode was controlled by a potentiostat (VersaSTAT 4, Princeton Applied Research, Oak Ridge, TN, USA) in the three-electrode cell and measured against the saturated silver–silver chloride electrode (Ag/AgCl/KCl$_{sat}$). A Ø 3.5 mm × 50 mm glassy carbon rod (Alfa Aesar) was used as the auxiliary electrode. The electrochemical deposition was performed at a potential slightly more positive than that of bulk lead reduction. The rate of cathodic reaction of PbSe formation was limited by the Se concentration. Since the PbSe formation requires electrons on p-ZnTe surface, the deposition was performed under the illumination of a light emission diode (8.7×10^{-4} W, 460 nm). The temperature of the electrolyte was stabilized at room (24°C) temperature. Potential of the p-ZnTe substrate as the working electrode was varied in time for controlling the PbSe film composition. Analysis of the film structure, chemical composition and crystalline properties was performed using a scanning electron microscope (Ultra 55, Carl Zeiss AG, Oberkochen, Germany) with a field electron emitter, equipped with EDX and EBSD units (Bruker Corporation, Billerica, MA, USA). Acceleration voltage was 30 kV. A composition ratio of Pb:Se = 1:1 was confirmed by EDX analysis of thicker (300–600 nm) PbSe films. The top transparent electrode and the second heterojunction were produced by sputtering 100 nm of n-ZnO on top of the p-ZnTe/PbSe structure. A gauze of Ø 135 μm stainless steel wires resulting in 530 × 530 μm² cells was used as a hard mask for structuring of top n-ZnO layer. The ZnO(99) Al(1)wt% target with 3N5 purity was purchased from FHR Analgebau GmbH, Ottendorf-Okrilla, Germany. The magnetron sputtering of ZnO, combined with the bonding of back silver contact to the massive copper plate, was done at an argon pressure of 2×10^{-3} mbar, at an RF power of 150 W and at a substrate temperature of 200°C.

Optoelectrical measurements

The individual cells were contacted by a spring loaded pin in all electrical measurements. The I-V measurements were done using Keithley 2401 source meter controlled by ReRa Tracer software (ReRa Solutions BV, Nijmegen, Niederlands). For the illuminated I-V measurements, the natural sunlight was focused by a 61 mm diameter glass lens with a focal length of 145 mm. The sunlight intensity was monitored by a calibrated reference solar cell type RS-00-1 (Fraunhofer ISE, Freiburg, Germany). Due to particular optical absorption in the atmosphere, related to the weather conditions, the spectral power density function continuously changes in time. Some part of this optical absorption that is due to N_2 and O_2 molecules is constant and is always present. Additional scattering may originate from eventual water droplets ("clouds"). On a sunny day, when the spectral power density is at its maximum, with the elevation of Sun of 41.8°, the "air mass" is 1.5 and the conditions are close to the "terrestrial 1 Sun" fixed in the Standard Test Conditions for solar cells. At these conditions, any calibrated solar cell will produce its rated short circuit current. If additional scattering due to moisture in the atmosphere is present, the short circuit current of the reference solar cell will be below the rating and the spectral power density function will randomly deviate from the standard test conditions. In our experiments we assured that the output of the reference solar cell was close to its rating, meaning that the spectral power density function is close to the standard one. The open circuit voltage versus temperature was measured using a digital voltmeter, whereas the temperature was controlled by a Peltier element and measured by a digital thermometer. The measurement was taken in an electrically shielded desiccator. Samples were illuminated by a 658 nm laser diode module, equipped with a focusing optics (Thorlabs, Dachau/Munich, Germany) and a silicon monitor diode. The laser module output power was calibrated according to the absolute thermal reference measurements.

Results and Discussion

We fabricated a double heterostructure according to the idea of Figure 2B on a (100) p-doped zinc telluride (ZnTe) substrate. ZnTe has zincblende structure type with a lattice constant of 0.6103 nm [15] and a band gap of 2.27 eV at 300 K [16]. A lead selenide (PbSe) film of 20 nm thickness, according to the Faraday law, was deposited by the electrochemical epitaxy [6, 17]. PbSe has the rock salt structure type, with a lattice constant of 0.6117 nm [18], and a band gap of 0.28 eV [19]. Figure 4 shows scanning electron microscopy (SEM) micrographs of the circular region where PbSe film was deposited. The region is visible at low magnification due to a mass difference of the constituting atoms in the substrate and in the film. The surface of the film remains microscopically rather smooth, so that the origin of the contrast is not related to an eventual change in the surface morphology. Pole figure, obtained by the electron backscattering diffraction (EBSD) shown in Figure 4B as the inset, confirms the epitaxial growth of the film.

The negative transparent contact was deposited by the magnetron sputtering of 100 nm ZnO doped with 1% Al_2O_3. ZnO layer having a band gap of 3.37 eV [19] was deposited through a mask, forming 0.24 mm² square areas (Fig. 5B). The areas in the middle part of the sample are located on top of the deposited PbSe layer, whereas ZnO was deposited directly on ZnTe at the circumference of ZnTe substrate. Both area types were illuminated by the solar radiation at various concentrations. The intensity of the solar radiation was monitored by short circuit current of a silicon reference solar cell. The measured current was 77–91% of the standard current corresponding to one Sun. The cell under investigation was contacted at the rear side by an evaporated silver layer and bonded at 200°C in vacuum/argon to a massive copper plate. During the I-V measurements, the temperature of the copper plate was in the range 49–55°C. At these test conditions, I-V characteristics shown in Figure 5C were measured.

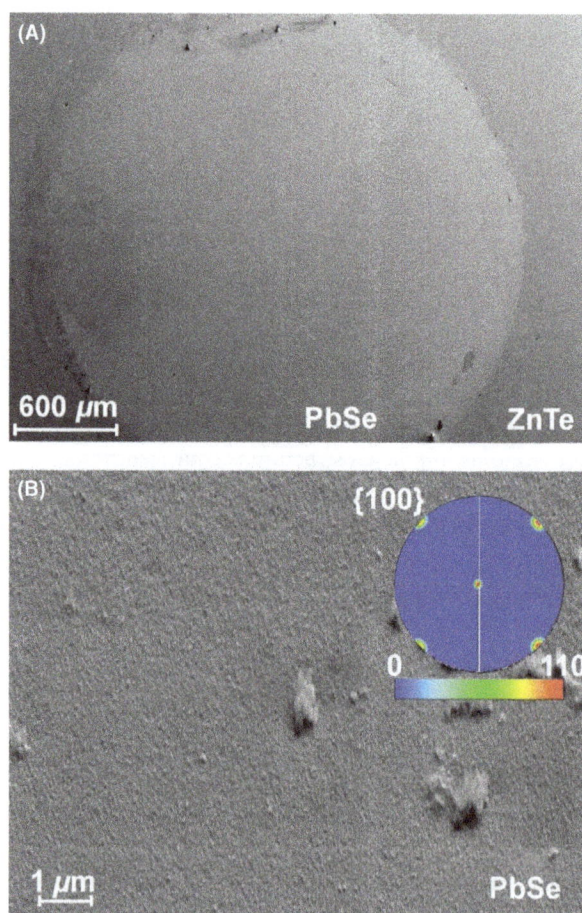

Figure 4. Scanning electron micrographs of 20 nm PbSe film on ZnTe. (A) Plan view of the PbSe thin film (circular region). (B) Tilted (70°) micrograph in the center of the covered region. Inset shows the pole figure obtained by the EBSD in the whole area of the micrograph B), confirming the epitaxial growth. The EBSD patterns were analyzed at 8153 pixels, resulting in 114 zero solutions (1.7%).

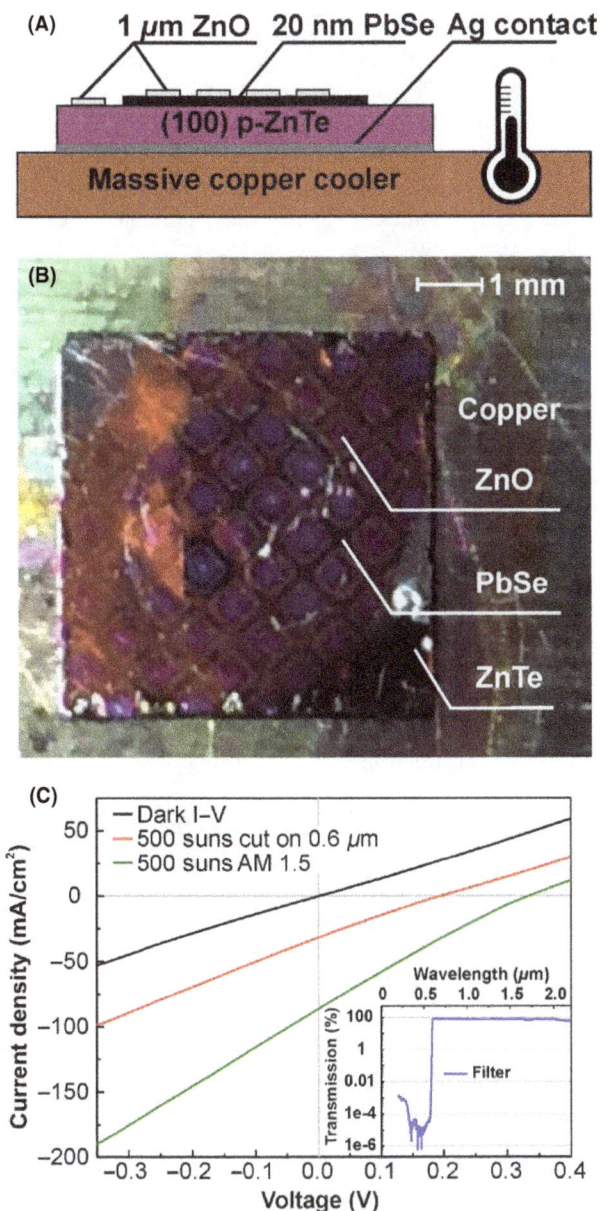

Figure 5. Hot carrier solar cell under test: (A) the layered structure of p-ZnTe/PbSe/n-ZnO double heterostructure bonded to copper cooler; (B) a plan view of the prototype: the circular region in the middle is the PbSe film; (C) the I-V characteristics of the hot carrier solar cell prototype in the dark, under concentrated (×500) and filtered concentrated (×500) sunlight; inset: the spectral characteristic of the long pass filter.

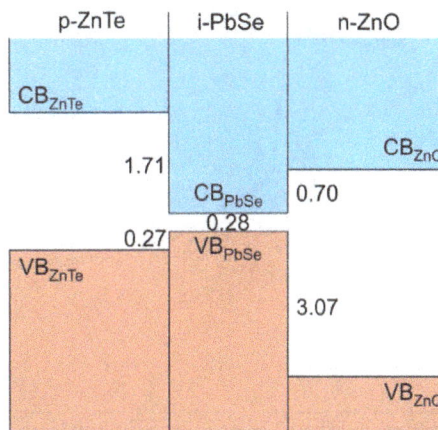

Figure 6. Band alignment in the double heterostructure ZnTe/PbSe/ZnO [19, 20]. The numbers shown near the interfaces are the band offset in electronvolts.

The band offsets at ZnTe/PbSe and PbSe/ZnO interfaces [19, 20] are shown in Figure 6. The band diagram implements the idea of the hot carrier solar cell in Figure 2B. Despite the small absorption in the thin PbSe film, there is a distinct improvement of the photo response in the structures with the PbSe absorber: while the structures with PbSe show open circuit voltage of 25 mV at 1 Sun and 180 mV at 15 Suns, the open circuit voltage of the ZnTe/

ZnO structure shows only 7 mV and 65 mV, respectively. Although a part of the solar spectrum is absorbed in ZnTe and results in some additional photo response, the major contribution originates from carriers in PbSe. Photocarriers originating from the ZnTe substrate may contribute to the photovoltage, if the prototype is excited in the spectral region at a wavelength below the absorption edge of ZnTe (520 nm). One of the I-V characteristics in Figure 5C was measured under concentrated solar illumination with the shortwave portion of the spectrum below 600 nm blocked by a filter. The solar cell continues to operate also at the filtered illumination conditions. We illuminated the solar cell with 1885 nm laser radiation and also measured a small photovoltage by using a lock-in amplifier. The open circuit voltage of the device under concentrated sunlight, 0.33 V, exceeds the bulk band gap of the PbSe material at room temperature, being 0.28 eV [19]. The quantum confinement in a 20-nm-wide quantum well is expected to result in a band gap evolution up to 0.318 eV [21]. This phenomenon can be explained by the diffusion of the hot carriers out of the absorber material prior to their cooling. The characteristic length L_c of hot carrier diffusion process is the hot carrier cooling length [22] with respect to the carrier cooling time τ_C, so that $L_C = \sqrt{D_C \tau_C}$, where $D_C = \mu_C kT/q$ is carrier diffusion coefficient and μ_C is carrier mobility. Assuming a temperature of 300 K, a carrier cooling time of 10 ps and a carrier mobility of $\mu_C = 1000$ cm^2/V s, one can obtain an L_C (PbSe) = 160 nm for both electrons and holes. Hot carrier cooling length is thus significantly larger than the thickness of the PbSe absorber in the prototype (20 nm). Therefore, the hot carriers are expected to leave the absorber before cooling down. The photovoltage due to hot carriers depends on the Seebeck coefficients in ZnTe and ZnO at their specific doping levels, and on the carrier temperature in the PbSe absorber.

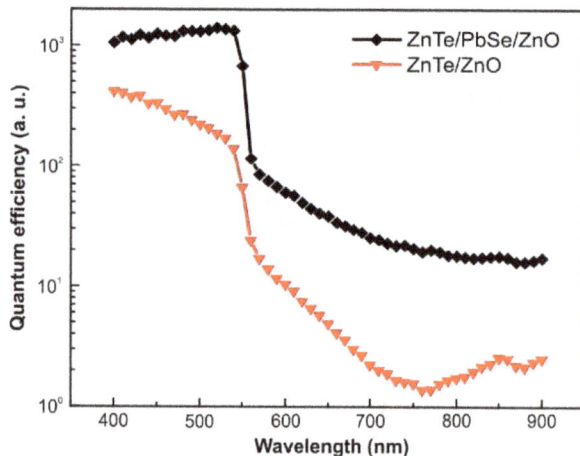

Figure 7. Quantum efficiency of the double heterostructure ZnTe/PbSe/ZnO (diamonds) and of the single heterostructure ZnTe/ZnO without the PbSe absorber layer (triangles) at low illumination intensity. The carrier collection in the 20-nm-thick absorber layer is demonstrated in the infrared region.

The upper graph in Figure 7 shows the quantum efficiency of complete double heterostructre ZnTe/PbSe/ZnO under small illumination intensity, whereas the lower graph shows the same for a reference heterostructure ZnTe/ZnO without the PbSe absorber layer. The absorption region of ZnTe is below 560 nm, resulting in additional strong photoresponse due to carrier collection in ZnTe. The lower photoresponse of ZnTe without PbSe may be related to sputtering damage of ZnTe during ZnO deposition. The structure with PbSe absorber shows at least one order of magnitude larger photoresponse in the infrared region, proving the existence and role of the absorber layer in the carrier collection, although the escape of cold carriers from the absorber is hindered by the band offsets.

The Seebeck coefficients were estimated in both ZnTe and ZnO. The ZnTe crystal of the same charge as used for the deposition was contacted by a copper foil at room temperature and by a hot metallic electrode at 550 K at its corner. A Seebeck coefficient of 430 μV/K was estimated. A ZnO film was sputtered on the soda lime glass at the identical conditions as during deposition of the heterostructures. Similarly measured Seebeck coefficient in ZnO was -67 μV/K. The difference of the Seebeck coefficients amounts to 500 μV/K. Under assumption that the open circuit voltage results only from the thermo power of the hot carriers generated in PbSe, the temperature of carriers in absorber reaches 600 °C at ×500 sunlight concentration. The lattice temperature in the double heterostructure at ×500 suns was roughly estimated from the 1D model of the transverse heat conductivity from the surface of the ZnTe wafer to the copper base

plate (Fig. 5A). Only a lattice thermal conductivity coefficient of 0.18 W cm^{-1} K^{-1} [23] was considered. The temperature difference between the faces of the wafer is expected to be 13 K at ×500 suns concentration. Therefore, it can be assumed that the maximum lattice temperature at the double heterostructure does not exceed 70°C being dramatically smaller than the estimated carrier temperature.

From the points of view of both the thermionic emission and the diffusion theories of charge transport in a junction, the relation between the open circuit voltage of a solar cell and its temperature is linear with the intercept at 0 K being the activation energy of the dominant charge transport mechanism in electronvolts [24]. For example, Figure 8A shows this relation for a conventional single crystalline silicon solar cell for two different photon fluences. Both the relations are linear in a narrow temperature range close to the room temperature and both of them can be extrapolated to a common intercept at about 0 K with its value close to the band gap of silicon in electronvolts. A similar measurement with the prototype hot carrier solar cell results also in a linear function, suggesting the thermionic nature of the current flow (Fig. 8B). However, the intercepts at 0 K strongly depend on the photon fluence. The magnitude of the intercepts is close to the smaller of the measured band offsets. The photon energy in this experiment was chosen smaller than the band gap in ZnTe, so that the carrier generation took place predominantly in PbSe layer. This experimental result may be considered from the kinetic point of view by consideration of different thermionic emission situations. If the carrier temperature is the same throughout the structure, the quasi-Fermi levels for electrons are the same in PbSe and in ZnO, whereas the quasi-Fermi levels for holes are the same in PbSe and in ZnTe (Fig. 8D left). The open circuit voltage, also extrapolated to 0 K, cannot exceed the band gap of PbSe in electronvolts (we expect no population inversion in PbSe through simple optical excitation in this material). At a larger illumination intensity, the carrier temperature in PbSe may become higher than both in ZnTe and ZnO, so that the three layers are out of thermodynamic equilibrium and the electrochemical potential (quasi-Fermi levels) exhibits an offset at the interfaces (Fig. 8D right). The potential barrier height for the current of the thermionic emission flowing across the barriers at the interfaces includes additional energy of the band offsets, so that it is larger than the band gap in PbSe. Therefore, the linear function V_{oc} versus temperature extrapolates to the voltages higher than the band gap in absorber in electronvolts. This is another evidence of the hot carrier filtering in the structure. On the other hand, the open circuit voltage represents a thermoelectric

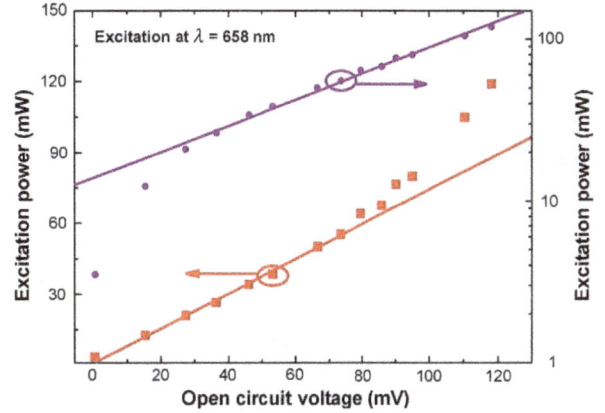

Figure 9. Suns-V_{OC} characteristics as a linear (lower) and as a semilogarithmic (upper) plots. The measurement data are the same. The optical excitation is at 658 nm, the temperature is 300 K.

voltage, showing linear dependence on the temperature difference according to the Seebeck coefficients:

$$V_{OC} = S\left(T_H - T_L\right). \quad (11)$$

The hot carrier temperature T_H can thus be calculated from the measured lattice temperature T_L, if the thermo-voltage V_{OC} and the difference of the Seebeck coefficients S are known. Figure 8C shows, that the hot carrier temperature depends linearly on the lattice temperature:

$$T_H = AT_L + B. \quad (12)$$

Here, A is the slope and B is the intercept of the linear dependence of T_H on T_L, both being functions of the illumination intensity. Substitution of equation (12) into (11) shows that V_{OC} still depends linearly on the lattice temperature, but with the slope being dependent on the illumination intensity:

$$V_{OC} = S(A-1)T_L + SB. \quad (13)$$

This behavior is indeed observed in experiment in Figure 8B.

Close to linear I-V characteristics like that in Figure 5C are a typical feature of thermoelectric converters in general. We performed *suns-V_{OC}* measurements [25] in order to exclude the influence of the series resistance and still obtained close to linear characteristic (Fig. 9 lower plot). A narrow gap semiconductor in the middle of a p-n diode represents an interface recombination path, dominating in the carrier transport. The ideality factor in a wide-gap heterojunction of Type II, dominated by the interface recombination, is expected to be [26] close to 2. An ideality factor of 2.3 may be deduced from a

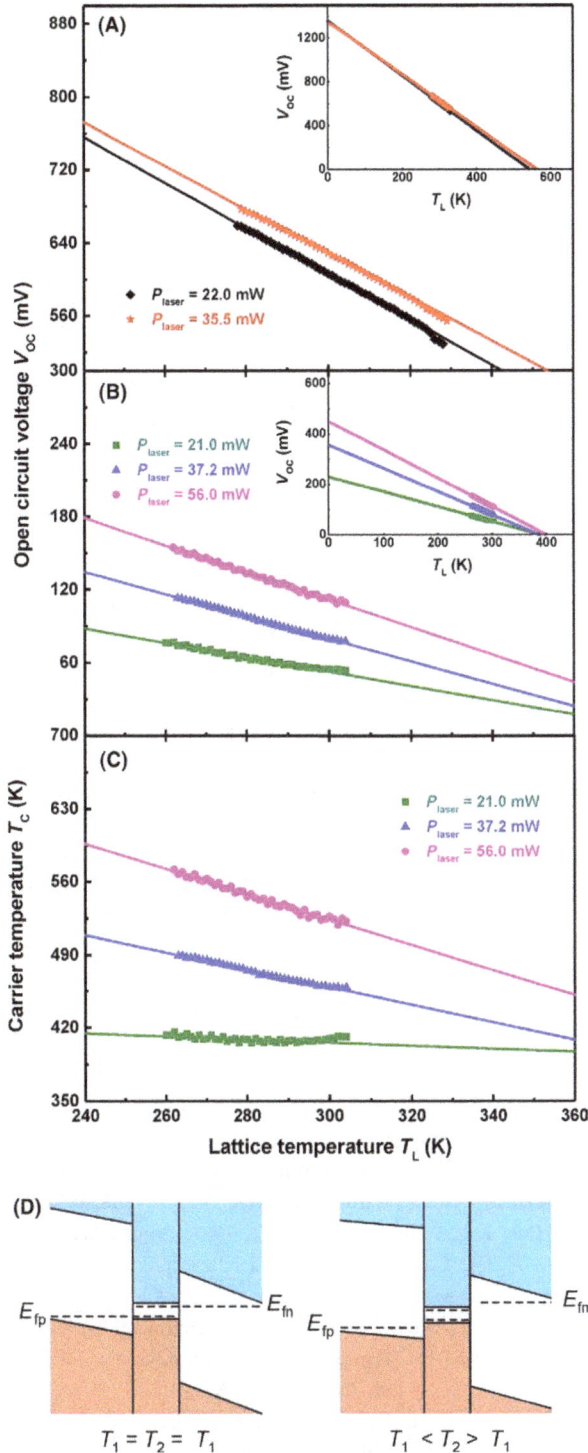

Figure 8. Open circuit voltage versus temperature due to the optical excitation at 658 nm for (A) conventional silicon solar cell; (B) prototype hot carrier solar cell; (C) carrier temperature versus the lattice temperature at various excitation power; (D) illustration of the origin of variable activation energy.

semi-logarithmic plot of the same dataset (Fig. 9 upper plot). This example illustrates two points of view on the same experiment.

Summary

In summary, we have demonstrated two equivalent approaches to describe operation and properties of hot carrier solar cells. While the common approach considers the structure as a diode where the thermionic emission takes place, an alternative approach of a thermoelectric device based on Seebeck effect is also valid. Theoretical consideration of the thermionic emission over a barrier in a heterostructure leads to a known Pisarenko relation for Seebeck coefficient. It follows from the thermoelectrical approach, that open circuit voltage of a hot carrier solar cell does not depend on interface properties of the heterostructure like band offsets, but is determined by the difference of bulk Seebeck coefficients in the energy selective contacts. A successful operation of the hot carrier solar cell prototype, based on the double heterojunction design, was demonstrated at the room temperature. The open circuit voltage of the prototype under concentrated sunlight exceeded the band gap in the bulk absorber material. The activation energy of the carrier transport is not constant, but depends on the excitation intensity. Surprisingly, the carrier temperature increases with decreasing lattice temperature at high excitation intensity. The shape of the I-V characteristics is close to linear, being common for thermoelectric devices.

Acknowledgments

We acknowledge endowed professorship of Carl Zeiss Foundation, financial support of Land Thüringen and European Commission through RFAproPV project and though Photograd network. We thank Ralf Linke (Innovent e. v. Jena, Germany) for x-ray photoelectron emission measurements for elucidation of the band diagram.

Conflict of Interest

None declared.

References

1. Ross, R. T., and A. J. Nozik. 1982. Efficiency of hot carrier solar energy converters. J. Appl. Phys. 53:3813.

2. Würfel, P. 1997. Solar energy conversion with hot electrons from impact ionization. Sol. Energy Mater. Sol. Cells 46:43.

3. Würfel, P., A. S. Brown, T. E. Humphrey, and M. A. Green. 2005. Particle conservation in the hot-carrier solar cell. Prog. Photovoltaics Res. Appl. 13:277–285.

4. Green, M. A. 2006. Third generation photovoltaics: advanced solar energy conversion. Springer-Verlag, Berlin Heidelberg, Germany.

5. Konovalov, I. 2013. Schichtenfolge zur photoelektrischen Umwandlung von Licht sowie Hot Carrier Solarzelle. German patent application DE 10 2013 105 462.5

6. Konovalov, I., V. Emelianov, and R. Linke. 2015. Hot carrier solar cell with semi infinite energy filtering. Sol. Energy 111:1–9.

7. Dimmock, J. A. R., S. Day, M. Kauer, K. Smith, and J. Heffernan. 2014. Demonstration of a hot-carrier photovoltaic cell. Prog. Photovoltaics Res. Appl. 22:151–160.

8. Dimmock, J. A. R., M. Kauer, P. N. Stavrinou, and N. J. Ekins-Daukes. 2015. A metallic hot carrier photovoltaic cell. In SPIE Physics, Simulation, and Photonic Engineering of Photovoltaic Devices, 9358, 935810, SPIE. https://doi.org/10.1117/12.2077573

9. Le Bris, A., and J.-F. Guillemoles. 2010. Hot carrier solar cells: achievable efficiency accounting for heat losses in the absorber and through contacts. Appl. Phys. Lett. 97:113506.

10. Mora-Sero, I., L. Bertoluzzi, V. Gonzalez-Pedro, S. Gimenez, F. Fabregat-Santiago, K. W. Kemp, E. H. Sargent, and J. Bisquert. 2013. Selective contacts drive charge extraction in quantum dot solids via asymmetry in carrier transfer kinetics. Nat. Commun. 4:2272.

11. Würfel, P. 2000. Physik der solarzellen (German Edition) Springer-Verlag, Berlin Heidelberg, Germany.

12. Sze, S. M., and K. N. Kwok. 2006. Physics of semiconductor devices, 3rd ed. John Wiley & Sons Inc, New York.

13. Bonch-Bruevich, V. L., and S. G. Kalashnikov. 1977. Semiconductor physics. Nauka, Moscow.

14. Ioffe, A. F. 1957. Semiconductor thermoelements and thermoelectric cooling. Infosearch. ISBN: 9780850860399

15. Hass, K. C., and D. Vanderbilt. 1987. Bond relaxation in Hg1−x Cd x Te and related alloys. J. Vac. Sci. Technol., A 5:3019.

16. Adachi, S.. 2009. Properties of semiconductor alloys: group-IV, III – V and II – VI semiconductors. John Wiley & Sons, Ltd, Chichester, UK.

17. Froment, M., L. Beaunier, H. Cachet, and A. Etcheberry. 2003. Role of cadmium on epitaxial growth of PbSe on InP single crystals. J. Electrochem. Soc. 150:C89.

18. Marinno, A. N., and K. L. Chopra. 1967. Polymorphism in some IV-VI compounds induced by high pressure and thin-film epitaxial growth. Appl. Phys. Lett. 10:282.

19. Cai, C. F., B. P. Zhang, R. F. Li, H. Z. Wu, T. N. Xu, W. H. Zhang, and J. F. Zhu. 2012. Band alignment determination of ZnO/PbSe heterostructure interfaces by

synchrotron radiation photoelectron spectroscopy. Europhys. Lett. 99:37010.

20. Konovalov, I., V. Emelianov, and R. Linke. 2016. Band alignment of type I at (100)ZnTe/PbSe interface. AIP Adv. 6:6.

21. Allan, G., and C. Delerue. 2004. Confinement effects in PbSe quantum wells and nanocrystals. Phys. Rev. B Condens. Matter. Mater. Phys. 70:1–9.

22. Kerner, B. S., and V. V. Osipov. 1976. Stratification of a heated electron-hole plasma. Sov. Phys. JETP 44:807–813.

23. Davami, K., A. Weathers, N. Kheirabi, B. Mortazavi, M. T. Pettes et al. 2013. Thermal conductivity of ZnTe nanowires. J. Appl. Phys. 114:134314.

24. Nadenau, V., U. Rau, A. Jasenek, and H. W. Schock. 2000. Electronic properties of CuGaSe2-based heterojunction solar cells. Part I. Transport analysis. J. Appl. Phys. 87:584–593.

25. Wolf, M., and H. Rauschenbach. 1963. Series resistance effects on solar cell measurements. Adv. Energy Convers. 3:455–479.

26. Grundmann, M., R. Karsthof, and H. von Wenckstern. 2014. Interface recombination current in type II heterostructure bipolar diodes. ACS Appl. Mater. Interfaces. 6:14785–14789.

8

Degradation of silicon wafers at high temperatures for epitaxial deposition

Thomas Rachow, Stefan Reber, Stefan Janz, Marius Knapp & Nena Milenkovic

Materials – Solar Cells and Technologies, Fraunhofer ISE, Heidenhofstrasse 2, 79110 Freiburg, Baden-Württemberg, Germany

Keywords
Annealing processes, epitaxial deposition, silicon wafers, solar cell performance

Correspondence
Thomas Rachow, Materials – Solar Cells and Technologies, Fraunhofer ISE, Heidenhofstrasse 2, 79110 Freiburg, Baden-Württemberg, Germany. E-mail: thomas.rachow@gmx.net

Funding Information
This work has been supported by the BMU in the EpiEm project under contract no. FKZ 0325199A.

Abstract

The material quality degradation of silicon wafers by metal impurities, various crystal defects as well as light and thermally induced mechanisms is very important for the solar cell performance and has been investigated by various groups. In this paper, the material degradation during epitaxial deposition at high temperatures above 1100°C will be discussed. Annealing experiments in hydrogen atmosphere are done with the laboratory rapid thermal chemical vapor deposition reactor to mimic the thermal process conditions for epitaxial growth of silicon from the gas phase. A general investigation of crystallographic and electronic properties of n- and p-type silicon wafers has been done between 950°C and 1150°C. A detailed sensitivity analysis of process parameters like cooling ramp, peak temperature, duration, and ambient gasses has been conducted. The degradation mechanism by metal impurities has been investigated by using silicon wafers with different diffusion barriers. Besides effective minority carrier lifetime, measurements by quasi steady state photo conductance, etch pit density, Raman spectroscopy, X-ray diffraction, and Fourier transform infrared spectroscopy measurements have been done. The presented results have been used to improve the deposition process of epitaxial thin-film solar cells, the production of silicon foils with a thickness <80 μm, and the fabrication of epitaxial multi junction solar cells with silicon bottom cell.

Introduction

The degradation of silicon material by metal impurities [1, 2], by light and thermally induced degradation [3], and by various crystal defects [4] like grain boundaries, crystal defects, and dislocations including vacancies, stacking faults, and interstitials is a critical factor for the resulting solar cell efficiency.

The described material degradation during the solar cell fabrication can be summarized as Shockley–Read–Hall recombination [5, 6] of excess carriers. Due to improvements in the fabrication processes, the material quality of mc, FZ, and Cz material could be significantly improved by magnetic assisted growth or defect annealing after ion implantation [7, 8].

On the other hand, the degradation of p-type and n-type wafers during the defect annealing [9, 10] and rapid thermal processing [11] in a commercial tube furnace has been reported. Furthermore, during the silicon deposition at temperatures above 1100°C by atmospheric pressure, chemical vapor deposition (APCVD) degradation of the silicon substrate was reported by [12, 13]. Since no conclusive explanation on the origin of this degradation has been presented, a close investigation of the degradation during the epitaxial deposition at high temperatures will be presented in this paper. Degradation is observed after epitaxial growth processes done in the laboratory rapid thermal chemical vapor deposition (RTCVD) reactor in a temperature range of 950°C and 1150°C. To simplify the investigation of degradation, a reference process using the same ambient gas without the trichlorosilane precursor has been used in order to be able to correlate degradation to the thermal process. Based on these experiments, the epitaxial silicon deposition process and the deposition parameters have been adjusted. Subsequently, these findings have been used to fabricate epitaxial p- and n-type emitters which results in an increase in cell efficiency from 18.6% to 23.7% [14].

Methods and Experimental Procedure

The high-temperature annealing processes were carried out in two RTCVD reactors [13, 15, 16]. The laboratory reactors operate at atmospheric pressure and consist of a quartz carrier inside a quartz furnace heated by halogen lamps. Process temperatures between 950°C and 1150°C as well as durations from 4 to 35 min have been varied depending on the experiment. The process times does not include the heating and cooling ramp if not stated otherwise. The heating and cooling ramps between 100 and 150 K min^{-1} are similar to the APCVD process and remain unchanged. Aside from hydrogen, the ambient gas for the silicon deposition process, argon, and nitrogen were used during the annealing. The reactor tubes as well as the quartz carrier are made of Heraeus HSQ 300 microelectronic-grade material which is also used in commercial diffusion furnaces. The reactor chamber and the carrier were cleaned by HCl etching, and the silicon wafers were chemically polished (CP etching) and cleaned by an RCA sequence prior to the experiment.

Depending on the experiment, n-type and p-type Cz and FZ silicon wafers have been used. The influence of the doping concentration on degradation was investigated by using 1 and 10 Ωcm wafers. In some cases, the wafers were coated with a diffusion barrier prior to the annealing process. In the first experiment, the diffusion barrier is a thermal silicon oxide of 105 or 275 nm thickness grown in a Centrotherm furnace at 950°C.

In a following experiment, an "ONO stack" of SiO_x and SiN_x by plasma-enhanced chemical vapor deposition (PECVD) was deposited on top of the thermal SiO_2 of 105 nm. After the annealing process, the diffusion barrier was etched away, the samples were cleaned by an RCA sequence, and passivated with SiN_x or Al_2O_3 by PECVD or atomic layer deposition (ALD), respectively.

The lifetime measurements have been conducted using a microwave photo conductance decay (MWPCD) setup WT-2000D by Semilab or a quasi steady-state photo conductance (QSSPC) setup by Sinton Instruments. A correction for the carrier diffusion at high lifetimes for MWPCD has been used according to the work of [17]. The QSSPC setup has also been used to determine the iron concentration based on the formation of interstitial iron Fe_i from iron boron complex Fe_B under illumination [18, 19]. The evaluation of the crystallographic quality by etch pit density (EPD) has been determined using a white light microscope after surface treatment via Secco etching [20].

Induced defects caused by the annealing process also lead to a lattice mismatch which results in lattice strains. These strains and the responsible defect atoms can be evaluated by μRaman spectroscopy [21] and Fourier transformed infrared spectroscopy (FTIR) [22]. For the FTIR measurements, a Bruker IFS 113 instrument was used to acquire spectra in the range of 400–2500 cm^{-1} with 6 cm^{-1} resolution. Additionally, X-ray diffraction (XRD) omega-2theta measurements [23, 24] have been done. The XRD tool used is a Philips X'Pert MRD with a CuK$_\alpha$ X-ray (λ = 0.154 nm) source.

Experimental Results and Discussion

Lifetime measurements after high-temperature annealing processes

A preliminary experiment was designated to determine the actual material degradation at 1000°C and 1050°C for 7 min as well as at 1150°C for 7 min and 35 min under hydrogen atmosphere. The heating-up and cooling took place with a ramp of 100 K min^{-1}. In Figure 1, the results of the three annealing temperatures for six different silicon wafers including Cz and FZ, n- and p-type as well as 1 and 10 Ωcm are summarized. In this case, all samples have been passivated by SiN_x deposited by PECVD and measured by QSSPC. The effective minority carrier lifetime degradation of almost two orders of magnitude to 1.5 μsec for p-type and 39 μsec for n-type FZ wafers with 1 Ωcm after a 7 min annealing process at 1150°C displays the magnitude of the degradation. The change of process time to 35 min shows further degradation in lifetime. A reduction of the annealing temperature to 1000°C lowers the degradation in lifetime to 253 μsec for n-type and 80 μsec for p-type FZ wafers with 1 Ωcm. Aside from the decrease in lifetimes for higher temperatures, it is apparent that the n-type FZ wafers show a slope with lower gradient. This effect can be correlated with a smaller capture cross section of certain defects in n-type silicon [6].

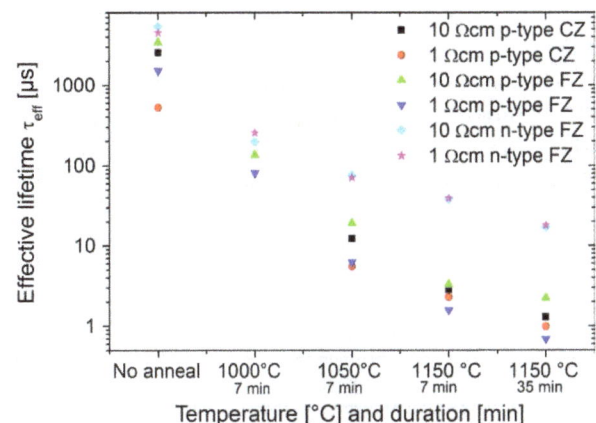

Figure 1. Effective lifetimes for six different silicon wafers processed for 7 min and 35 min, respectively, at three different temperatures.

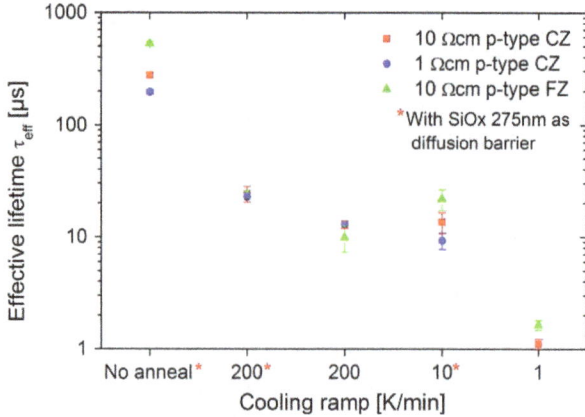

Figure 2. Lifetime degradation depending on cooling ramp for 200, 10, and 1 K min^{-1} with and without diffusion barrier processed at RTCVD 100.

The second experiment examines not only the thermal load of different annealing processes but also the influence of the cooling ramp on the material quality. Each sample was heated up with a ramp of 150 K min^{-1} from 25°C to 1120°C. After a 4 min anneal at 1120°C, the samples were cooled down with varied cooling ramps from 1120°C to 750°C. The minority-carrier lifetime results are shown in Figure 2. There is an overall decrease of lifetime in the case of 10 Ωcm p-type FZ material from about 500 μsec to below 30 μsec, which is just slightly different for the other materials investigated. Although untreated FZ-material should have lifetime values in the range of ms, in this case, the influence of a low passivation quality is apparent. Since the overall effect of lifetime degradation is still observable, the process-related problems during the passivation will not be discussed.

The lifetime degradation by reducing the cooling rate from 200 to 10 K min^{-1} and 1 K min^{-1} is related to a change in thermal load and duration of the cooling process. The process time for a cool down from 1120°C to 750°C increases from 2 to 370 min. Furthermore, this experiment investigates the effect of a 275 nm thermal SiO$_2$ diffusion barrier. A few samples had this SiO$_2$ layer, whereas the others had no diffusion barrier at all. Despite an increased lifetime from 13 to 24 μsec (+84%) at 200 K min^{-1} for samples with diffusion barrier, the significant degradation is not prevented. This result can be

explained by the high thermal load and the insufficient quality of thermal SiO$_2$ as diffusion barrier. Additional investigations about SiO$_2$ as diffusion barrier have been done by [25].

Unfortunately, the presented experiments only prove a negative correlation between the thermal budget applied and the minority carrier lifetime, but not if the degradation is caused by the diffusion of impurities or by the generation of crystal defects.

Influence of diffused metal impurities on material degradation

As part of the second experiment, it was shown that 275 nm thermal silicon dioxide did not prevent the lifetime degradation. This leads to the conclusion that SiO$_2$ does not work as a proper diffusion barrier for this experiment under the assumption that metal impurities are responsible for the lifetime degradation. In equation (1), the general definition of the diffusion coefficient $D_{impurity}$ of metal impurities [26, 27] depending on the distance between atoms α_0, number of vacant lattice sites N, jump frequency ω, and energy barrier E_B is shown.

$$D_{impurity} = \alpha_0^2 N \omega * e^{-\frac{E_B}{RT}} \tag{1}$$

With the relation between $D_{impurity}$ and diffusion time t (annealing time), one obtains the diffusion length L

$$L = \sqrt{D_{impurity} * t} \tag{2}$$

In Table 1, the diffusivity depending on the temperature and the corresponding diffusion length of three different metal impurities for a 4 min annealing process are listed. In case of iron, the diffusivity changes from 1.6×10^{-6} m^2 sec^{-1} at 900°C to 5.1×10^{-6} m^2 sec^{-1} at 1150°C [28, 29]. The corresponding diffusion length for a 4 min annealing process increases from 193 to 349 μm. In case of nickel, the diffusion length at 1150°C is >1000 μm. These values highlight the difference and the critical impact of the diffusion of metal impurities.

Since a lifetime degradation has been observed for samples with SiO$_2$ as a diffusion barrier, an additional experiment with and without SiO$_2$-ONO-stack as diffusion barrier were annealed in hydrogen for 4 min at 950°C,

Table 1. Diffusion coefficient by [28, 29] and calculated diffusion length of specific metal impurities in c-Si material at 900°C, 1000°C, and 1150°C.

Temperature	D_{Fe} [cm² sec^{-1}]	L_{Fe} [μm]	D_{Ti} [cm² sec^{-1}]	L_{Ti} [μm]	D_{Ni} [cm² sec^{-1}]	L_{Ni} [μm]
900°C	1.56×10^{-6}	193	2.96×10^{-10}	2.67	1.91×10^{-5}	678
1000°C	2.64×10^{-6}	252	1.19×10^{-9}	5.34	2.76×10^{-5}	813
1150°C	5.08×10^{-6}	349	6.64×10^{-9}	12.6	4.33×10^{-5}	1020

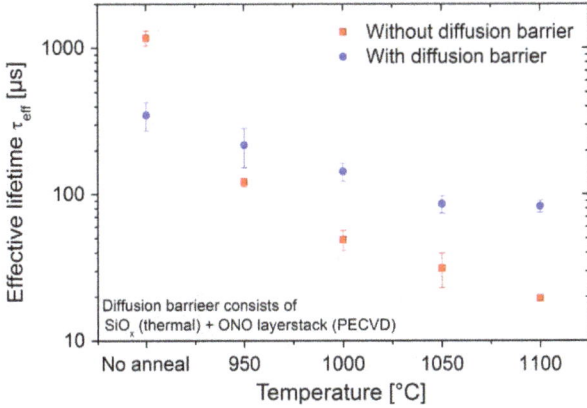

Figure 3. Effective lifetime for different annealing temperatures regarding samples with and without ONO-stack as diffusion barrier.

1000°C, 1050°C, and 1100°C with a temperature ramp of 150°C min^{-1}. The SiO$_2$-ONO diffusion barrier consists of thermal SiO$_2$ (105 nm) and a layer system of SiO$_x$ (1 μm), SiN$_x$ (100 nm), and SiO$_x$ (1 μm) deposited by PECVD.

In Figure 3, the lifetime measurements for samples with and without ONO-stack are shown. Again, the figure shows the mean values of three equally processed samples including errors bars. At 1100°C, the lifetime of the 1 Ωcm, p-type FZ sample is reduced to 83 μsec with and to 20 μsec without barrier. The correlation between annealing temperatures and degradation shows a higher gradient for the samples without diffusion barrier. However, despite SiO$_2$-ONO-stack, there is still a significant decline in effective minority carrier lifetime.

After annealing, the iron content of the samples was estimated by measurement of FeB complexes in the samples via QSSPC [30] assuming that the metal impurities are the main influence on the lifetime degradation. The samples were measured both after 24 h storage in darkness and after light soaking. The difference in lifetime due to the formation of interstitial iron Fe$_i$ by dissociation of FeB complexes gives information about the iron content in the sample [31] with a measurement error of about 20 %. The lifetime measurements are injection independent for high FeB concentrations and strongly differ in their shape due to increased SRH-lifetime by Fe$_i$ after illumination. In Table 2, the iron content of samples with and without

SiO$_2$-ONO-stack at four temperatures between 950°C and 1100°C are shown. It is important to note that there is a rise in iron content for samples with and without diffusion layer, even if they were not annealed. The iron content of samples with SiO$_2$-ONO-stack, is increased from below the detection limit ($\approx 1.0 \times 10^8$ 1 per cm^3, without barrier) to approximately 2.1×10^{10} 1 per cm^3. This process-related effect due to the deposition of the SiO$_2$-ONO-stack plays a minor role for annealed samples because the iron concentration in annealed samples of 10^{11} impurities per cm^3 is one order of magnitude higher. The difference in iron concentration depending on the annealing temperature is limited, but in case of the samples without ONO stack, the concentration increases from 1.1×10^{11} 1 per cm^3 at 950°C to 4.1×10^{11} 1 per cm^3 at 1100°C.

Corresponding calculations of the Shockley–Read–Hall lifetime depending on the iron concentration have been done to evaluate the measured concentration. The SRH lifetime has been calculated by equation (3) according to [7] using defect energy E_t, the carrier concentration in thermal equilibrium n_0 and p_0 and the capture cross-section σ_p and σ_n for holes and electrons.

$$\tau_{SRH} := \frac{\tau_{n0}\left(p_0 + p_1 + \Delta n\right) + \tau_{p0}(n_0 + n_1 + \Delta n)}{p_0 + n_0 + \Delta n} \quad (3)$$

Based on [32] the electron and hole lifetime in thermal equilibrium and the carrier concentration under illumination have been calculated using equation (4) and (5).

$$\tau_{p0} := \left(N_t \sigma_p v_{th}\right)^{-1} \quad \& \quad \tau_{n0} := \left(N_t \sigma_n v_{th}\right)^{-1} \quad (4)$$

$$n_1 := N_L e^{\left(\frac{E_t - E_L}{k_B T}\right)} \quad \& \quad p_1 := N_V e^{\left(\frac{E_V - E_t}{k_B T}\right)} \quad (5)$$

The valence band $E_V = -5.2$ eV and the conduction band $E_C = -4.1$ eV in crystalline silicon as well as data for different defect centers can be found in [33]. In case of iron impurities,

$$E_t = E_V + 0.38\,\text{eV} = -4.82\,\text{eV} \quad \sigma_p = 0.68 \cdot 10^{-16}\,\text{cm}^2$$
$$\sigma_n = 4 \cdot 10^{-14}\,\text{cm}^2 \quad (6)$$

have been used for the calculations. The radiative lifetime τ_{rad} and Auger lifetime τ_{Auger} have been determined for a 1 Ωcm, p-type, FZ wafer according to [7]. The calculated effective lifetimes τ_{eff} are shown in Figure 4.

Table 2. Iron impurity concentration in silicon wafers after high-temperature annealing process. An overall error of N_{Fe} is assumed as 20%.

Type of sample/Temperature	No process	950°C	1000°C	1050°C	1100°C
N_{Fe} [1 per cm³] with ONO	2.1×10^{10}	1.4×10^{11}	2.2×10^{11}	3.3×10^{11}	2.8×10^{11}
N_{Fe} [1 per cm³] without ONO	1.0×10^{8}	1.1×10^{11}	2.8×10^{11}	2.7×10^{11}	4.1×10^{11}

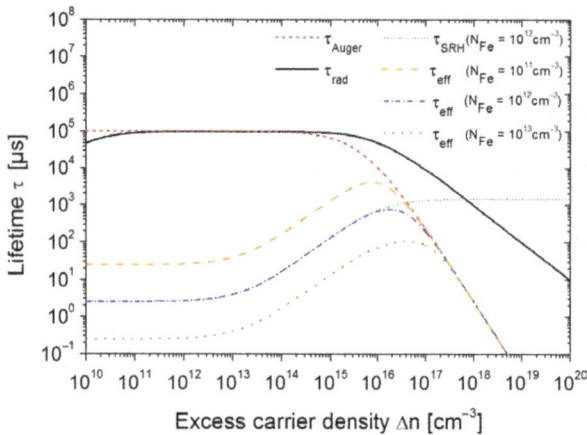

Figure 4. Calculated radiative, auger, and SRH-lifetimes for three different iron concentrations and the resulting effective lifetimes.

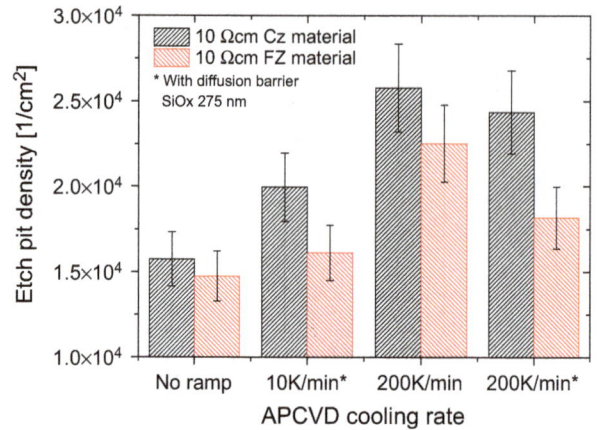

Figure 5. Etch pit density measurements of samples with and without diffusion barrier after annealing and two different cooling ramps.

The detrimental effect on the SRH-lifetime τ_{SRH} at low injection density is shown for an iron concentration of $N_{Fe} = 1 \times 10^{12}$ cm^{-3}. The impact of the metal impurity concentration is depicted by the effective lifetimes for three different iron concentrations. The differences in τ_{eff} between $N_{Fe} = 1 \times 10^{10}$ cm^{-3} and $N_{Fe} = 1 \times 10^{12}$ cm^{-3} are approximately two orders of magnitude. These calculations are in agreement with the measured iron concentration and the corresponding lifetime values.

Nevertheless, additional investigations of the degradation effect including different metal impurities are necessary. Therefore, deep level transient spectroscopy (DLTS) [31] and total reflection X-ray fluorescence (TXRF) [27] measurements will be conducted.

However, the degradation by metal impurities alone does not explain the difference between the silicon wafers with and without diffusion barrier. This suggests that there is an additional degradation mechanism. In [9], the impurity diffusion into n-type material during the defect annealing after ion implantation in a diffusion furnace was ruled out due to the fact that the sub-bandgap energy level corresponding to the SRH lifetime did not change.

An explanation of the additional degradation in lifetime is the appearance and propagation of thermal defects induced into the silicon crystal [3]. The reason for a reduced degradation of samples with dielectric layer at higher temperatures could be the reduction of stress and strain on the sample surface [34].

Formation and migration of crystal defects

The formation of crystal defects can be influenced by peak temperature, thermal load, and by the heating and cooling ramp as suggested by [3]. Since the temperature ramps are always associated with the thermal load, a complete separation is very difficult. In theory, a very high temperature gradient is supposed to cause the "freezing" of crystal defects by crossing a specific temperature. The results would be point defects like interstitials or vacancies, dislocations, and stress in the crystalline silicon wafer.

The determination of the crystal defect formation at high temperatures has been evaluated by EPD measurements. In Figure 5, the EPD measurements of the first experiment show a minor increase in defect density of 18–12% by increasing the cooling ramp from 10 to 200 K min^{-1}. The direct comparison of etch pit densities on samples (200 K min^{-1} cooling ramp) annealed with SiO$_2$ of 1.8×10^4 1 per cm² and without SiO$_2$ of 2.2×10^4 1 per cm² in case of FZ material shows the indirect impact of the diffusion barrier. This difference can be explained by the reduced stress on the silicon surface [34] and the positive effect of a diffusion barrier which prevents silicon etching during hydrogen anneal. These results prove that the difference in etch pit density for an annealing process at high temperatures is negligible as one of the main reasons for the material degradation. However, additional experiments are required to have enough statistics for a substantial analysis because the measurement error is about 25–50%. Supplementary EPD measurements of the samples without diffusion barrier show similar etch pit densities after an additional etching of 10 μm by chemical polishing. These measurements demonstrate that there are no additional crystalline defects in the near-surface region. Furthermore, a comparison between the Cz samples without annealing (1.5×10^4 1 per cm²) and the annealed Cz samples (2.2×10^4 1 per cm²) show no relevant increase. Since no significant effect in lifetime degradation can be found by various cooling ramps (see Fig. 2), the standard cooling process of 150 K min^{-1} will remain unchanged.

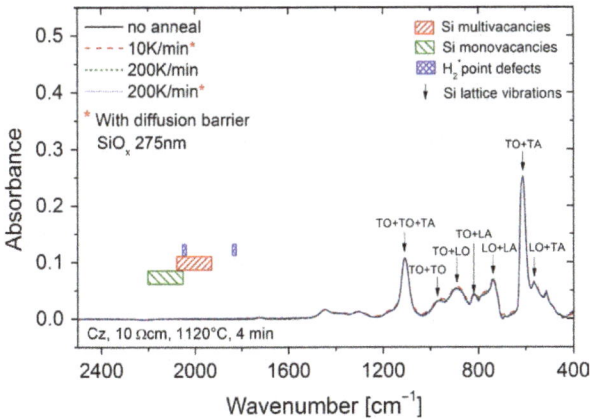

Figure 6. Fourier transformed infrared spectroscopy measurements of the silicon wafers after different annealing processes with and without diffusion barrier. The labeled features correspond to transverse acoustic (TA), transverse optic (TO), longitudinal acoustic (LA), and longitudinal optic (LO).

In [35] FTIR spectra have been measured to characterize the crystallographic defects before and after annealing. Based on these experiments, FTIR measurements on FZ and Cz material after annealing using the RTCVD reactor have been done. The measurements in Figure 6 show no additional peak due to crystal defects between 1800 and 2200 nm [35]. Despite the vibrations of the silicon lattice below 1200 nm, no absorbance by vacancies or hydrogen point defects has been detected. Supplementary X-ray diffraction (XRD) omega-2theta measurements [23, 24] of a reference and an annealed wafer at 1100°C have been made. Despite an increase in FWHM from 30.3 to 47.37 arcsec, no significant change in crystal quality is identified. The μRaman measurements have not shown a significant increase in crystal defect density as well because neither an additional defect peak nor an increasing FWHM of the Si peak Lorentz fit could be determined.

Process parameters and reactor properties

Since no significant crystallographic degradation has been found, metal impurities seem to be the limiting degradation effect. Therefore, additional experiments have been done to explain the origin of metal impurities The lifetime measurements of 1 Ωcm, p-type, FZ wafers after the different high-temperature oxidation processes (105 nm thermal SiO_2 at 950°C, 275 nm at 1050°C) using a diffusion furnace show no significant degradation. Since the same quartz material is used in a commercially available diffusion furnace (validated by INAA measurements at the University of Mainz) and the gasses used have a certified quality of 6N with an additional purification process of the hydrogen by a palladium cell, there is no apparent

reason for lifetime degradation in the used RTCVD reactor. The determination of absolute impurity concentration by SIMS or GDMS has been difficult because the concentrations are below the detection limit [30].

The major difference between both annealing processes is the peak temperature of 1150°C during the APCVD process. The duration is similar because the defect annealing, as post processing after the ion implantation, can take up to 60 min. Another difference between a diffusion furnace and the RTCVD reactors is the ambient gasses which are used. At 1120°C, the lifetime of 56 ± 9 μsec for an annealing in Ar atmosphere decreases to values below 10 μsec in hydrogen. The difference in lifetime between Ar and H flow suggest an increased effective diffusion coefficient on the silicon surface in hydrogen atmosphere. This difference in effective diffusion constant results in an increased transport of impurities from the quartz carrier.

Analysis and Conclusions

In summary, it can be stated that the degradation in material quality from $\tau_{eff} > 1$ msec to values <10 μsec during an APCVD process is dominated by the diffusion of metal impurities. However, the decrease in lifetime cannot be attributed to metal impurities alone, which suggests an additional crystallographic degradation mechanism. In [3], it is suggested that the dangling silicon bonds at the silicon surface propagate as vacancies into the silicon bulk material at high temperatures. However, if the dangling bonds are saturated by the deposition of a dielectric passivation layer, the thermal degradation by 1 D crystallographic dislocation can be reduced [3]. This hypothesis is supported by various groups reporting on the degradation of p-type and n-type wafers during the defect annealing [9, 10] and rapid thermal processing [11] in a commercial tube furnace. In [2], a discrepancy between DLTS measurements and measured lifetime is shown. Although a different conclusion is drawn, these findings could also suggest an additional degradation mechanism. In case of [9], the impurity diffusion into n-type material during the defect annealing after ion implantation in a diffusion furnace was ruled out as degradation mechanism due to the fact that the sub-band gap energy level corresponding to the SRH lifetime measurements did not change. The results presented in this paper show that the formation of crystal defects is not a dominating factor.

Summary

In this work, an investigation of minority carrier lifetime degradation in silicon wafers during a high-temperature annealing processes in a reactor for epitaxial deposition is presented. The variation of peak temperature and total annealing time shows the significance of these process parameters. Various theories on the formation and influence of thermal defects have been investigated by a thorough electrical and crystallographic characterization. In conclusion, it can be stated the the main effect on lifetime degradation results from metal impurities diffusing into the samples, since a diffusion barrier like an ONO-stack increases the lifetime significantly. In order to obtain the iron content in the samples, specific measurements via QSSPC were done and discovered a positive correlation between the amount of iron atoms in the sample and the annealing temperature in the process. Since comparable annealing experiments in other furnace tubes with the same certified quality of process gasses and quartz material show no degradation in lifetime, the sample placement in the carrier or on a susceptors as well as the ambient gasses have been evaluated. Based on these results, the epitaxial deposition by APCVD in general has been optimized. These findings have been used to optimize the deposition of epitaxial p- and n-type emitters which results in an increase in cell efficiency from 18.6% to 23.7% [14]. Furthermore, these experiments contribute to the fundamental understanding of the different degradation mechanism.

Acknowledgments

The authors express their gratitude to Harald Lautenschlager, Mira Kwaitkowska, Elke Gust, Kai Schillinger, Marion Drießen, Stefan Lindekugel, and Michaela Winterhalder at ISE for their support and input in many valuable discussions. This work has been supported by the BMU in the EpiEm project under contract no. FKZ 0325199A.

Conflict of Interest

None declared.

References

1. Schubert, M. C. et al. 2012. *Modeling distribution and impact of efficiency limiting metallic impurities in silicon solar cells.* Proceedings of the 38th IEEE Photovoltaic Specialists Conference, Austin, Texas, USA, Pp. 286–291.
2. Eichhammer, W. et al. 1989. On the origin of rapid thermal process induced recombination centers in silicon. J. Appl. Phys. 66:3857–3865.
3. Sah, C. T., and C. T. Wang. 1975. Experiments on the origin of process–induced recombination centers in silicon. J. Appl. Phys. 46:1767.
4. Schwuttke, G. H. 1982. *Crystal defect problems.* Proceedings of the 16th IEEE Photovoltaic Specialists Conference, San Diego, CA, USA, Pp. 327–332.
5. Shockley, W., and W. T. J. Read. 1952. Statistics of the recombinations of holes and electrons. Phys. Rev. 87:835–842.
6. Macdonald, D. H., and L. J. Geerligs. 2004. *Recombination activity of iron and other transition metals in p- and n-type crystalline silicon.* Proceedings of the 19th European Photovoltaic Solar Energy Conference, Paris, France. P. 492.
7. Rein, S. 2005. Lifetime spectroscopy – a method of defect characterization in silicon for photovoltaic applications. [Dissertation], Fakultät für Physik, Universität Konstanz, Konstanz.
8. Hermle, M. et al. 2011. *N-type silicon solar cells with implanted emitter.* Proceedings of the 26th European Photovoltaic Solar Energy Conference and Exhibition, Hamburg, Germany, Pp. 875-8,
9. Mäckel, H., and K. Varner. 2012. On the determination of the emitter saturation current density from lifetime measurements of silicon devices. Prog. Photovolt. 21:850–866.
10. Glunz, S. W. et al. 2001. Minority carrier lifetime degradation in boron-doped Czochralski silicon. J. Appl. Phys. 90:2397–2404.
11. Poggi, A. et al. 1994. Rapid thermal annealing of p-type silicon. J. Electrochem. Soc. 141:754.
12. Kiefer, F. 2009. *Untersuchungen zu epitaktischen Emittern an Silizium-Solarzellen und kristallinen Silizium-Dünnschicht-Solarzellen.* Diploma, Fakultät für Mathematik und Physik, Albert-Ludwigs-Universität, Freiburg.
13. Schmich, E. K. 2008. High-temperature CVD processes for crystalline silicon thin-film and wafer solar cells. [Dr. rer. nat. Dissertation], Fachbereich Physik, Universität Konstanz, Konstanz.
14. Rachow, T. 2014. Deposition and characterisation of crystalline silicon. [Dr. rer. nat. Dissertation], Fachbereich Physik, Universität Konstanz, Konstanz.
15. Faller and, F. R., and A. Hurrle. 1999. High-temperature CVD for crystalline-silicon thin-film solar cells. IEEE Trans. Electron Devices 46:2048–2054.
16. Bau, S. 2003. High-temperature CVD silicon films for crystalline silicon thin-film solar cells. [Dissertation], Fakultät für Physik, Universität Konstanz, Freiburg.
17. Semilab. 2013. npv-Workshop. Cambery, France.
18. Lauer, K. et al. 2008. *Effect of iron-boron pairs on crystalline silicon solar cells.* Proceedings of the 23rd European Photovoltaic Solar Energy Conference, Valencia, Spain, Pp. 1660–1663.
19. Palais, O. et al. 2002. Minority carrier lifetime scan maps applied to iron concentration mapping in silicon wafers. Mater. Sci. Eng., B 91–92:216–219.

20. Secco d'Aragona, F.. 1972. Dislocation etch for (100) planes in silicon. Solid State Sci. Technol. 119:948–951.

21. Gundel, P. 2011. Neue mikroskopische Opto-Spektroskopie-Messtechniken für die Photovoltaik. [Ph.D], Technische Fakultät Freiburg.

22. Boyle, R.. 2008. FT-IR measurement of interstitial oxygen and substitutional carbon in silicon wafers. Thermo Fisher Scientific, Madison, WI.

23. Erdtmann, M., and T. A. Langdo. 2006. The crystallographic properties of strained silicon measured by X-ray diffraction. J. Mater. Sci.: Mater. Electron. 17:137–147.

24. Klug, H. P., and L. E. Alexander, eds. 1974. X-ray diffraction procedures. John Wiley & Sons, New York, NY.

25. Reber, S. 2000. Electrical confinement for the crystalline silicon thin-film solar cell on foreign substrate. [Ph.D], Fachbereich Physik, Johannes Gutenberg-Universität, Mainz.

26. Bentzen, A. 2006. Phosphorus diffusion and gettering in silicon solar cells. [Dissertation], Department of Physics, University of Oslo, Oslo.

27. Kitagawa, H. 2000. Diffusion and electrical propertoies of 3d transition-metal impurities in silicon. Solid State Phenom. 71:51–72.

28. Weber, E. R. 1983. Transition metals in silicon. Appl. Phys. A Solids Surf. A30:1–22.

29. Graff, K. 1995. Metal impurities in silicon-device fabrication, 1st ed. Springer, Berlin.

30. Habenicht, H. et al. 2008. *Defect transformation in intentionally contaminated FZ silicon during low temperature annealing.* Proceedings of the 23rd European Photovoltaic Solar Energy Conference, Valencia, Spain. Pp. 1933–1937.

31. Habenicht, H. et al. 2007. *Out-diffusion of metal from grain boundaries in multicrystalline silicon during thermal processing.* Proceedings of the 22nd European Photovoltaic Solar Energy Conference Milan, Italy. Pp. 1519–1523.

32. Würfel, P., ed. 1995. Physik der solarzellen. Spektrum Akademischer, Heidelberg.

33. Istratov, A. A. et al. 1999. Iron and its complexes in silicon. Appl. Phys. A 69:13–44.

34. Bearda, T. et al. , 2013. *Low-temperaure emitter passivation for solar cells bonded to glass.* Proceedings of the 28th European Photovoltaic Solar Energy Conference, Paris.

35. Moriceau, H. et al. 2012. Smart Cut™: Review on an attractive process for innovative substrate elaboration. Nucl. Instrum. Methods Phys. Res. B 277:84–92.

9

A study of the annual performance of bifacial photovoltaic modules in the case of vertical facade integration

Bruno Soria[1], Eric Gerritsen[1], Paul Lefillastre [1] & Jean-Emmanuel Broquin[2]

[1]Photovoltaic Modules Laboratory, CEA-INES, 50 avenue du Lac Léman, Le Bourget-du-Lac F-73375, France
[2]CNRS, IMEP-LACH, Grenoble Alpes University, Grenoble F-38000, France

Keywords
Bifacial, facade, half-cells, indoor, N-type, outdoor, ray-tracing, textured glass

Correspondence
Bruno Soria, CEA-INES, Photovoltaic Modules Laboratory, 50 avenue du Lac Léman, Le Bourget-du-Lac F-73375, France.
Tel: +33 6 67 77 60 59;
E-mail: brunosoria@ntymail.com

Funding Information
Study funded by CEA-INES.

Abstract

Despite the apparent benefits of bifacial modules, their application still suffers from a lack of visibility on the performance gain that they can actually provide. In this work, we consider the specific application of vertically oriented bifacial modules, notably for facade integration. We have developed a methodology to evaluate the annual electrical performance of bifacial modules based on three tools. First, a double illumination characterization setup is used in a solar simulator for comparing module architectures. Then, a reduced scale outdoor test bench allows us to evaluate bifacial module performance in a variety of configurations. Finally, a ray-tracing model validated with short-term outdoor data leads to the determination of the annual performance gain. This methodology allowed us to find optimal performance according to the most important parameters of application and module. Specifically, a module architecture using half-cut cells, a parallel cell interconnection, and textured glasses have been analyzed with respect to their influence on the resistive losses which increase in dual side illumination as well as to their influence on the effect of non-uniform and diffuse irradiance on the backside of the module. This work enabled us to give directions for innovative full-size module architectures.

Context and Aim of the Study

Bifacial photovoltaic (PV) cells which are able to convert solar radiation from both sides have been developed since the 1960s, but their terrestrial application only started around the 1980s [1, 2]. Even though some modeling work is performed at the megawatt scale [3], only few large bifacial plants have been set up so far, probably due to a lack of confidence from investors in the actual gain of bifacial technology. Indeed, actual commercial energy prediction softwares do not take into account the complex opto-geometrical environment of the modules (generating non-uniform and diffuse backside albedo for instance), and the experimental tests performed so far have often been short term, at small scale or in specific configurations only. New methodologies which consider the complex opto-geometrical environment of bifacial modules are however in development in the research community [4, 5].

Vertical facade integration of PV modules attracts interest in the building integrated photovoltaics (BIPV) community because PV facades have a large collection area without additional footprint and their vertical installation makes them less sensitive to snow, dust, bird droppings, or tree leaves. However, they suffer from non-optimal orientation and module heating due to integration onto a wall. Using bifacial modules in a double skin configuration can minimize these drawbacks. Figure 1 shows how the gap between module and inner wall allows backside albedo and natural or forced ventilation to give additional power to the module. Additionally, there are seasonal advantages like a higher PV production and heating inside the building during winter as well as vertical sun shading during summer.

The aim of this article is to evaluate and to maximize the annual performance of a bifacial module in a double skin configuration compared to a monofacial one. For this purpose, we developed a methodology based on indoor characterization with a double illumination setup in solar simulator, on outdoor characterization with an adjustable test bench of the application, and on ray-tracing simulation with an optical model of the application. The study has been performed on

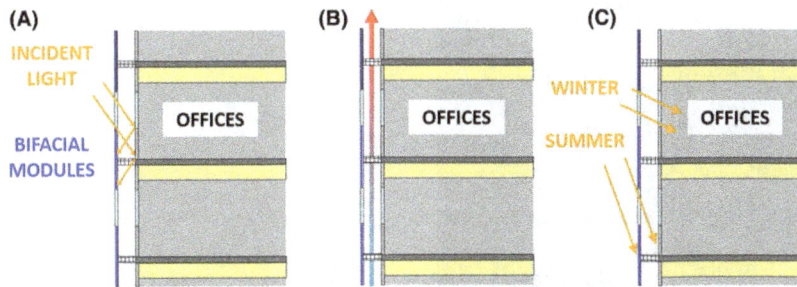

Figure 1. Advantages of using bifacial modules on a vertical facade in a double skin configuration: a backside albedo (A), natural or forced ventilation (B), and seasonal advantages (C).

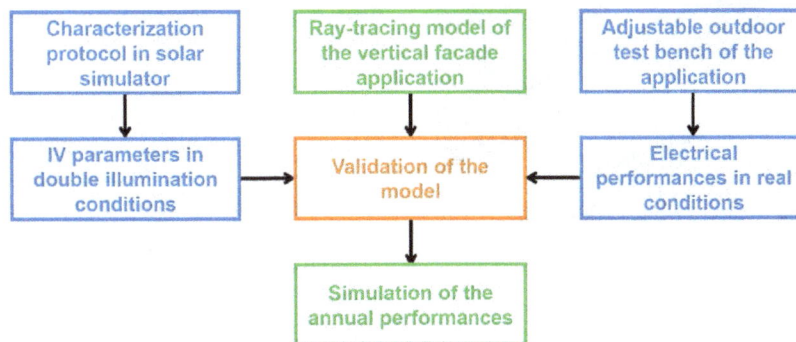

Figure 2. Schematics of the methodology used for evaluating the annual performance of bifacial modules in a vertical facade application (experimental part in blue, simulation part in green, and validation in orange).

a reduced scale application using mini-modules. The bifacial gain is dependent on the opto-geometrical parameters of the application (double skin distance, reflective properties of the inner wall) and on the module architecture (electrical and optical parameters). Therefore, these two categories of parameters have been studied separately. The results enabled us to propose some design rules of the application and perspectives for the optimal bifacial module architecture.

Methodology of the Study

We used mini-modules with four N-type bifacial cells connected in series, encapsulated between two glasses in EVA-HLT (ethylene vinyl acetate - high-light transmission from STR). The type of cells used in this study has a back to front power ratio of 0.9 and has been reported in detail in [6]. Figure 2 shows the different steps implemented for the simulation of the annual performance and the following subsections will describe the three tools used in this methodology.

Double illumination characterization setup

Standard characterization protocols for bifacial modules in solar simulators usually consist in measuring each side of the module independently using a black mask on the opposite side. Here, we adapted a double illumination setup existing at the cell level [7–9] for our mini-modules. Figure 3 shows the setup we designed. The bifacial module is placed between two aluminum mirrors (Alanod MIRO® 4200GP, Ennepetal, Germany) with an angle between mirror and module of $\psi = 44.1°$ optimized with a ray-tracing model of our PASAN solar simulator performed with TracePro® [10]. Metallic grid filters with several pitches are used for attenuating light on the backside of the module in order to mimic different albedo conditions.

With this setup, we used a protocol with nine main steps depending on the PASAN intensity filters and on our metallic filters (see Table 1). We performed a reproducibility study in order to quantify the 3σ relative uncertainty of every current–voltage (IV) parameter for each step. Figure 4 shows that this uncertainty is comprised in the $\pm 3\%$ range as in Ohtsuka's study [7]. The uncertainty is high at step 1 probably because the voltage range of the IV sweeping (kept equal for each step of our protocol) is not fully adapted for low irradiance measurements. It is also high at steps 7, 8, and 9 because the manual change of filters can slightly modify the alignment between the PASAN source and the module. Finally, step 5 shows the lowest uncertainty as this is the calibration

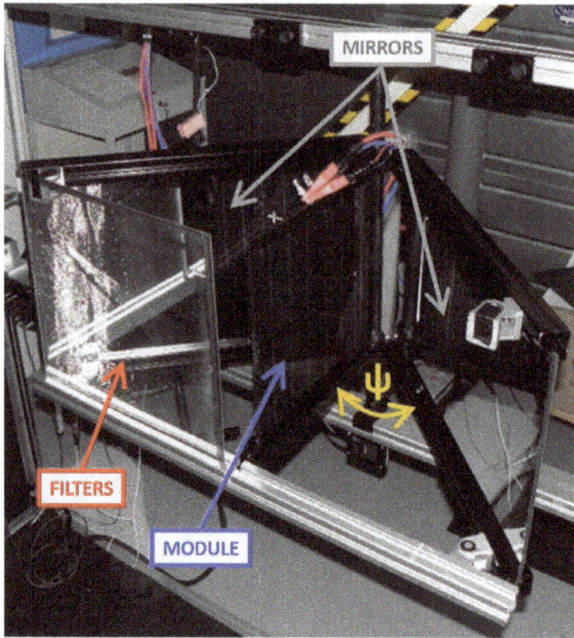

Figure 3. Picture of the double illumination setup positioned in our PASAN solar simulator.

step: here, the front short circuit current measured with our setup ($I_{sc\text{-}front}$) is compared to the I_{sc} measured in standard test conditions (STC) with a black mask on the back ($I_{sc\text{-}front\text{-}STC}$) and the PASAN light source intensity is adjusted in order to compensate mirrors absorption (about 5%).

Figure 5 shows the spectral intensity measured with a spectrometer in the standard STC measurement case (in blue) and pondered with the spectral reflectivity of the aluminum mirror with our calibrated setup (in red). This shows that the spectral intensity of the PASAN source belongs to the A+ class (deviation compared to AM1.5G

Table 1. The nine main steps of our protocol. Additional steps consist in measuring the backside of the module (black mask on the frontside) in order to measure the optical transmission of the metallic filters.

Protocol step number	Irradiance on the reference cell (below our setup)	Percentage of light on the back of the module
1	100 W/m²	0% (mask)
2	200 W/m²	0% (mask)
3	400 W/m²	0% (mask)
4	700 W/m²	0% (mask)
5	1000 W/m²	0% (mask)
6	1000 W/m²	23% (filter)
7	1000 W/m²	47% (filter)
8	1000 W/m²	78% (filter)
9	1000 W/m²	100% (no filter)

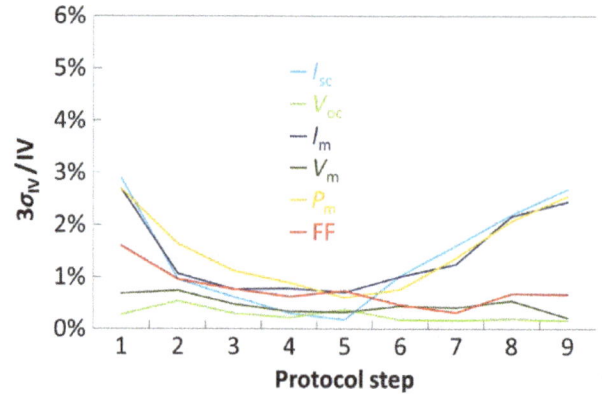

Figure 4. 3σ relative uncertainty measured for each step of the protocol as well as each IV parameter (σ being the standard deviation).

below ±12.5%) as long as the calibration step which compensates the mirror absorption is performed.

Using our setup, we ensured that all the IV parameters follow the same tendency independently of the illumination mode (frontside only, backside only, double side) and identified the resistive losses which increase when the module receives higher irradiance. Figure 6A shows that the power in double illumination ($P_{m\text{-}bi}$) gets lower than the linear tendency, which is the sum of frontside and backside powers ($P_{m\text{-}front} + P_{m\text{-}back}$) as the total irradiance on the module $I_{sc\text{-}bi}/I_{sc\text{-}front\text{-}STC}$ increases. Figure 6B details this behavior plotting the gain between $P_{m\text{-}bi}$ and ($P_{m\text{-}front} + P_{m\text{-}back}$). This shows that the resistive losses reach −5% at 2 sun. Note that for front characterization at low irradiances, the gain is actually positive because of the logarithmic behavior of the voltage (see the V_m parameter evolution in blue).

Outdoor test bench

Our outdoor study has been performed on an adjustable reduced scale test bench inspired by a real building facade. From November 2012 until March 2013, we studied the influence of the opto-geometrical parameters of the application, namely the diffuse reflector on the inner wall and its distance to the module. Figure 7A shows the bifacial module in the middle, a monofacial reference on the right (bifacial cells from the same batch encapsulated with EVA-HLT between a glass and a white backsheet), and a reflector behind. In this article, the measure for length will be quantified in terms of number of c: $c = 36$ cm being the length of our modules, $3c \times 2c$ the area of the reflector, and $3c \times 3c$ the aperture of the opto-geometrical environment. From April 2013 until December 2013, we studied the influence of different module architectures. Figure 7B shows two bifacial modules (a standard architecture on the left and novel one on

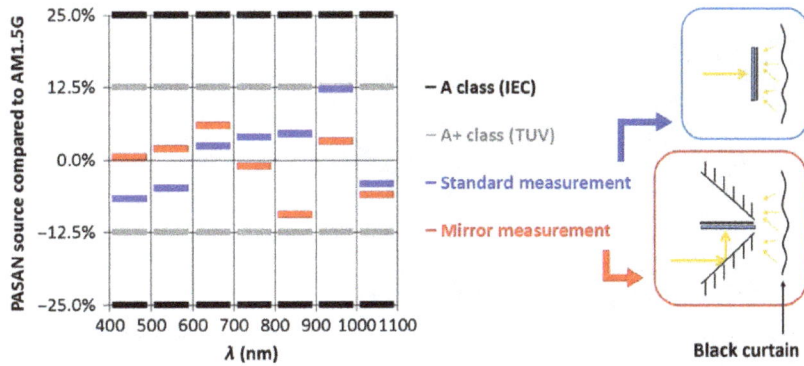

Figure 5. Spectral repartition of the PASAN source intensity compared to AM1.5G reference. The blue schematics shows a standard STC measurement and the red one shows a measurement with our double illumination setup. Black curtain reflections are also shown behind the setup.

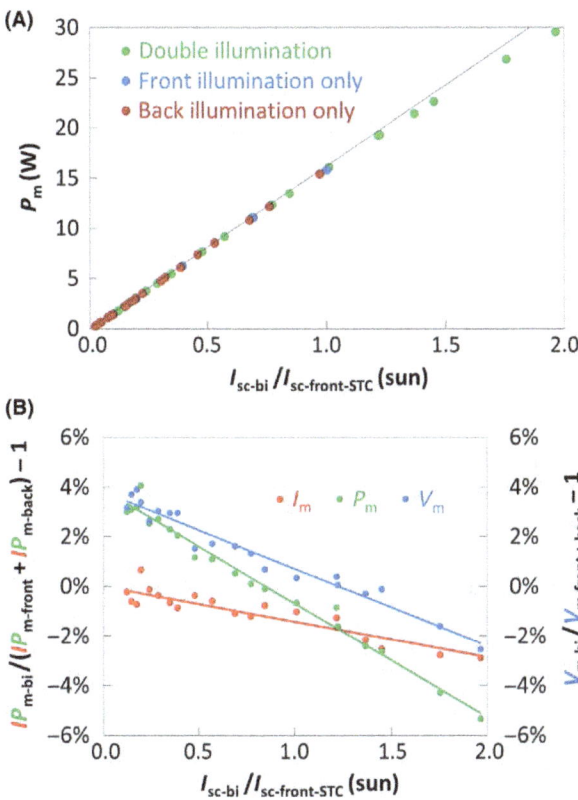

Figure 6. (A) Power P_m measured for different illumination modes and linear tendency showing the sum of frontside and backside powers ($P_{m\text{-}front}$ and $P_{m\text{-}back}$). (B) Gain between the double illumination power $P_{m\text{-}bi}$ and ($P_{m\text{-}front}$ + $P_{m\text{-}back}$) and its separation into the current contribution I_m and the voltage contribution V_m.

the right), the monofacial reference below, and a reflector behind. The whole setup is equipped with frontside and backside irradiance sensors (in yellow and orange) as well as calibrated temperature sensors on the back of the modules. These are T-type sensors glued with epoxy and protected from light using aluminum adhesive. The IV

parameters have been followed over the year with a Daystar MT5 multi-tracer. These data have been used to compare the temperatures measured with thermocouples and the ones calculated using outdoor and indoor IV parameters. Equation 1 shows the calculation of the temperature T as a function of the open circuit voltage V_{oc} and the voltage temperature coefficient β ($-0.31\%/°C$, assumed independent of the irradiance G [11]). $V_{oc}(T, G)$ is directly measured outdoor and $V_{oc}(25, G)$ is calculated from $I_{sc}(T, G)$ also measured outdoor. Note that we assume $I_{sc}(25, G) = I_{sc}(T, G)$, that is to say a zero current temperature coefficient, otherwise the equation would become non-linear.

$$T = 25 + \frac{V_{OC}(T, G) - V_{OC}(25, G)}{\beta \cdot V_{OC}(25, 1000)} \quad (1)$$

Figure 8 shows the energy yield of the monofacial module in terms of kilowatt-hour per kilowatt-peak over the period of test. Minimum production appears near summer solstice due to non-optimal orientation of the module with respect to solar positions (high azimuth angles on the morning and evening and high elevation angles around noon), and near winter solstice due to short duration of irradiation over the day. This results in a maximum production achieved around spring and autumn solstices.

Annual performance simulation

Ray-tracing model

We used the ray-tracing software TracePro® to create a simple optical model of our test bench (see Figure 9A). The solar disk sources can be set for any position in the sky with solid angle of the sun not being taken into account. Moreover, an isotropic semi-hemispherical background source is defined with an importance

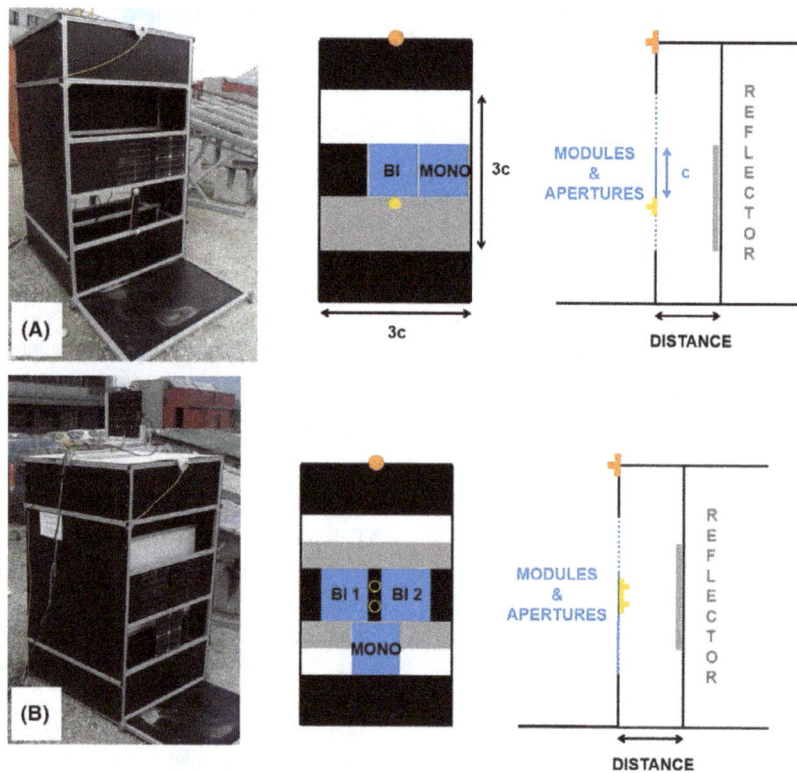

Figure 7. Picture and schematics of our reduced scale test bench for studying the opto-geometrical parameters (A) and the module parameters (B). Filled orange circle denotes the front pyranometer CMP3, while the filled and empty yellow circles denote the SPlite pyranometers for front and rear, respectively. All pyranometers are installed vertically and front pyranometers face south.

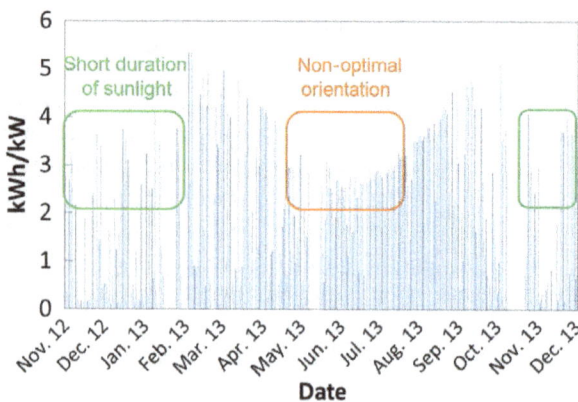

Figure 8. Daily performance of the monofacial reference module over the whole period of experimental tests.

sampling applied on the aperture area $3c \times 3c$. We limit the model to the opto-geometry of the system and do not take into account spectral effects. Additionally, we did not take into account the ground albedo in front of our test bench in order to be able to use the importance sampling tool. Also, the reflective losses, semi-transparency of the module as well as global irradiances below 10 W/m² have not been integrated in

the model. Note that Reich went down to 8 W/m² [12] and Sprenger to 5 W/m² [13]. With this model, we can set any diffuse to global ratio and any sun position in order to simulate the spatial and angular distribution of hourly irradiance on the backside of the bifacial module. We defined a typical day for each season of the year as shown on Figure 9B. The input data is taken from PVsyst database [14] and averaged over ±1.5 month around each solstice. Note that the number of rays have been chosen in order to ensure sufficient uniformity of the irradiance as well as limited time of simulation.

Optical to electrical conversion at 25°C

Once the model is designed, we use indoor flash test measurements (front and back independently measured) for converting irradiance data into electrical power data at 25°C. The spatial distribution of the irradiance is taken into account associating indoor STC measurements (Fig. 10A) with the least irradiated cell in the outdoor test bench simulations (Fig. 10B). The conversion is done using a logarithmic relationship between efficiency and irradiance ($\eta = a_1 \ln(G) + b_1$). The angular distribution

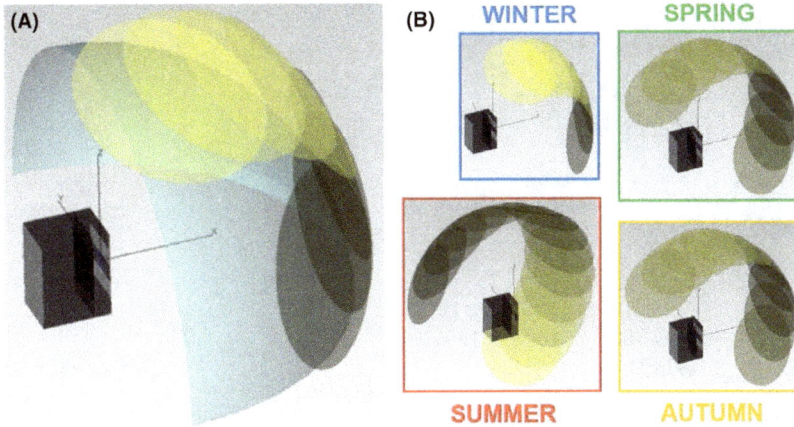

Figure 9. (A) Optical model of the test bench with direct (solar disks) and diffuse (isotropic sky) sources irradiating the test bench. (B) Solar positions for each typical day.

is taken into account associating indoor angle measurements (Fig. 10C) with the incidence angles of light rays in the outdoor test bench simulations (Fig. 10D). The conversion is done using a cosinus relationship between power and angle ($P_m = c_1 \cos(\theta)$). These conversions are critical hypothesis as we associate irradiance data from input meteorological data based on the measurement device spectral and angular responses with irradiance data based on the monitor cell spectral response at normal incidence and AM1.5G indoor spectrum.

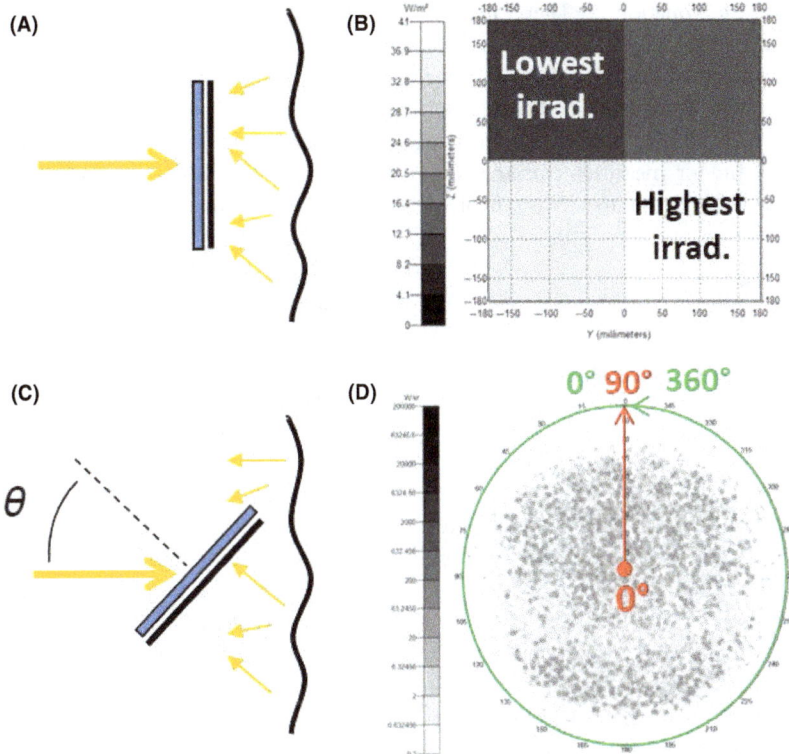

Figure 10. Standard indoor measurement (A) correlated with an example of spatial distribution of outdoor irradiance data (B) each of the four pixels corresponds to a cell, the darker the less irradiated. Indoor angle measurements (C) correlated with an example of angular distribution of outdoor irradiance data (D): red straight axis from 0° to 90° shows the incidence angle of a ray and green circular axis from 0° to 360° maps the direction of incidence of a ray, the darker the more irradiated in this case.

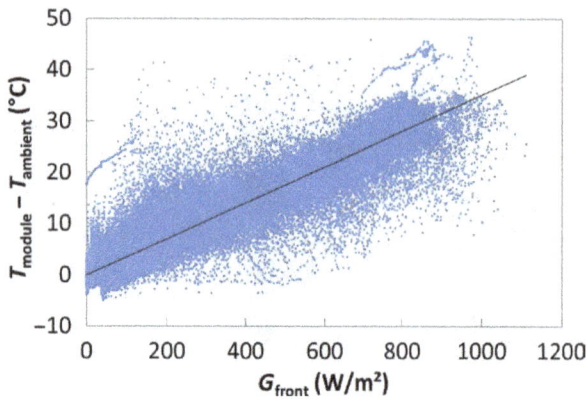

Figure 11. Temperature difference between monofacial module and ambient as a function of the front irradiance in minute data over the year of test, and its linear approximation.

Bifacial power at real temperature

In order to get the electrical power at real operating temperatures, we use a linear model based on outdoor data obtained for the monofacial module. Figure 11 shows the temperature difference $T_{module} - T_{ambient}$ as a function of irradiance G_{front} over the whole year in minute data. The linear approximation lead to a coefficient $\kappa_{mono} = 0.035°C/(W/m^2)$ in Equation 2. Therefore, the module temperature can be evaluated using the ambient temperature and irradiance measured on site, and the real powers can be calculated. These front and back powers are summed to give the bifacial power. We ensured with double illumination indoor measurements that this hypothesis is valid for the level of total irradiance involved in our application (<1.04 sun).

$$T_{module} = T_{ambient} + \kappa \cdot G_{front} \qquad (2)$$

Influence of Opto-geometrical Parameters

Experimental data

Between November 2012 and March 2013, several opto-geometrical configurations have been tested one after each other varying the module–reflector distance (0.5c, c and 2c; with the module height c = 36 cm) and the inner reflector (black or white). In this test, we used a module of standard architecture (glass–glass structure, full-size cells in series) with the configuration of the test bench shown on Figure 7A. Figure 12 shows the daily electrical energy gain brought by the use of a bifacial module instead of a monofacial one for four configurations. The variability in the data points is due to the period of test and the meteorological conditions, which were more or less cloudy.

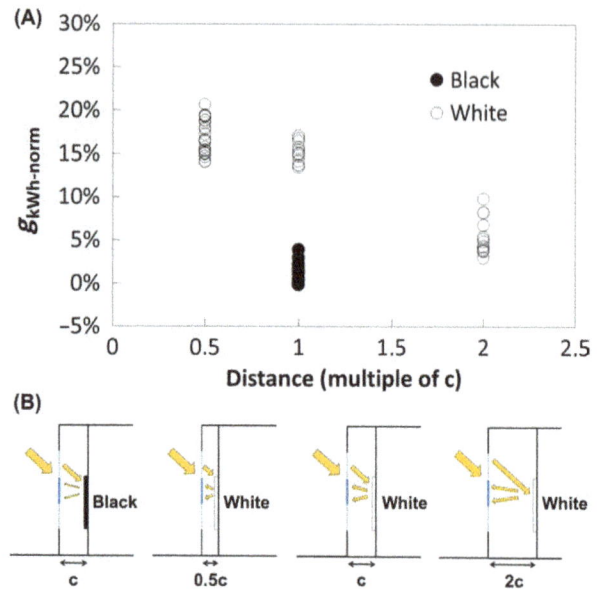

Figure 12. Daily bifacial electrical energy gain compared to monofacial (powers normalized to the front STC for each module) for different opto-geometrical configurations. The module reflector distance is shown in terms of module height, c = 36 cm.

At a distance c, the use of a white reflector (weighted reflectivity with respect to the solar spectrum, R ~80.4%) gives an energy gain compared to monofacial module ($g_{kWh-norm}$) 13% higher than the one obtained with a black reflector (R ~4.9%). At a smaller distance (0.5c), maximal gains appear to be comprised between 15% and 20%.

Additionally, the temperature difference between bifacial and monofacial modules is shown in Figure 13 for the four configurations. At low irradiances, a measurement artefact due to the combination of data measured with different IV tracers is visible (temperatures calculated with Equation 1). For higher irradiances, all the configurations follow the same linear tendency, which reaches −2°C at a front irradiance of 1000 W/m². The main reason is that the bifacial module has twice as much surface compared to the monofacial one to release heat by thermal radiation. Using these data, we define a new coefficient $\kappa_{bi} = 0.033°C/(W/m^2)$ for the bifacial module, which will be used for simulating its thermal advantage over the year. It must be noticed that both the temperatures measured with thermocouples and calculated with V_{oc} gave the same linear trend (except at low irradiances) after removing thermal conduction contributions, which are different for a thermocouple glued on a 0.17 mm backsheet compared to a 3 mm backside glass [15].

Validation of the model

A sunny and a cloudy day have been used to compare the simulation and experimental results for winter period as well as for summer period. The model sources (see Fig. 9) are defined using direct irradiance measured with a sun tracker on our site and diffuse irradiance on the module plane calculated using global irradiance measured with the front pyranometer (filled orange circle on Fig. 7). The backside irradiance is simulated and converted in electrical power data as explained in section Optical to electrical conversion at 25°C. These data are temperature corrected using experimental temperature monitoring. Figure 14 shows the comparison for these 4 days. We use the determination coefficient R^2 and the Student test (see Equation 3 and Equation 4 – N being the number of data points over the day) to compare the experimental and simulated temporal series (x^i_{exp} and x^i_{sim}). Each R^2 is above 95.1% and the Student test allows us to say that simulations estimate measurements with a confidence interval of 99.9% [16].

$$R^2 = 1 - \frac{\sum_{i=1}^{N}(x^i_{sim} - x^i_{exp})^2}{\sum_{i=1}^{N}(x^i_{exp} - x^{ave}_{exp})^2} \qquad (3)$$

$$t_s = \sqrt{\frac{(N-2) \cdot R^2}{1 - R^2}} \qquad (4)$$

Annual performance

In the previous sections, we obtained experimental data in different configurations of our test bench at specific periods of the year. The model being validated, we can now extrapolate our results over the whole year. Figure 15 shows the bifacial module energy production simulated in watt-hour for each typical day as well as over the year (cumulating each typical day) according to the distance between module and reflector in the white reflector case. As in the monofacial case, we can see that the energy production is maximum near spring and autumn solstices independently of the distance (see Fig. 8). The annual energy production (black curve) clearly shows a maximum between distance 0.5c and 0.75c. Above these distances, the decrease is due to side effects of our test bench, namely the black side walls between the module and the reflector, which absorb the light diffused by the white reflector and prevent the ambient diffuse light from reaching the backside of the module. Below these distances, the decrease is due to current limitation by the least irradiated cell in the module. Indeed, Figure 16 shows the non-uniformity of backside irradiance (daily average of hourly data points calculated with Equation 5) according to the distance for each typical day as well as over the year. The black curve on the chart shows a strong increase in non-uniformity at short distances. This is mainly due to the module shade as well as the black side walls

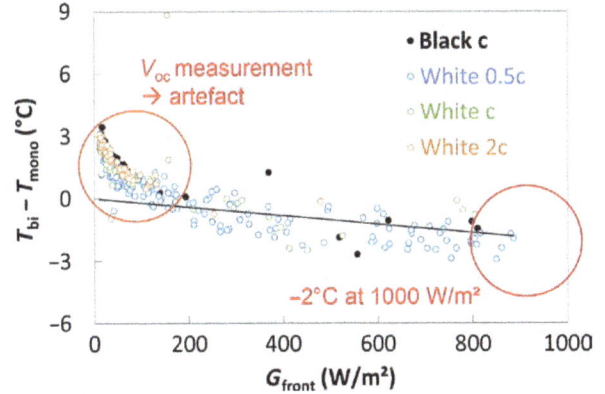

Figure 13. Temperature difference between bifacial and monofacial module as a function of front irradiance for the four configurations shown in Figure 12 (hourly data points). The linear tendency without taking into account the low irradiance artefact concerns the four configurations all together.

shade both projected onto the reflector (depending on the position of the sun).

$$NU = \frac{G_{max} - G_{min}}{G_{max} + G_{min}} \qquad (5)$$

Influence of Module Architecture

Experimental data

Between April 2013 and December 2013, innovative module architectures have been tested one after each other with the configuration of the test bench shown on Figure 7B. We saw before that bifacial modules suffer from resistive losses due to higher currents, and the backside irradiance is often non-uniform and diffuse. In order to address these issues, we tested two different architectures. On the one hand, a module has been made with two series of half-cut cells connected in parallel in order to reduce the resistive losses and to be less sensitive to non-uniform irradiance (Fig. 17A). We ensured that this architecture has the same $I_{sc\text{-}front\text{-}STC}$ and $V_{oc\text{-}front\text{-}STC}$ as a standard architecture (four full size cells in series). On the other hand, we used a textured glass on both sides for collecting high incidence angles of diffuse radiation (Fig. 17B). The following subsections describe the experimental tests (indoor and outdoor) performed for finding empirical models, which will allow us later to identify the potential of these novel module architectures at the annual scale.

Influence of using half-cells

In order to separate the influence of half-cells from the influence of parallel interconnection, we had to measure

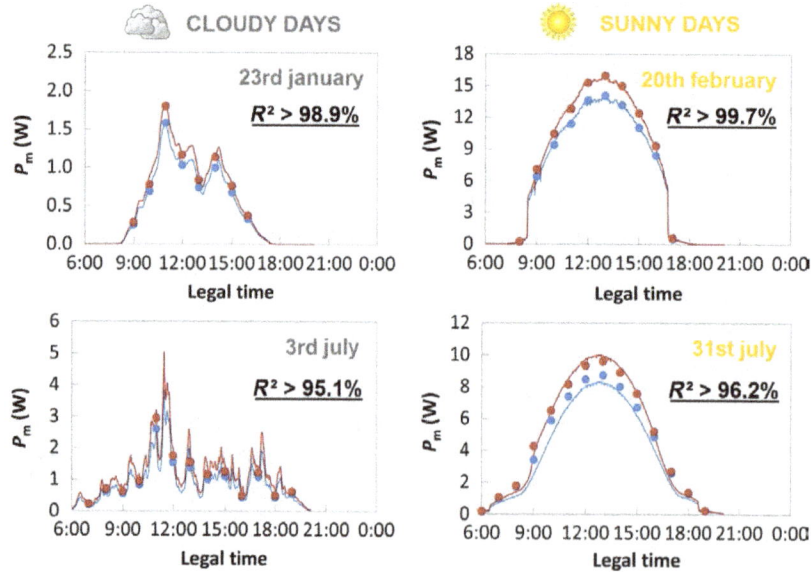

Figure 14. Comparison of simulated (dots) and experimental (curves) data for bifacial (red) and monofacial (blue) modules against legal time, with associated R^2. Cloudy days on the left, sunny days on the right, winter days on the top, and summer days on the bottom.

the module under uniform illumination. This was possible with our indoor double illumination setup (Fig. 18A) and with our outdoor test bench masking the backside of the module (Fig. 18B). Both charts show the gain brought by the novel architecture compared to the standard one for parameters I_m, V_m, and P_m. The indoor and outdoor measurements are in agreement showing a linear increase in P_m and V_m with total irradiance on the module: we observe a negative gain below 0.25 sun and ~3% at 1 sun. The indoor bifacial characterization shows up to 7% gain at 2 sun. Indeed, as the total irradiance increases on the module, the resistive losses increase on the standard architecture and the use of half-cells brings more and more advantage (higher power due to higher voltage). Note that the negative gain at low irradiances might be due to shunt currents induced by laser cutting of our N-type cells.

From the indoor experimental data, we used a linear relationship (green line shown on Fig. 18A) between the power gain g_{Pm} and the total irradiance incident on the module $I_{sc-bi}/I_{sc-front-STC}$ shown in Equation 6 for simulating the contribution of half-cells at the annual scale.

$$g_{Pm} = 4.6\% \cdot I_{sc-bi}/I_{sc-front-STC} - 1.4\% \qquad (6)$$

Influence of parallel interconnection

The bifacial gain is critically dependent on the uniformity of irradiance on the module backside (examples in [4, 17]). For vertical facades, the section Annual performance showed that the backside irradiance can be strongly non-uniform if the module–reflector distance is very small.

In this case, the upper and lower cells do not receive the same amount of radiation due to the module shade projection onto the back reflector. Therefore, we investigated the advantage of using two strings of cells connected in parallel in the module architecture in order to be less sensitive to non-uniform irradiance.

First, we artificially created strong non-uniformities on the backside of the module placed in the double illumination setup. Table 2 shows the gain between the novel architecture and the standard one for parameters I_m, V_m, and P_m in three different backside configurations: no mask, a $c \times 0.25c$ mask, and a $c \times 0.5c$ mask (no mask on the

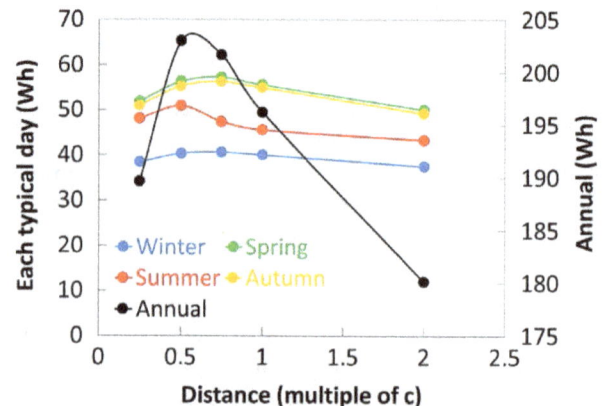

Figure 15. Simulated daily energy production for each typical day (winter in blue, spring in green, summer in red, and autumn in yellow) and over the year (black) as a function of the distance between module and reflector.

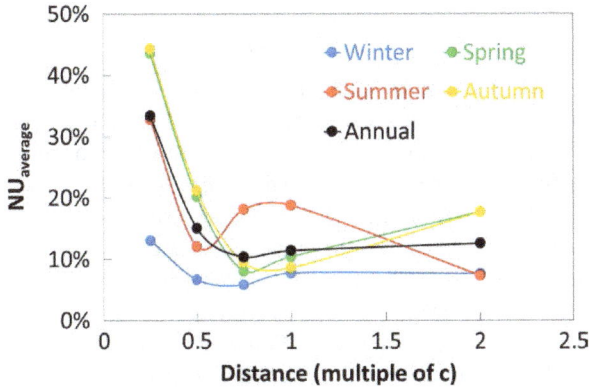

Figure 16. Simulated daily averaged non-uniformity of backside irradiance for each typical day (winter in blue, spring in green, summer in red, and autumn in yellow) and over the year (black) as a function of the distance between module and reflector.

frontside). For comparison, the gain taking into account resistive losses only (calculation with eq. 6) is written between brackets. The configuration $c \times 0.25c$ shows a strong current gain (12%) due to the interconnection in parallel since the shaded string does not limit the unshaded string in the novel architecture. In terms of voltage, the shaded string slightly limits the unshaded string: a 6% gain due to resistive losses becomes only a 1.4% gain with the parallel interconnection in addition. This still results in a positive effect on the power (13.5% gain). If one of the two strings is fully masked (configuration $c \times 0.5c$), this phenomenon is increased with a higher current gain (+41%) and a voltage loss (−2.2%).

Then, we compared two different configurations of our outdoor test bench by varying the distance between module

and reflector: a more uniform irradiance and a less uniform irradiance on the backside of the module as shown on Figure 16. Figure 19 shows that the power gain is about 0.5% higher for the less uniform configuration. This confirms the slight advantage of using a parallel interconnection in outdoor conditions for our specific case. This is the result of about 1% gain for I_m (no current limitation by the least irradiated string) and about 0.5% loss for V_m (voltage limitation by the least irradiated string). Note that both curves have a similar slope due to the resistive loss effect as explained previously.

In order to be able to simulate the maximal contribution brought by the parallel interconnection on the annual scale, we expressed the I_{sc} gain using the spatial distribution of irradiance on each side. The following equations have been used to get Equation 7:

1. $I_{sc\text{-}bi} = I_{sc\text{-}front} + I_{sc\text{-}back}$ for both modules. This is verified at −0.5% in average, while Ohtsuka had +0.4% [7] and Ezquer −1.8% [8].
2. $I_{sc\text{-}front} = I_{sc\text{-}front\text{-}STC} \times G_{front}$ for both modules (uniform irradiance on their frontside). G is the global irradiance incident on one face in terms of number of sun.
3. $I_{sc\text{-}back} = \min(I_{sc\text{-}back\text{-}string1}, I_{sc\text{-}back\text{-}string2}) = I_{sc\text{-}back\text{-}STC} \times \min(G_{back\text{-}string1}, G_{back\text{-}string2})$ for the standard architecture.
4. $I_{sc\text{-}back} = I_{sc\text{-}back\text{-}string1} + I_{sc\text{-}back\text{-}string2} = 0.5 \times I_{sc\text{-}back\text{-}STC} \times (G_{back\text{-}string1} + G_{back\text{-}string2})$ for the novel architecture.
5. $I_{sc\text{-}back\text{-}STC} = r \times I_{sc\text{-}front\text{-}STC}$ for both modules, which have the same I_{sc} in STC conditions. r is the bifacial ratio.
6. the numbers of sun G_{front}, $G_{back\text{-}string1}$ and $G_{back\text{-}string2}$ are linked through the non-uniformity of irradiance NU

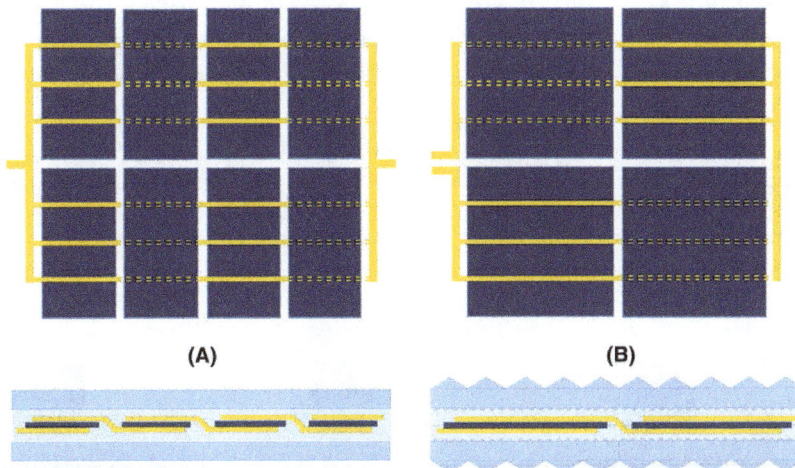

Figure 17. Schematics of the two different architectures tested: half-cells and parallel strings with flat glasses (A), and four full-size cells in series with textured glasses (B). The modules have been made with cells from the same batch and will be compared to a reference module made of four full-size cells in series encapsulated between flat glasses.

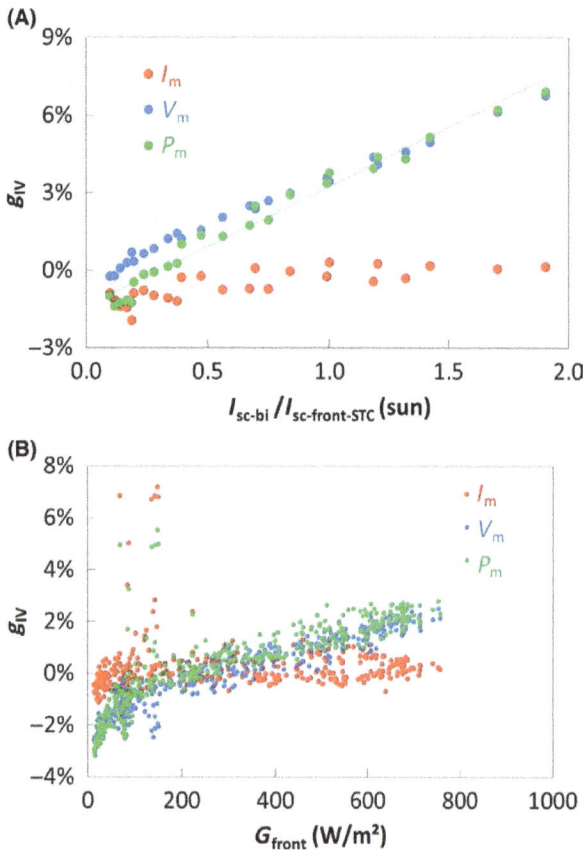

Figure 18. Gain of the novel architecture compared to a standard module for the parameters I_m, V_m, and P_m obtained with indoor measurements (A – bifacial illumination characterization) and outdoor measurement (B – front characterization only in hourly data).

between the upper string and the lower string of the backside (see eq. 5), and through the effective albedos $A_{min} = \min (G_{back\text{-}string1}, G_{back\text{-}string2})/G_{front}$ and $A_{ave} = 0.5 \times (G_{back\text{-}string1} + G_{back\text{-}string2})/G_{front}$.

$$g_{Isc} = NU \cdot \frac{r \cdot A_{ave}}{1 + r \cdot A_{min}} \qquad (7)$$

Table 3 shows the comparison between calculation and measurements with our indoor setup. Our calculations are validated experimentally in this case of strong non-uniformities artificially created (masks $c \times 0.25c$ and $c \times 0.5c$), so we applied this model using outdoor I_{sc} data. We observed that the measurement values were lower than the calculated values for unknown reasons (chart not represented). Therefore, we decided to add an empirical coefficient of 0.4 in Equation 7 in order to fit outdoor observations. Finally, we associate the resulting I_{sc} gain to the upper limit of the P_m gain in Equation 8, since the shaded string can suffer from reverse currents. Equation 8 will therefore allow us to evaluate the maximum power gain due to non-uniform irradiance even though the real power gain will be always lower.

$$g_{Pm\text{-}max} = 0.4 \cdot NU \cdot \frac{r \cdot A_{ave}}{1 + r \cdot A_{min}} \qquad (8)$$

Influence of linear textured glasses

Diffuse radiation plays an important role in bifacial applications particularly on the backside of the module. We compared a bifacial module with a unidirectional textured glass from the company AGC on both sides, to a bifacial module with standard flat glasses. Such a unidirectional texturation can have the advantage of self-cleaning with rain. In order to identify the effect of the glass only, we measured the gain between a monofacial module with the front linear textured glass and a monofacial module with the front flat glass as a function of the incidence angle of light θ (see Fig. 10C). Figure 20 shows the results for two orientations of the texturation: parallel and orthogonal to the rotation axis. For angles below 40°, the gain is about 2–3% due to a light trapping effect. Above 60°, the orthogonal orientation shows a gain increase due

Table 2. Gains of the novel architecture compared to standard one for parameters I_m, V_m and P_m for different shading configurations on the backside of the modules placed in our double illumination setup (see the image on the left corner). The measured gains (in bold) are compared with calculated gains taking into account resistive losses only (between brackets).

	No mask	Mask c × 0.25c	Mask c × 0.5c
g_{Im}	**−0.3%** (0.3%)	**12%** (0%)	**41%** (−0.4%)
g_{Vm}	**6%** (6.2%)	**1.4%** (4.6%)	**−2.2%** (3.1%)
g_{Pm}	**5.7%** (6.5%)	**13.5%** (4.6%)	**37.9%** (2.7%)

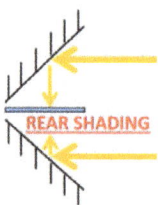

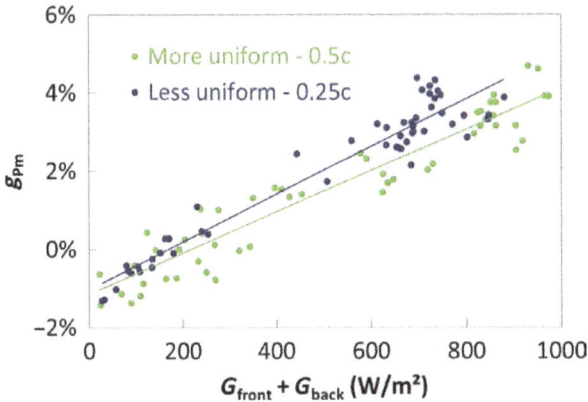

Figure 19. Hourly data point of the power gain g_{Pm} of the novel architecture compared to standard one plotted against the total irradiance on the module $G_{front} + G_{back}$ for two different configurations of the test bench: smaller distance (less uniform irradiance on the backside of the module) and larger distance (more uniform irradiance).

to the divergence of the PASAN source as well as parasitic reflections on the black curtain behind the module. For this range of angles, a clear increase in the gain confirms the positive effect of textured glass for the parallel orientation.

In order to be able to simulate the influence of textured glasses at the annual scale on our application, the angular distribution of light on each face of the module has been simulated hourly for each typical day (see Fig. 10D as an example). For each side of the module, the optical flux is integrated over ±45° around the horizontal direction on the one hand, and around the vertical direction on the other hand. We associated this simulated data (normalized with the total flux incident on the surface) with the indoor characterization data shown in Figure 20 in order to obtain the power gain brought by the use of a textured glass for each hourly data point and each side of the module in two different orientations.

Validation of the model

The empirical correlations obtained in section Experimental data have been used to simulate the

Table 3. Comparison between calculated and measured g_{Isc} in the case of indoor artificially created non-uniformities on the backside of both modules (masks c × 0.25c and c × 0.5c).

Mask on backside	NU	A_{min}	A_{ave}	g_{Isc} calc.	g_{Isc} meas.
c × 0.25c	33.3%	0.5	0.75	15.2%	13.1%
c × 0.5c	100%	0	0.5	43.6%	44.3%

evolution of P_m over two summer days. We chose a sunny day giving non-uniform irradiance on the backside of the module (NU comprised between 31% and 60%) for the half-cells and parallel interconnection architecture, and a cloudy day giving diffuse irradiance on both sides (diffuse to global ratio above 99.7%) for the linear textured glasses architecture. The comparison of temporal series (simulated and measured) gives R^2 above 98.8% for the half-cells and parallel interconnection architecture and R^2 above 94.3% for textured glasses architecture (see eq. 3). Knowing the number of hourly data points for each day, the Student test allows us to say that the model estimate the reality with a confidence interval of 99.9% (see eq. 4 and [16]).

Figure 21 shows the resulting gain from experimental and simulated data for these two days. The upper chart (sunny day) shows a strong loss for the novel architecture compared to standard one (four full size cells in series) around 9:00 am and a strong gain around 4:30 pm. Indeed, in this configuration, the standard module is on the west side meaning that its backside receive more radiation when the sun rises, and the novel module is on the east side meaning that its backside receives more radiation when the sun goes down. By contrast, the lower chart (cloudy day) does not show this asymmetrical production as the optical flux is completely diffuse on the backside of the module.

Annual performance

Half-cells and parallel interconnection architecture

Figure 22 shows the annual simulation of the gain brought by the novel architecture compared to standard one with the influence of half-cells only and with the influence of the parallel interconnection in addition. We observe that the gain due to less resistive losses (half-cells influence) is limited independently of the distance (below 0.4%) as the total irradiance on our south-oriented vertical bifacial module is not very high (1.04 sun maximum, 0.32 sun in average). This type of architecture could be suitable in applications where the module receives more radiation, which implies more resistive losses. We also observe that the maximal gain due to non-uniform irradiance (parallel interconnection influence) is obviously higher for short distances when the non-uniformity of backside irradiance increases. Therefore, this architecture could be suitable for applications which suffer from high non-uniform irradiance on the module like in compact bifacial solar plants for example. Note that the real annual gain

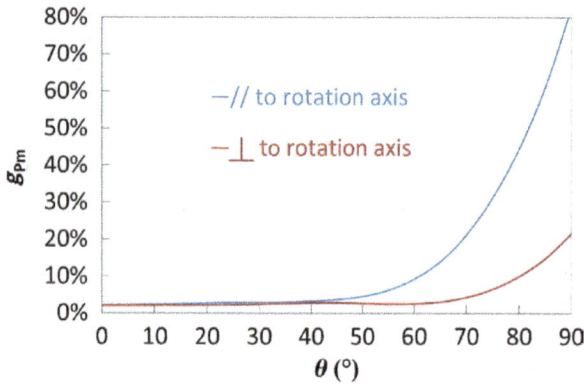

Figure 20. Power gain for a monofacial module with a unidirectional textured glass compared to flat glass as a function of the incidence angle of light (0° corresponds to normal incidence) for two different orientations of the textured glass.

considering both effects would be comprised between the dotted and plain curve since the plain red curve describes the maximal g_{Pm} (see eq. 8). However, we cannot quantify this with our simple model.

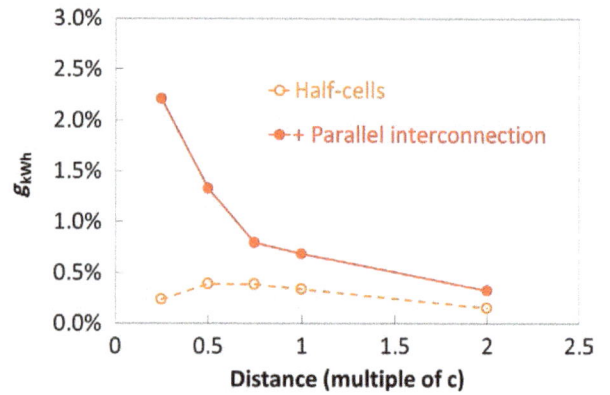

Figure 21. (A) Simulated and measured power gain of the half-cells and parallel interconnection architecture compared to standard one on the 31/07/13. (B) Simulated and measured power gain of the textured glasses architecture compared to standard one on the 03/07/13.

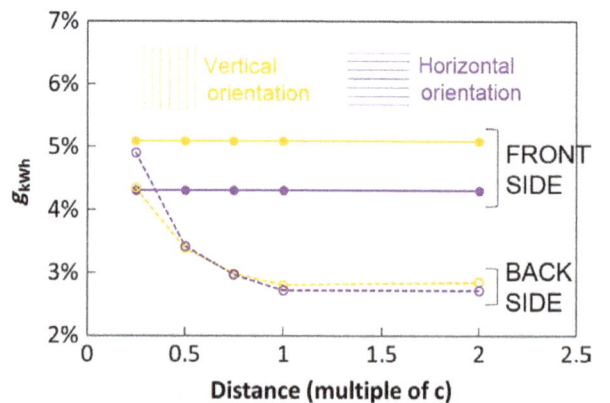

Figure 22. Simulation of the annual gain of the novel architecture compared to standard one varying the distance between module and reflector. The red plain curve shows the parallel interconnection influence in addition to the half-cells influence shown by the orange dotted curve.

Textured glasses architecture

Figure 23 shows the annual simulation of the gain brought by the novel architecture compared to standard one in different orientation of the linear textured glass and for each side of the module. We observe that the backside gain does not depend a lot on the texturation orientation as the white backside reflector is quasi-lambertian, and it increases for shorter distances as the proportion of high incidence angles increases. In a real facade application, the backside gain would be certainly higher since the black inner side walls of our test bench absorb a part of radiations. We also observe that the frontside gain is independent of the distance and lower for the horizontal orientation. This gain would be certainly higher if the ground albedo had been taken into account in our model since more radiation would be incident from the ground.

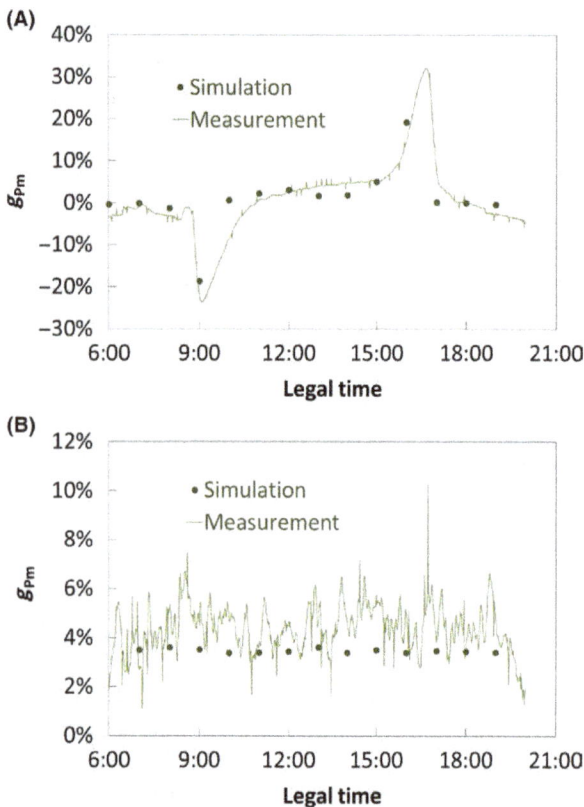

Figure 23. Simulation of the annual gain of the novel architecture compared to standard one varying the distance between module and reflector.

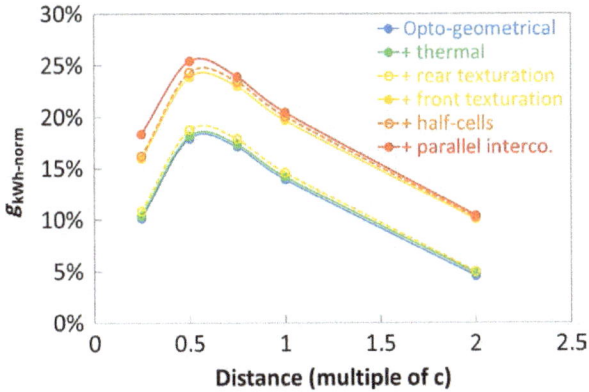

Figure 24. Simulation of the annual gain of a bifacial module compared to a monofacial one as a function of the distance between module and reflector. From the bottom to the top, each curve shows an additional contribution to the gain.

Bifacial gain considering all contributions

Finally, Figure 24 shows the annual gain of a bifacial module compared to a monofacial one in our reduced scale vertical facade. First, the largest part of the gain is due to the opto-geometrical environment (18.1% for the optimal distance – see section Annual performance). Then, a slight gain is due to the thermal advantage of the bifacial module (0.3% maximum – see section Experimental data). The use of textured glasses in their vertical orientation adds a large gain (up to 5.8%) mainly due to the front glass (up to 5.1%). In comparison, the gain brought by the half-cells and parallel interconnection architecture appears to be limited in our vertical facade application.

Perspective at Real Scale

Full-size module architecture

The electrical and optical architecture of bifacial modules need to be optimized regarding the issues of resistive losses, non-uniform and diffuse backside irradiance. Based on the work described here, we propose the module architectures shown in Figure 25.

On the electrical part, the module has two parallel blocks of 60 half-cut cells (Fig. 25A), three parallel blocks of 60 cells cut in three (Fig. 25B), or four parallel blocks of 60 cells cut in four (Fig. 25C). Possible positions of bypass diodes are shown on the images. The two blocks module seems to be the most suitable in practice, since the three blocks module requires electrical insulation between the yellow and brown busbars, and the four blocks module could imply complex soldering process due to the small size of the cells. However, with three or four

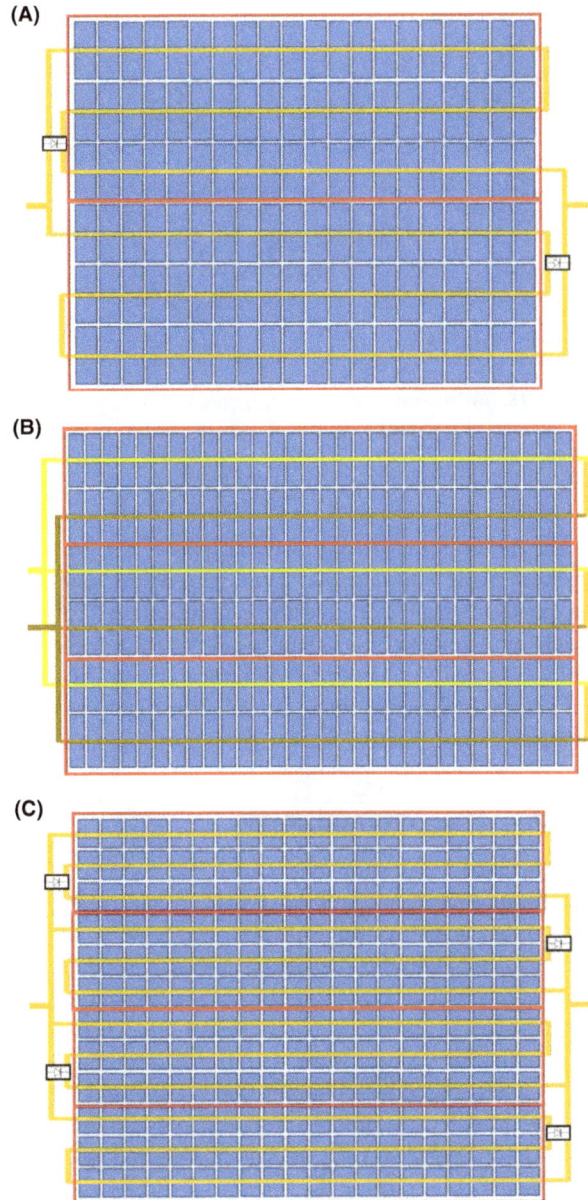

Figure 25. Module electrical architectures equivalent to a 6 × 10 full-size cells module. The division factor of a cell is equal to the number of blocks in parallel shown with red rectangles: factor two (A), factor three (B), and factor four (C).

parallel blocks (smaller cells), the robustness to non-uniform irradiance would be higher and the resistive losses would be lower compared with the two blocks module. Finally, the two and four parallel blocks modules seem to be suitable for vertical facade applications as their positive and negative outputs are positioned on both sides of the module. Therefore, the neighboring modules can be more easily connected in series and resistive losses are minimized at the system level.

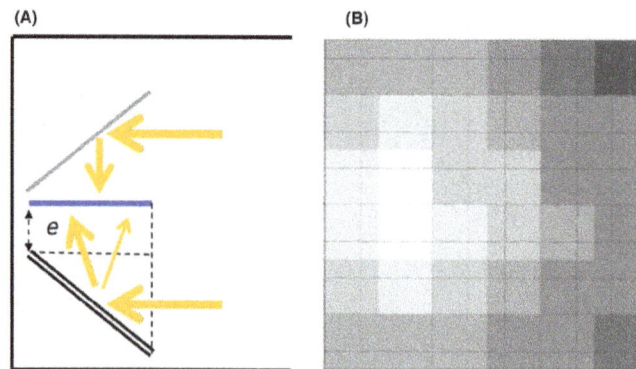

Figure 26. (A) Schematics of the adapted double illumination setup (aluminum mirror on the front, white outdoor reflector on the back). (B) Ray-tracing simulation of the distribution of irradiance on the backside of a 6 × 6 cells module (instead of a 6 × 10 cells in portrait orientation) with the adapted double illumination setup (white pixels show the most irradiated cells).

On the optical part, the textured glasses can be used on both sides knowing that most part of the gain will be due to the front glass. Ideally, the texturation of glasses would be optimized for several types of bifacial applications: tilted face to the equator or vertical face to east–west on a horizontal white ground or white roof, and vertically integrated onto a white facade.

Further experimental tests

The potential of the module architectures proposed in the previous section must be evaluated experimentally. This must be performed with both an outdoor and an indoor setup. On the one hand, a real size adjustable facade would allow evaluating the optimal module–reflector distance in order to compare it to the $0.5c$–$0.75c$ found on our reduced scale test bench. On the other hand, a real size double illumination setup could be designed in order to simulate real outdoor environment conditions. Figure 26A shows that a white reflector (the one used outdoor on a flat roof or vertical facade for example) can be used instead of the back mirror in the double illumination setup. In this case, the radiation incident on the backside of the module would be non-uniform (Fig. 26B), diffuse, and would overall take into account the spectral reflectivity of the reflector as in a real outdoor environment.

Finally, we designed an optimal real-size standard double illumination setup (mirrors on both sides of the module) with our ray-tracing model of the PASAN solar simulator. We found that a portrait orientation of the module is suitable otherwise the mirrors length would not fit inside the PASAN. We also found an optimal angle between mirror and module of 43° ± 1°. That leads to a similar uniformity of irradiance, a similar global irradiance and a normal incidence in average for our setup and for a standard STC test (considering a perfect mirror and for a distance $e = 5$ cm – see parameter e on Fig. 26A). For this angle, increasing parameter e does not improve the uniformity of irradiance. Note that an experimental validation of this setup is required since the use of real mirrors (with a part of diffuse reflection, dust, micro-scratches, and less clean areas) could make the non-uniformity of irradiance too high for the characterization standards (NU < 2% for class A [18]). The protocol described in section Double illumination characterization setup must be used and the uncertainties must be evaluated again in the real size case. Such a study could provide additional information to the actual discussions about defining a standard for indoor characterization of bifacial modules in solar simulators.

Conclusion

In this article, we developed a methodology for evaluating the annual performance of bifacial modules integrated on a vertical facade at a reduced scale. First, a protocol using a double illumination setup in a solar simulator has been used. We found that the use of aluminum mirrors does not impact the spectral class of measurement as long as the PASAN source intensity is adjusted. Also, the uncertainty of measurement has been quantified and does not exceed ±3%. Additionally, we performed outdoor tests in real conditions, which allowed us to validate our ray-tracing simulations for both sunny and cloudy days in winter as well as in summer: we found $R^2 > 94.3\%$ with Student test giving a high confidence interval. With these three tools, we were able to quantify the resistive losses which are due to high total irradiance on both sides of the module, the non-uniformity of irradiance on the backside which is due to the module shade projected onto the back reflector, and its diffuse character which is due to the use of quasi-lambertian reflectors.

Regarding the issues above mentioned and using our methodology, we varied opto-geometrical parameters of the application (particularly the distance between module and reflector and the type of reflector) as well as electrical and optical aspects of the module architecture (particularly the cell interconnection and the glasses) in order to maximize the annual performance. The annual electrical energy gain of a bifacial module compared to monofacial could reach 25% with all the cumulated contributions for a module–reflector distance ranging between $0.5c$ and $0.75c$, c being the module height. The highest contribution comes from the opto-geometrical environment of the module namely the irradiance incident on its back, which gives a 18% maximum gain. Another large contribution comes from the use of a linear textured glass vertically oriented on the frontside, which gives a 5% maximum gain.

Based on our tests and simulations for a reduced scale application, we proposed a full-size module architecture which could be more robust to non-uniform irradiance on the backside with less resistive losses and more harnessing of high incidence angles. The next research step will be to evaluate the optimal module–reflector distance in the case of a real-size vertical facade, and to quantify the advantage of the proposed module architecture in several kinds of bifacial applications. This could be achieved with more general simulation tools taking into account the complex opto-geometrical environment of the bifacial module. Finally, the optimized applications will have to be long term tested at the real size power plant scale in order to give confidence to investors.

Acknowledgments

We thank AGC Glass Europe for providing the textured glass samples and for technical discussions. Bruno Soria's work has been supported by a PhD grant from the Commissariat à l'Energie Atomique et aux Energies Alternatives (CEA). The resulting thesis is available online in French [19]. This article is dedicated to the memory of Yves Delesse who actively participated to the instrumentation of the test bench applied in this study.

Conflict of Interest

None declared.

References

1. Cuevas, A. 2005. The early history of bifacial solar cells. Pp. 801–805 *in* Proceedings 20th European PV solar energy conference, Barcelona.

2. Luque, A., and S. Hegedus. 2003. Handbook of photovoltaic science and engineering. Wiley & Sons, UK.

3. Obara, S., D. Konno, Y. Utsugi, and J. Morel. 2014. Analysis of output power and capacity reduction in electrical storage facilities by peak shift control of PV system with bifacial modules. Appl. Energy 128:35–48.

4. Yusufoglu, U. A., T. M. Pletzer, L. J. Koduvelikulathu, C. Comparotto, R. Kopecek, and H. Kurz. 2015. Analysis of the annual performance of bifacial modules and optimization methods. IEEE J. Photovolt. 5:320–328.

5. Lo, C. K., Y. S. Lim, and F. A. Rahman. 2015. New integrated simulation tool for the optimum design of bifacial solar panel with reflectors on a specific site. Renew. Energy 81:293–307.

6. Cabal, R., Y. Veschetti, V. Sanzone, S. Manuel, S. Gall, F. Barbier, et al. 2013. Industrial process leading to 19.8% on N-Type Cz silicon. Energy Procedia. 33:11–17.

7. Ohtsuka, H., M. Sakamoto, M. Koyama, K. Tsutsui, T. Uematsu, and Y. Yazawa. 2001. Characteristics of bifacial solar cells under bifacial illumination with various intensity levels. Prog. Photovolt. Res. Appl. 9:1–13.

8. Ezquer, M., I. Petrina, J. M. Cuadra, A. R. Lagunas, and F. Cener-Ciemat. 2009. Design of a special set-up for the IV characterization of bifacial photovoltaic solar cells. Presented at 24th European Photovoltaic Solar Energy Conference and Exhibition, Hamburg.

9. Edler, A. 2012. Flasher setup for bifacial measurements. Presented at BIFIPV workshop, Konstanz.

10. Lambda Research Corporation. TracePro: Software for designing optical systems. Available at http://www.lambdares.com/.

11. King, D. L., J. A. Kratochvil, and W. E. Boyson. 1997. Temperature coefficients for PV modules and arrays: measurement methods, difficulties, and results. Photovoltaic Specialists Conference, 1997., Conference Record of the Twenty-Sixth IEEE:1183–1186.

12. Reich, N. H., W. G. J. H. M. van Sark, W. C. Turkenburg, and W. C. Sinke. 2010. Using CAD software to simulate PV energy yield – The case of product integrated photovoltaic operated under indoor solar irradiation. Sol. Energy 84:1526–1537.

13. Sprenger, W. 2013. Electricity yield simulation of complex BIPV systems. Delft University of Technology, Fraunhofer-Verlag. ISBN 978-3-8396-0606-3, 158 pages.

14. PVsyst. Software for designing PV plants. Available at www.pvsyst.com.

15. Soria, B., and E. Gerritsen. 2013. Vertical Facade Integration of Bifacial PV Modules: Outdoor Testing and Optical Modelling. Presented at 28th European Photovoltaic Solar Energy Conference and Exhibition, Paris.

16. Muneer, T. 2004. Solar radiation and daylight models: (with software available from companion web site), 2nd ed. Elsevier Butterworth Heinemann, Oxford; Burlington, MA.

17. Kreinin, L., N. Bordin, A. Karsenty, A. Drori, D. Grobgeld, and N. Eisenberg. 2010. PV module power gain due to bifacial design. Photovoltaic Specialists Conference (PVSC), 2010 35th IEEE:002171–002175.

18. International Electrotechnical Commission. 2007. Norme IEC 60904-9 Ed 2.0 Photovoltaic devices - Part 9: Solar simulator performance requirements. October, 2007.

19. Soria, B. 2014. Etude des performances électriques annuelles de modules photovoltaïques bifaces. Cas particulier? Modules bifaces intégrés en façade verticale. Université de Grenoble, https://tel.archives-ouvertes.fr/tel-01126959/document.

10

Stable ultrathin surfactant-free surface-engineered silicon nanocrystal solar cells deposited at room temperature

Vladimir Švrček[1] ⓘ Calum McDonald[1,2], Mickael Lozac'h[1], Takeshi Tayagaki[1], Tomoyuki Koganezawa[3], Tetsuhiko Miyadera[1], Davide Mariotti[2] & Koji Matsubara[1]

[1]Research Center for Photovoltaics, National Institute of Advanced Industrial Science and Technology (AIST), Central 2, Umezono 1-1-1, Tsukuba, 305-8568, Japan
[2]Nanotechnology and Integrated Bio-Engineering Centre (NIBEC), Ulster University, Coleraine, UK
[3]Japan Synchrotron Radiation Research Institute (JASRI), 1-1-1, Kouto, Sayo-cho, Sayo-gun, Hyogo 679-5198, Japan

Keywords
Silicon nanocrystals, surfactant-free surface engineering, ultrathin film solar cells

Correspondence
Vladimir Švrček, Research Center for Photovoltaics, National Institute of Advanced Industrial Science and Technology (AIST), Central 2, Umezono 1-1-1, Tsukuba 305-8568, Japan.
E-mail: vladimir.svrcek@aist.go.jp

Funding Information
The GIWAXS measurement was performed at SPring-8 BL46XU with the approval of the Japan Synchrotron Radiation Research Institute (JASRI, proposal nos. 2015B1600 and 2015B1891). This work was partially supported by EPSRC (EP/K022237/1 and EP/M024938/1).

Abstract

We present a scalable technology at room temperature for the fabrication of ultrathin films based on surfactant-free surface-engineered silicon nanocrystals (SiNCs). Environmentally friendly pulsed fsec laser induced surface engineering of SiNCs and vacuum low-angle spray deposition is used to produce ultrathin films. Surface engineering of SiNCs improved stability and dispersibility of SiNCs by allowing thin (30 nm thickness) and exceptionally smooth (mean square roughness corresponds to 0.32 nm) film deposition at room temperature. The quality of the SiNC thin films is confirmed by ultrafast photoluminescence measurements and by applying such films for solar cells. We demonstrate that films produced with this approach yield good and stable devices. The methodology developed here is highly relevant for a very wide range of applications where the formation of high-quality ultrathin films of quantum dots with controllable thickness and smoothness is required.

Introduction

Colloidal semiconductor nanoscale crystals, with quantum confinement, offer a powerful platform for a new generation of device engineering [1, 2]. Colloidal nanocrystals may be tailored in size, shape, and composition and their surfaces functionalized by diverse chemical approaches [3–7]. The quantum mechanical coupling of over hundreds to thousands of atoms varies the electronic structure, whereby the optical and magnetic properties of materials can be tuned leading to new phenomena [8, 9]. The energy band gap can be willingly tuned to emit and absorb photons in the broader spectral region in comparison to the bulk [1, 10]. In the last few years, the research on colloidal nanocrystals has moved from fundamental research to first applications in biology, optoelectronics, and photovoltaics [11–15].

In particular, sunlight to electricity conversion (e.g., photovoltaics) has undergone a spectacular drop in cost as a result of breakthroughs in material science, engineering, design, and manufacturing. The silicon-based solar cell market showed an exponential growth rate in the last 20 years [16]. It is believed that further significant improvements with bigger impact in solar cell efficiency are possible through so-called third-generation photovoltaic device architectures where the use of silicon-based quantum dots or silicon nanocrystals (SiNCs) is particularly appealing [17]. Clearly opportunities exist through the precise control of the nanocrystal size which determines quantum confinement properties, through accurate control of the surface characteristics, and also through control of the distance between NCs. Desired surface properties can be achieved via surface engineering (with and without mechanical stress on surface), in particular, without the use of surfactants [18] and where the very large surface area to volume ratio can impinge on the overall NC behavior when it is accurately engineered without surfactants. Control of the NC size and surface properties determines the energy band gap of silicon to an optimal value (i.e., 1.6 eV), and allows precise design of absorbing layers for tandem solar cells [17, 19, 20]. Furthermore, quantum confinement can extensively improve efficiency (>40%) through enhanced inverse Auger recombination and the generation of multiple excitons (MEG) [21], while the large surface area can contribute not only to tuning the electronic properties of the nanocrystals, but also to controlling their interactions with the host matrix constituting a new class of hybrid quantum–material composites [22, 23].

In contrast to other semiconductor nanocrystals prepared by solution chemistry [24, 25], intrinsic and doped SiNCs can be synthesized simply and cheaply from low-cost materials (e.g., polycrystalline wafers) by an electrochemical etching approach [26]. As SiNCs can be in colloidal dispersion, solar cells or any optoelectronic device (e.g., light-emitting diodes) can be fabricated by various low-cost deposition techniques including simple one-step solution coating and spray procedures [27]. Indeed, various solar cell structures ranging from sensitized solar cells on mesoscopic semiconducting TiO_2 to the planar and bulk heterojunction architecture can be designed and produced [28]. However, to date, the impact of the SiNC size distribution at quantum confinement sizes on the film formation, and thus on the solar cell performance, has not been solved. Due to the large SiNC size distribution, a noncontinuous and rough film is usually obtained, where pinholes introduce shunting pathways limiting the solar cell performance [29]. On the other hand, maintaining a low temperature during deposition is also a crucial factor to avoid recrystallization of the SiNCs and to allow for device fabrication which involves temperature-sensitive materials (e.g., glass, polymer) [30].

Over the recent years, it has become necessary to deploy thin-film deposition methods for SiNCs suitably applicable to large area deposition, as wafers for the manufacture of optoelectronic devices and/or solar panels have become larger in size [31]. Furthermore, the development of devices with multilevel interconnects requires the surface of insulation films to be planarized, therefore surface

planarization techniques to form planar SiNC films can also become a necessity [32,33]. In an effort to satisfy these requirements, in general, for nanocrystal-based thin-film formation techniques at low temperatures, the aggregation of nanocrystals can be avoided by surface engineering [18]. In addition, plasma-induced surfactant-free surface chemistries on SiNCs showed encouraging results in terms of improved quality and stability of the SiNCs [18]. Surfactant-free femtosecond (fsec) laser-induced surface-engineered SiNCs also proved to be highly effective in promoting the separation of SiNCs [34], which largely contribute to preserve the individuality of the SiNCs during integration into thin films.

In this work, we employ pulsed fsec laser-induced surface engineering of colloidal SiNCs (produced by electrochemical etching, ~3 nm average diameter) together with low-angle spray deposition in vacuum to produce ultrathin films (<30 nm) of SiNCs for solar cells. We demonstrate that surface engineering of SiNCs induced by fsec laser allows the fabrication of smooth photoluminescent thin films with mean square roughness (Rms) corresponds to 0.32 nm. Photoluminescence dynamic studies confirm that the characteristic of the SiNCs are preserved throughout the surface engineering and film preparation steps. We show that the methodology developed here is highly relevant for a very wide range of applications where high-quality films of nanocrystals are required, and as an example it is demonstrated here for fabricating PV devices.

Experimental Details

Synthesis and Surface engineering of SiNCs

SiNCs were produced by electrochemical etching of Si wafer in HF followed by mechanical removal [26]. More specifically, SiNCs were produced by electrochemical etching of p-type Si wafers (HF:ethanol 1;4, constant current 20 mA/cm^2) for 1 h, which yields SiNC agglomerates. The fsec laser treatment is used for both SiNC deagglomeration and simultaneous surface passivation [34]. In particular, 5 mg of SiNC powder (i.e., SiNC agglomerates) from the electrochemical etching process was transferred into 5 mL ethanol. The fsec laser irradiation [34] used a wavelength of 400 nm and a pulse width 100 fsec. A barium borate (BBO) crystal was used to select 400 nm. During the irradiation, the glass container was rotated. The process was conducted at room temperature and for 30 min. The laser beam was shaped and focused onto a spot (2 mm in diameter) on the liquid surface by an optical lens with a focal length of 250 mm. The average laser power was set to be approximately 30 mW while using the repetition rate of 1 kHz. In order to check the efficiency of the laser-based surface engineering process with simplified experimental conditions and larger absorption profile of chosen wavelength we first compare with our previous works [34]. Similar to our previous studies, we clearly confirmed the trends in stability and an increase in the PL intensity after fsec laser processing after simplified experimental conditions.

Low-angle spray deposition

The colloids of SiNCs treated by the laser process were then deposited using the deposition system depicted in the schematic diagram of Figure 1 without any further step or filtration. This apparatus consists of a colloidal solution tank, stopper valve placed after the tank, and a spray nozzle. The IOTA ONE (Parker Hannifin Corporation, East Pine Brook, NJ, USA) valve driver for high-speed solenoid valves was used to control automatically the valve. The valve was used to perform pulsed deposition by stop/start action and by controlling the time interval. The pulse duration for the valve was set at 0.5 sec open and 1 sec closed, and repeated 30 times until the 5 mL colloidal SiNCs/ethanol dispersion was fully consumed. During deposition, the pressure in the chamber increased from 2×10^{-5} Torr to 10^{-2} Torr. The actual sample preparation was done at room temperature on quartz or glass/indium-doped tin oxide (ITO)/compact-TiO_2/mesoporous-TiO_2 substrates resulting in a thickness of about 30 nm.

Fabrication of solar cell devices

Figure 2 represents schematics cross-section of solar cell structure. Glass substrates with patterned indium-tin-oxide (ITO) were cleaned by O_2 plasma. The TiO_2 compact blocking layer was formed by dissolving titanium (IV)

isopropoxide and triethanolamine in ethanol, stirred for 2 h at 40°C, and then left for 24 h. The solution was spin coated at 5000 revolution per minute (rpm) for 30 sec and then annealed at 400°C in a furnace for 2 h. The mesoporous TiO_2 layers were deposited by spin coating (2000 rpm for 60 sec) a solution of commercial dyesol 18-NRT titania nanoparticle paste dissolved in ethanol in a 1:2 ratio of paste to ethanol. The films were then annealed again in a furnace at 400°C for 2 h. Then by low-spray angle deposition at room temperature, 30-nm thick film of fsec laser surface-engineered SiNCs was deposited (Fig. 2B). The hole transport layer was prepared by dissolving 0.207 g of 2,2',7,7'-Tetrakis[N,N-di(4-methoxyphenyl)amino]-9,9'-spirobifluorene (Spiro-MeOTAD, Sigma-Aldrich, St. Louis, MO, USA) in 1 mL chlorobenzene and deposited by spin coating at 3000 rpm for 30 sec. Silver metal contacts were deposited by thermal evaporation using a shadow mask. The resulting device active area was 0.04 cm².

Power conversion efficiency

Solar simulated AM 1.5G sunlight was generated using Wacom Electric Co. solar simulator (JIS, IEC standard conforming, CLASS AAA, Wacom, Taito-Ku, Tokyo, Japan) calibrated to give 100 mW/cm² using a silicon reference cell. The electrical data were recorded using a Keithley 2400 source meter (Tektronix, Inc., Beaverton, OR, USA). Devices were measured in the range −0.1 to 0.6 V with scan rates between 150 and 1500 mV/sec.

External quantum efficiency

The solar cell external quantum efficiency (EQE) characteristics were measured under illumination using an Air Mass 1.5 Global (AM1.5G) with light intensity calibrated to 100 mW/cm² by using a reference cell for a-Si solar cells.

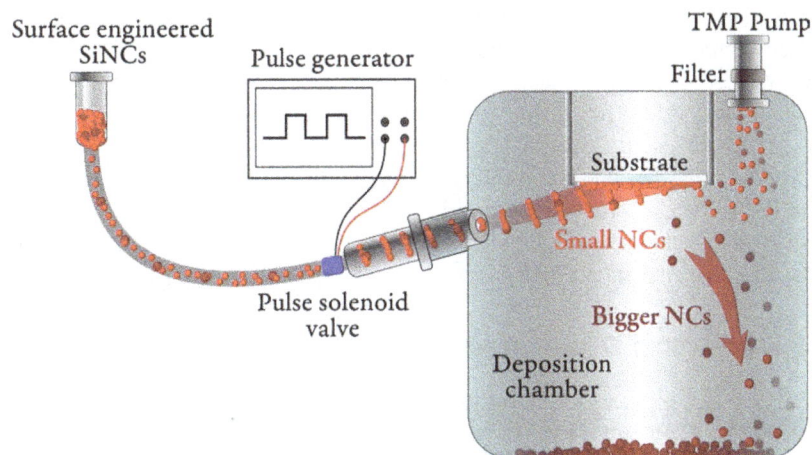

Figure 1. Schematic diagram of the deposition system used to produce ultrathin surface-engineered silicon nanocrystal (SiNC) films; the system uses a homemade pulsed spray system where the speed of the solenoid duration range was controlled automatically by an IOTA ONE (Parker) valve driver.

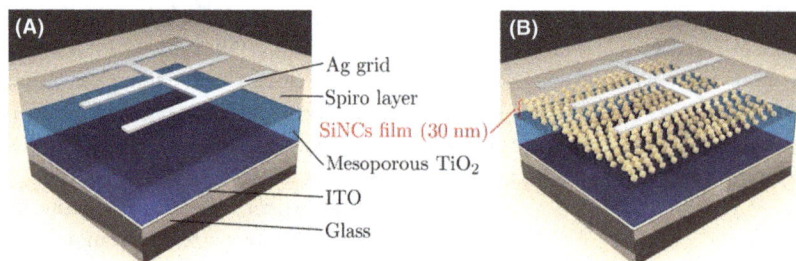

Figure 2. (A) Schematics showing cross-section of reference solar cell structure and (B) solar cells with ultrathin film of surface-engineered silicon nanocrystals (SiNC).

Photoluminescence and PL decay measurements

The photoluminescence (PL) was measured by a spectrometer (Spectrofluorometer, Horiba Jobin Yvon, Palaiseau, France) at room temperature with excitation at 375 nm. Time-resolved PL spectra were measured using a streak camera with the second harmonics of a Ti:sapphire laser (400 nm) as an excitation light source.

XRD analysis

The X-ray diffraction (XRD) measurements of thin films were performed with a synchrotron radiation (SPring-8) at the beam line BL46XU. The film was mounted on the gonio stage (HUBER X-ray diffractometer) to allow XRD analysis. GIWAXS data were acquired using 12.398 keV X-rays (λ = 1.000 Å) at an incident angle of 0.25° and diffracted X-rays were captured at 1.0 sec intervals by a two-dimensional detector (PILATUS 300K, DECTRIS, Baden-Daettwil, Switzerland) located at a distance of L = 174.5 mm from the sample.

Thickness and surface analysis

The atomic force microscopy (AFM) measurements (Nano/Navi, E-SWEEP SII Nanotechnology) on SiNC ultrathin films on quartz substrate were performed either with contact or with tapping mode. Dektak-XT (Bruker, Kanagawa, Japan) and AFM was used to evaluate the thickness of the deposited thin film on a quartz substrate and a corresponding thickness of 30 nm from the step size was evaluated. The AFM image analysis was carried out using commercial software procedures to determine the surface roughness.

Experimental Results and Discussion

Luminescence properties of ultrathin SiNCs films

The steady-state photoluminescence and time-resolved photoluminescence have been recognized to be useful tools to understand charge extraction in thin film solar cells; therefore, we first looked into the charge dynamics of surface-engineered SiNCs directly in colloidal solution and after SiNC ultrathin film deposition on quartz substrate

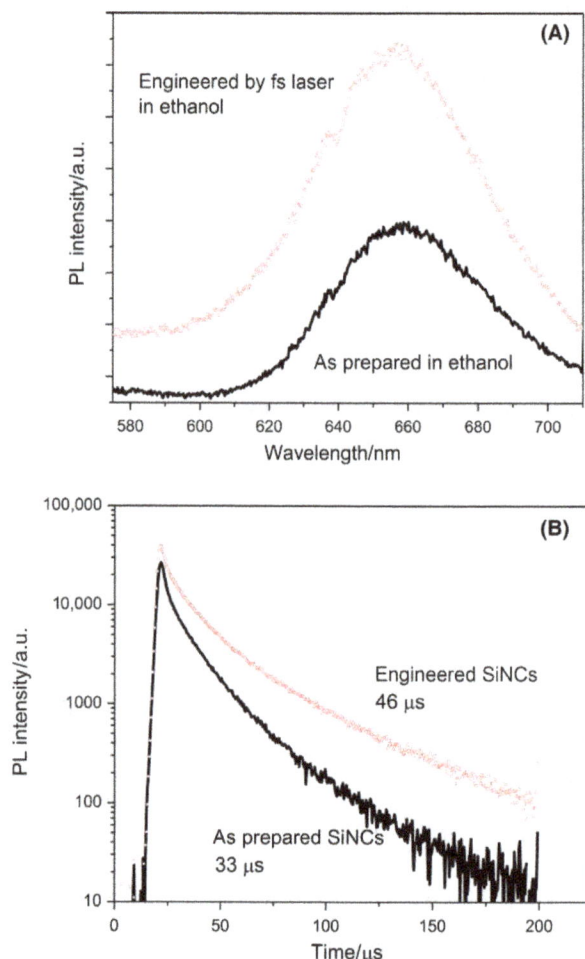

Figure 3. (A) Time-gated photoluminescence (PL) shows a broad spectra spanning from 600 to 750 nm for as-prepared SiNCs dispersed in ethanol (black line) and fsec laser processed SiNCs (red symbols) also in ethanol, both measured directly in colloid with excitation at 400 nm. (B) PL decays corresponding to PL emission in (A).

by low-angle deposition techniques. Figure 3A shows the time-gated PL spectra (400 nm excitation) for as-prepared SiNCs dispersed in ethanol and for SiNCs processed by fsec laser in ethanol, where both samples were dispersed in ethanol for same time duration. The time-gated PL is broad, spanning from 600 to 750 nm; however, it evolves on the μsec timescale (Fig. 3B) from a band centered around 660 nm. The PL peak exhibits an increase in intensity for surface-engineered SiNCs. Corresponding PL decays on the μsec timescale are shown in Figure 3B. The PL decay time is extended by a few tens of microsecond in the fsec laser-processed sample compared to as-prepared SiNCs, which underline higher quality SiNCs after laser processing. The μsec transition can originate from the relaxation of the excited electrons in the SiNC core to the lower states, or from trapping by defect states on the surface which are reduced under surface engineering. As a result, the PL peak exhibits an increase in intensity for surface-engineered SiNCs. The PL decay is in the microseconds range which suggests that the nature of the transition is most likely indirect. The decay constants were determined by fitting the experimental data with a single exponential function in the 100–200 nsec region. Our results indicate that the PL decay constant after surface engineering increased from 33 to 46 μsec.

Next, surface-engineered SiNCs were used to fabricate ultrathin films by low-angle spray deposition on quartz substrates at room temperature. Figure 4 shows a typical AFM image of the deposited SiNCs (exposed to air). Such SiNC thin films exhibit a very smooth surface morphology. The analyzed surface has an area of 200 × 200 nm^2 and the evaluated Rms corresponds to 0.32 nm.

The samples were then analyzed by grazing incidence wide-angle X-ray scattering (GIWAXS). Figure 5A shows GIWAXS pattern of SiNCs for ultrathin films represented in reciprocal lattice space. The XRD broad spectrum (Fig. 5B) of the SiNCs film evaluated from the GIWAXS pattern confirms the presence of the SiNCs in the ultrathin films, where the broadening is ascribed to the quantum confinement size of the SiNCs whereby the general expression for the line broadening is given by Williamson and Hall [35]. The characteristic diffraction lines are broadened and their width differs by variation of grazing incidence angle indicating the variation of nanocrystals sizes [36].

Figure 6A shows typical time-gated PL spectra of the 30-nm thick film on quartz substrate. The PL maxima (Fig. 6B) is located at 630 nm, which is slightly shifted in comparison to the PL emission of the corresponding colloid (Fig. 3A). This 30 nm blue shift is most likely due to the deposition approach, which selectively deposit the smallest SiNCs (see Fig. 1). However, the PL dynamics of the SiNCs in the films is the same as for the colloids (Fig. 3B) with a decay constant of about 43 μsec (Fig. 6C).

Although the exact nature is still debated, PL in SiNCs is often associated to quantum confinement [37, 38]. In addition, several works point out and attribute the luminescence emission to the variable structures of the samples and the many possible defect states in SiNCs mostly localized at the SiNC surface [39]. Figure 7A shows typical time-gated PL spectrum of the 30-nm thick film on quartz substrate at excitation 10^{-1} I_0, ($I_0 \wedge 1$ kJ/cm^2), while Figure 7B shows the PL spectra as a function of the wavelength at different excitation intensities

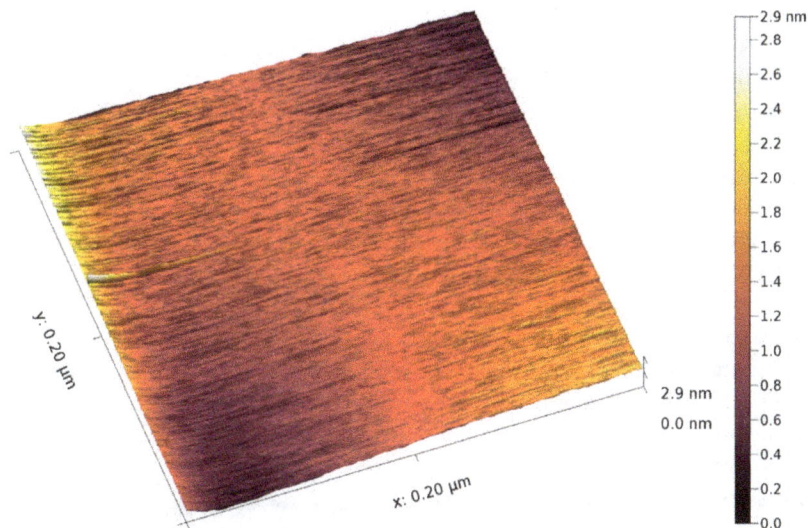

Figure 4. Typical AFM image of the SiNCs thin film with thickness 30 nm deposited at low angle and room temperature. The analyzed surface has an area of 200 × 200 nm^2 and the evaluated mean square roughness (Rms) corresponds to 0.32 nm.

corresponding to I_0, $10^{-0.3} I_0$, $10^{-0.5} I_0$, $10^{-1} I_0$, $10^{-2} I_0$. The inset in Figure 7B plots the PL intensity as a function of excitation intensity and Figure 7C presents respective PL decays whereby the legend summarizes the corresponding decay constants. It is well accepted that the size of the SiNCs is the key factor to determining the ratio of zero-phonon transitions to phonon-assisted transitions [40, 41]. On the other hand, the state-filling effect is closely linked to the recombination rate and consequently high recombination rate leads to a shortening of the lifetime at decreased excitations intensities. However, we observe no significant changes in the decay constants, thus suggesting that the state-filling effect is not significant [42]. Furthermore, the PL results suggest that in surface-engineered SiNCs the state-filling effect can effectively prevent hot carriers from being thermalized quickly. Due to the indirect gap of crystalline silicon, a thermalization relaxation time of SiNCs is about 1000 times slower than common carrier cooling rates in bulk semiconductors [43]. Since Auger recombination is proportional to the cube of the carrier density and, initially, a large population of carriers is pumped at higher excitation intensities (comparable to solar irradiation), we should expect Auger recombination to occur at an accelerated rate, possibly eased also by a lower defect state density of surface-engineered SiNCs that present reduced non-radiative recombination paths.

Photovoltaic applications of ultrathin films based on doped SiNCs

The PL studies suggest that such thin films with high quality of highly packed SiNCs could be advantageous for carrier multiplication in SiNCs and photovoltaic applications [44, 45]. In order to test the photovoltaic properties of the ultrathin SiNCs films, we fabricated solar cells. The photovoltaic properties of the typical solar cell based on the ultrathin film of SiNCs deposited at similar conditions are shown in Figure 8. Silver (Ag) was used as the top electrode and ITO as the bottom electrode. Figure 8A and B show the current density–voltage (J–V) curves of the SiNC solar cell device (ITO/compact-TiO$_2$/mesoporous-TiO$_2$/SiNCs/spiro-OMeTAD/Ag) and the control solar cell device (ITO/compact-TiO$_2$/mesoporous-TiO$_2$/spiro-OMeTAD/Ag) in the dark and under illumination. As can be seen from the results, the control device exhibits a negligible PV effect (Fig. 8B), while a considerable PV output emerges after the introduction of an ultrathin surface-engineered SiNCs layer (Fig. 8A). As it is shown in Figure 8A, the SiNCs solar cell has an open-circuit voltage (V_{oc}) and short-circuit current density (J_{sc}) of 0.42 V and 112 mA/cm^2, respectively, with a corresponding power conversion efficiency of 0.0096% and a fill factor

of 20%. We also measured the EQE of the corresponding cells, which are compared in Figure 8C. From the overall EQE it is evident that a broad range of incident light from 300 to 600 nm can be harvested to generate electric power, thereby validating the contribution of the ultrathin film made of surface-engineered SiNCs.

The inset in Figure 8C shows the schematic band diagram of the device with SiNCs. The band alignment without a potential barrier ensures an efficient charge transport to their respective ohmic contacts. In our case the charge carriers are separated most likely between TiO$_2$ (electrons) and SiNCs (holes). The TiO$_2$ layer acts as a selective carrier transport that provides an excellent hole blocking layer [46, 47]. The photogenerated hole carriers in SiNCs are separated efficiently by using p-type doped SiNCs. Electron and hole separation between TiO$_2$ and

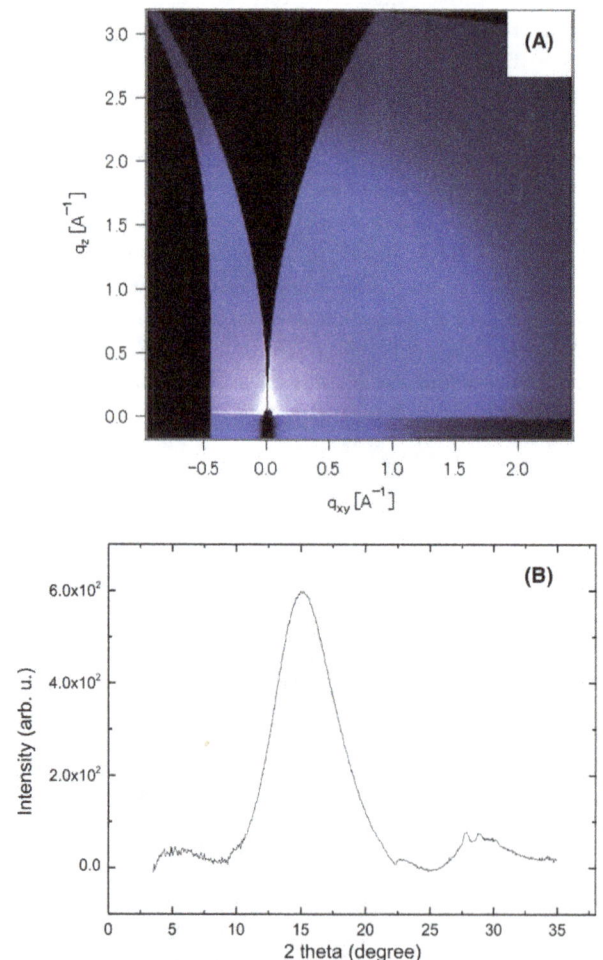

Figure 5. (A) Grazing incident wide-angle X-ray scattering (GIWAXS) patterns of surface-engineered SiNCs deposited by low-angle spray deposition, represented in reciprocal lattice space. (B) XRD spectrum corresponding to figure (A).

SiNCs is crucial for the open-circuit voltage (Voc) by preventing charge recombination and enabling an efficient extraction even at low electric field (Fig. 8A). Another crucial factor concerns the nanocrystal surface chemistry. We believe that the charge carrier separation at SiNCs interface is deeply related to the fabrication process by laser engineering that allows a refinement of the nanocrystal surface chemistry. Indeed, nonradiative recombination at dangling bond sites may be responsible for the emission quenching due to poorly passivated SiNCs. Therefore, the high intensity from the PL spectra of our films as shown in Figure 6B underlines a low concentration of dangling

bonds that indirectly supports the effective charge separation present in our devices.

The improved PL decay due to surface-engineered SiNCs may be closely related to a lower interfacial barrier and a more efficient charge carrier extraction as discussed above, thus contributing to an enhancement of photovoltaic properties, such as reduced hysteresis (Fig. 9). In Figure 9A we show J–V curves for the best-performing solar cell devices in forward and reverse scans with the scan rate of 150 mV/sec. The champion ultrathin SiNC device achieved, in the reverse scan, a power conversion efficiency of 0.016% with a short-circuit current density

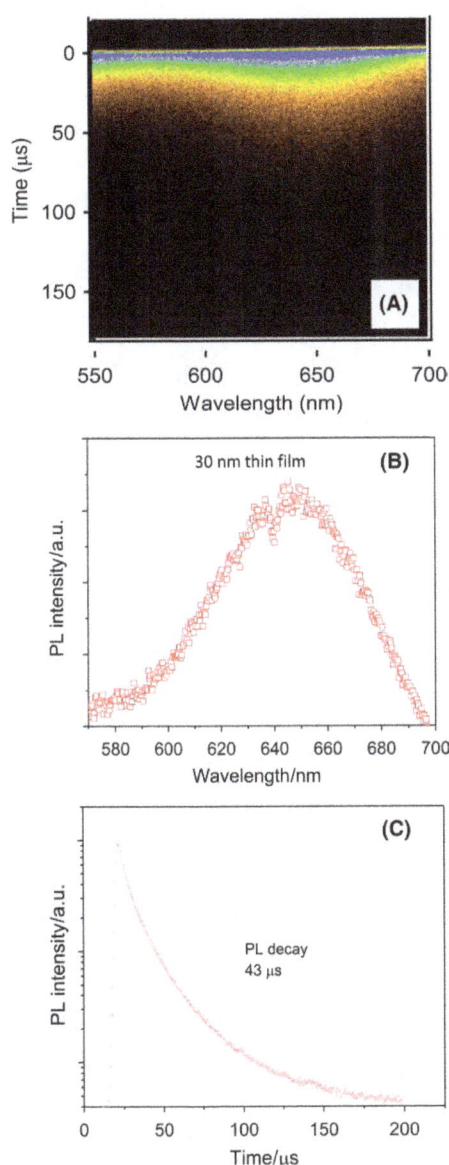

Figure 6. (A) Time-gated PL spectra of the 30-nm thick film on quartz substrate. (B) PL spectrum at room temperature. (C) PL decay corresponding to (B).

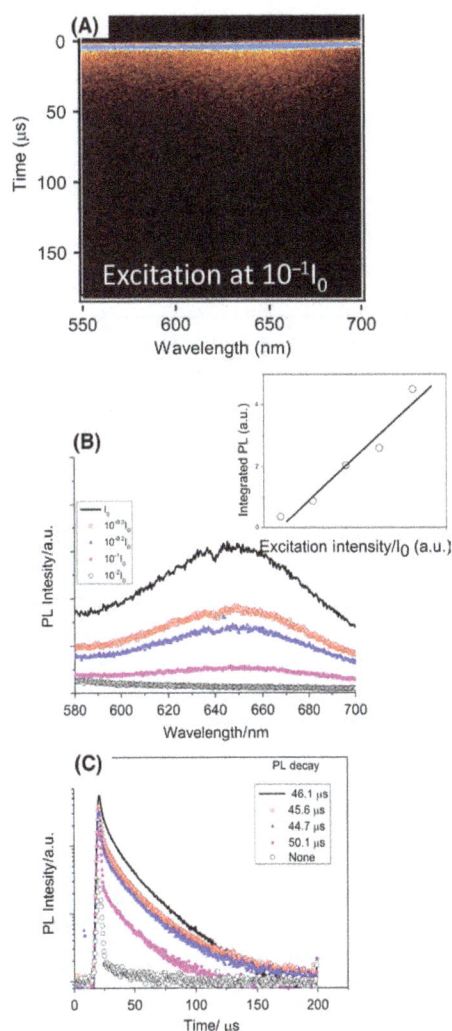

Figure 7. (A) Typical time-gated photoluminescence (PL) spectrum of the 30-nm thick film on quartz substrate at excitation $10^{-1}I_0$. (B) The PL spectra as a function of the wavelength at different excitation intensities: I_0, $10^{-0.3}\,I_0$, $10^{-0.5}\,I_0$, $10^{-1}\,I_0$, $10^{-2}\,I_0$. The inset in (B) plots PL intensity as a function of excitation intensity. (C) PL decays whereby the legend summarizes evaluated decay time constants.

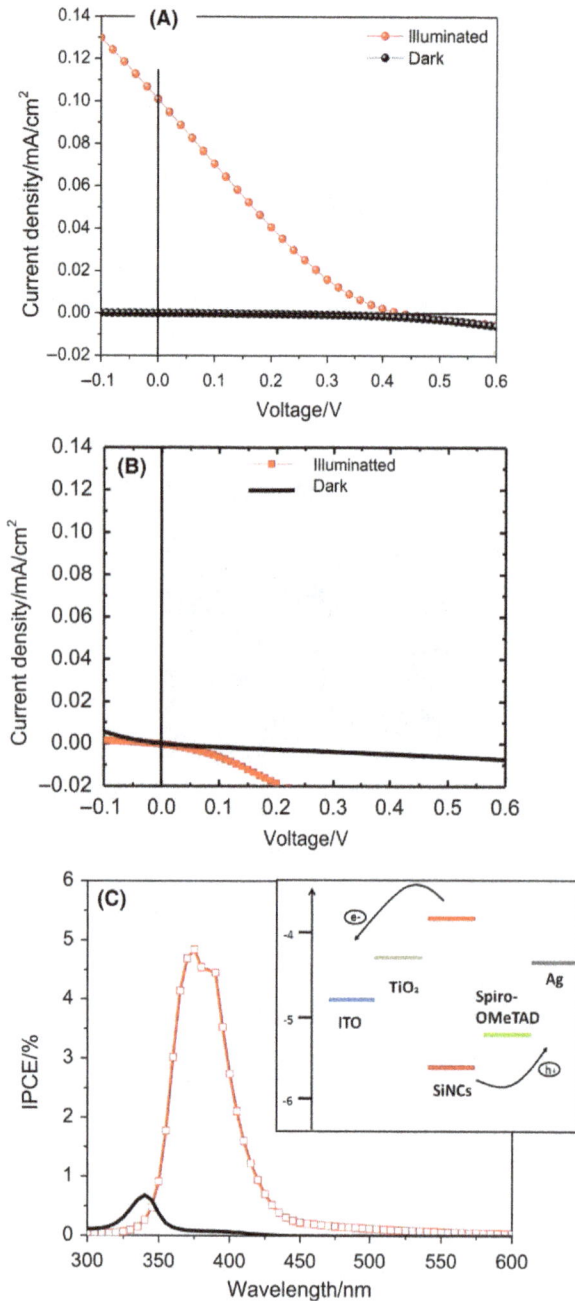

Figure 8. (A) Current density–voltage characteristic for a SiNC solar cell (red symbols) and (B) for the same structure without SiNCs. (C) External quantum efficiency (EQE) as a function of the wavelength for corresponding solar cells is plotted. Inset in (C) shows the schematic band diagram of the solar cell with SiNCs.

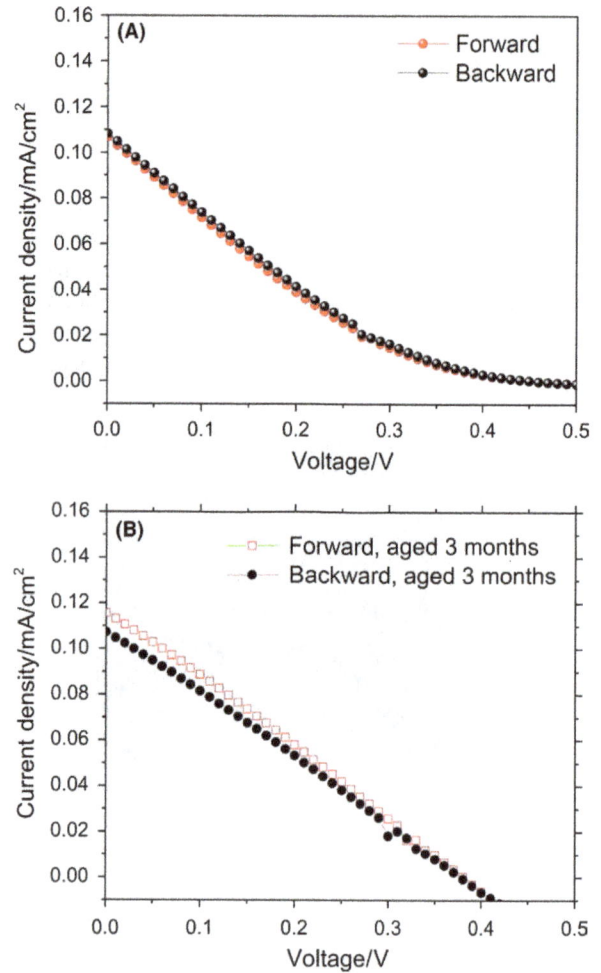

Figure 9. (A) Current density–voltage characteristic for silicon nanocrystal solar cells. (B) Current density–voltage characteristic of the devices after 3 months stored in air and ambient conditions.

both J–V curves of solar cell in forward and reverse scans that confirms also a low hysteresis is maintained after aging.

Conclusions

The PL and photovoltaic properties of silicon nanocrystals (SiNCs) 30-nm thick and smooth ultrathin films have been investigated. A technological approach to fabricate SiNC thin films is proposed and based on the combination of surface engineering by fsec laser irradiation and low-angle SiNCs deposition. The longer lifetime in the PL emission band is likely attributed to the nature of the surface chemistry, which has been largely improved by the laser-based process. We demonstrate that the PL decay properties are preserved during the room-temperature deposition. The surface engineering step together with the production of thin films has allowed the

(J_{SC}) of 0.12 mA/cm^2, open-circuit voltage (V_{OC}) of 0.43 V. Device stability is another major concern with regard to practical applications. Therefore, we recorded the device performance after 3 months with devices stored in air and ambient conditions. The solar cells are very stable at room temperature as illustrated in Figure 9B. We show fabrication of prototype transparent solar cells. The low-temperature deposition approach provides great flexibility, whereby the concepts discussed in this work can be also expanded to other ultrathin nanocrystal solar cells based on different material systems at low cost.

Acknowledgments

The GIWAXS measurement was performed at SPring-8 BL46XU with the approval of the Japan Synchrotron Radiation Research Institute (JASRI, proposal nos. 2015B1600 and 2015B1891). This work was partially supported by EPSRC (EP/K022237/1 and EP/M024938/1). The authors also acknowledge the support of the EU COST Action TD1208.

Conflict of Interest

None declared.

References

1. Rogach, A. 2008. Semiconductor nanocrystal quantum dots: synthesis, assembly, spectroscopy and applications. Springer, Wien.

2. Delerue, C. 2016. Nanocrystal solids: Order and progress. Nat. Mater. 15:498–499.

3. Shevchenko, E. V., M. I. Bodnarchuk, M. V. Kovalenko, D. V. Talapin, R. K. Smith, S. Aloni et al. 2008. Nanoparticles with different functionalities and their periodic structures. Adv. Mater. 20:4323–4329.

4. Talapin, D. V., J. H. Nelson, E. V. Shevchenko, S. Aloni, B. Sadtler, and A. P. Alivisatos. 2007. Seeded Growth of Highly Luminescent CdSe/CdS Nano-Heterostructures with Rod and Tetrapod Morphologies. Nano Lett. 7:2951–2959.

5. Xu, W., P. Jain, B. Beberwyck, and A. P. Alivisatos. 2012. Probing Redox Photocatalysis of Trapped Electrons and Holes on Single Sb-doped Titania Nanorod Surfaces. J. Am. Chem. Soc. 134:3946–3949.

6. Liu, W. Y., A. Y. Chang, R. D. Schaller, and D. V. Talapin. 2012. Colloidal InSb nanocrystals. J. Am. Chem. Soc. 134:20258–20261.

7. Kagan, C. R., and C. B. Murray. 2015. Charge transport in strongly coupled quantum dot solids. Nat. Nanotechnol. 10:1013–1026.

8. Talapin, D. V., and C. B. Murray. 2005. PbSe Nanocrystal Solids for n- and p-Channel Thin Film Field-Effect Transistors. Science 310:86–89.

9. Klimov, V. I., S. A. Ivanov, J. Nanda, M. Achermann, I. Bezel, J. A. McGuire et al. 2007. Single-exciton optical gain in semiconductor nanocrystals. Nature 447:441–446.

10. Murray, C. B., D. J. Norris, and M. G. Bawendi. 1993. Synthesis and characterization of nearly monodisperse CdE (E = sulfur, selenium, tellurium) semiconductor nanocrystallites. J. Am. Chem. Soc. 115:8706–8715.

11. Kovalenko, M. V., L. Manna, A. Cabot, Z. Hens, D. V. Talapin, C. R. Kagan et al. 2015. Prospects of Nanoscience with Nanocrystals. ASC Nano 9:1012–1057.

12. Kamat, P. V. 2012. Boosting the Efficiency of Quantum Dot Sensitized Solar Cells through Modulation of Interfacial Charge Transfer. Acc. Chem. Res. 45:1906–1915.

13. Shirasaki, Y., G. J. Supran, M. G. Bawendi, and V. Bulovic. 2013. Emergence of Colloidal Quantum-Dot Light-Emitting Technologies. Nat. Photonics 7:13–23.

14. Peteiro-Cartelle, J., M. Rodríguez Pedreira, F. Zhang, P. Rivera-Gil, L. L. Mercato, and W. J. Parak. 2009. How colloidal nano- and microparticles could contribute to medicine - a personal perspective both from the eyes of physicians and materials scientists. Nanomedicine 4:967–979.

15. Wang, Y., X. Li, J. Song, L. Xiao, H. Zeng, and H. Sun. 2015. All-Inorganic Colloidal Perovskite Quantum Dots: A New Class of Lasing Materials with Favorable Characteristics. Adv. Mater. 27:7101–7108.

16. Fraunhofer, I. S. E. 2015. Current and Future Cost of Photovoltaics—Long-term Scenarios for Market Development, System Prices and LCOE of Utility-Scale PV Systems" (PDF). Available at http://www.agora-energiewende.org/ [accessed on 01 February 2015]

17. Green, M. A. 2003. Third generation photovoltaics. Springer, Berlin.

18. Mariotti, D., V. Švrcek, W. J. Hamilton, M. Schmidt, and M. Kondo. 2012. Silicon Nanocrystals in Liquid Media: Optical Properties and Surface Stabilization by Microplasma-Induced Non-Equilibrium. Adv. Funct. Mater. 22:954–964.

19. Miles, R. 2006. Photovoltaic solar cells: Choice of materials and production methods. Vacuum 80:1090–1097.

20. Luque, A., and S. Hegedus, eds. 2003. Handbook of photovoltaic science and engineering. John Wiley & Sons, Ltd., Chichester, UK.

21. Nozik, J. 2008. Multiple exciton generation in semiconductor quantum dots. Chem. Phys. Lett. 457:3–11.

22. Dirin, D. N., S. Dreyfuss, M. I. Bodnarchuk, G. Nedelcu, P. Papagiorgis, G. Itskos et al. 2014. Lead Halide Perovskites and Other Metal Halide Complexes as Inorganic Capping Ligands for Colloidal Nanocrystals. J. Am. Chem. Soc. 136:6550–6553.

23. Svrcek, V., T. Yamanari, D. Mariotti, S. Mitra, T. Velusamy, and K. Matsubara. 2015. A silicon nanocrystal/polymer nanocomposite as a down-conversion layer in organic and hybrid solar cells. Nanoscale 7:11566–11574.

24. Kopping, J. T., and T. E. Patten. 2008. Identification of Acidic Phosphorus-Containing Ligands Involved in the

Surface Chemistry of CdSe Nanoparticles Prepared in Tri-N-octylphosphine Oxide Solvents. J. Am. Chem. Soc. 130:5689–5698.

25. Hassinen, A., I. Moreels, K. De Nolf, P. F. Smet, J. C. Martins, and Z. Hens. 2012. Short-chain alcohols strip X-type ligands and quench the luminescence of PbSe and CdSe quantum dots, acetonitrile does not. J. Am. Chem. Soc. 134:20705.

26. Svrcek, V., A. Slaoui, and J. C. Muller. 2004. Ex situ prepared Si nanocrystals embedded in silica glass: Formation and characterization. J. Appl. Phys. 95:3158–3164.

27. Purkait, T. K., M. Iqbal, M. Amirul Islam, M. H. Mobarok, C. M. Gonzalez, L. Hadidi et al. 2016. Alkoxy-Terminated Si Surfaces: A New Reactive Platform for the Functionalization and Derivatization of Silicon Quantum Dots. J. Am. Chem. Soc. 138:7114–7120.

28. Xin, X., J. Wang, W. Han, M. Ye, and Z. Lin. 2012. Dye-sensitized solar cells based on a nanoparticle/nanotube bilayer structure and their equivalent circuit analysis. Nanoscale 4:964–971.

29. Velusamy, T., S. Mitra, M. L. Macias-Montero, V. Svrcek, and D. Mariotti. 2015. Varying Surface Chemistries for p-Doped and n-Doped Silicon Nanocrystals and Impact on Photovoltaic Devices. ACS Appl. Mater. Interfaces. 7:28207–28214.

30. Ito, M., C. Koch, V. Švrcek, M. B. Schubert, and J. H. Werner. 2001. Silicon thin film solar cells deposited under 80°C. Thin Solid Films 383:129–131.

31. Mason, T. L., A. Straub, D. Inns, D. Song, A. Aberle, and G. Armin. 2005. A Novel method for the development of Solar Energy. Appl. Phys. Lett. 86:V172108–172112.

32. Doering, R., and Y. Nishi, eds. 2007. Handbook of semiconductor manufacturing technology, 2nd ed. CRC Press, Boca Raton, FL.

33. Zhuang, Z., F. Huang, Z. Lin, and H. Zhang. 2012. Aggregation-Induced Fast Crystal Growth of SnO_2 Nanocrystals. J. Am. Chem. Soc. 134:16228–16234.

34. Svrcek, V., D. Mariotti, U. Cvelbar, G. Filipič, M. Lozac'h, C. McDonald et al. 2016. Environmentally Friendly Processing Technology for Engineering Silicon Nanocrystals in Water with Laser Pulses. J. Phys. Chem. C 120:18822–18830.

35. Williamson, J. K., and W. H. Hall. 1953. X-ray line broadening from filed aluminium and wolfram Die verbreiterung der roentgen interferenzl inien von aluminium- und wolfram spaenen. Acta Metall. 1:22–31.

36. Juraic, K., D. Garcin, B. Santic, D. Meljanac, N. Zoric, A. Gajovic et al. 2010. GISAXS and GIWAXS analysis of amorphous–nanocrystalline silicon thin films. Nucl. Instrum. Methods Phys. Res. B 268:259–262.

37. Garrido, G. C., B. Pellegrino, P. Ferre, R. Moreno, J. Morante, L. Pavesi et al. 2003. Size dependence of lifetime and absorption cross section of Si nanocrystals embedded in SiO_2. Appl. Phys. Lett. 82:1595(1-3).

38. Wolkin, M. V., J. Jorne, P. M. Fauchet, G. Allan, and C. Delerue. 1999. Electronic States And Luminescence In Porous Silicon Quantum Dots: The Role Of Oxygen. Phys. Rev. Lett. 82:197–200.

39. Lin, G. R., C. J. Lin, C. K. Lin, L. Chou, and Y. Chueh. 2005. Characteristics of constrained ferroelectricity in $PbZrO_3$/$BaZrO_3$$PbZrO_3$/$BaZrO_3$ superlattice films. J. Appl. Phys. 97:034105–034112.

40. Kovalev, D., H. Heckler, M. Ben-Chorin, G. Polisski, M. Schwartzkopff, and F. Koch. 1998. Breakdown of k-conservation rule in Si nanocrystals. Phys. Rev. Lett. 81:2803–2807.

41. Hybertsen, M. S. 1994. Excitons in Si nanocrystals. Phys. Rev. Lett. 72:1514–1518.

42. Park, Y. M., Y. J. Park, K. M. Kim, J. C. Shin, J. D. Song, J. I. Lee et al. 2004. State filling phenomena in modulation-doped InAs quantum dots. J. Cryst. Growth 271:385–390.

43. Zhang, P., Y. Feng, X. Wen, W. Cao, R. Anthony, U. Kortshagen et al. 2016. Generation of hot carrier population in colloidal silicon quantum dots for high-efficiency photovoltaics. Sol. Energy Mater. Sol. Cells 145:391–396.

44. Beard, M. C., K. P. Knutsen, P. Yu, J. M. Luther, Q. Song, W. K. Metzger et al. 2007. Multiple Exciton Generation in Colloidal Silicon Nanocrystals. Nano Lett. 7:2506–2512.

45. Govoni, M., I. Marri, and S. Ossicini. 2012. Carrier multiplication between interacting nanocrystals for fostering silicon-based photovoltaics. Nat. Photonics 6:672–679.

46. Weickert, J., R. B. Dunbar, H. C. Hesse, W. Wiedemann, and L. Schmidt-Mende. 2011. Nanostructured Organic and Hybrid Solar Cells. Adv. Mater. 23:1810–1828.

47. Chen, L-M., Z. Hong, G. Li, and Y. Yang. 2009. Recent Progress in Polymer Solar Cells: Manipulation of Polymer: Fullerene Morphology and the Formation of Efficient Inverted Polymer Solar Cells. Adv. Mater. 21:1434–1449.

11

Status and future strategies for Concentrating Solar Power in China

Jun Wang[1], Song Yang[1], Chuan Jiang[1], Yaoming Zhang[1] & Peter D. Lund[1,2]

[1]Key Laboratory of Solar Energy Science and Technology in Jiangsu Province, School of Energy and Environment, Southeast University, No. 2 Si Pai Lou, Nanjing 210096, China
[2]School of Science, Aalto University, P. O. Box 15100, FI-00076 Aalto (Espoo), Finland

Keywords
13th 5-year plan, China, Concentrating Solar Power, innovation, policy

Correspondence
Jun Wang, Key Laboratory of Solar Energy Science and Technology in Jiangsu Province, School of Energy and Environment, Southeast University, No. 2 Si Pai Lou, Nanjing 210096, China. E-mail: 101010980@seu.edu.cn

Funding Information
No funding information provided.

Abstract

China is the world leader in several areas of clean energy, but not in Concentrating Solar Power (CSP). Our analysis provides an interesting viewpoint to China's possible role in helping with the market breakthrough of CSP. We present a short overview of the state-of-the-art of CSP including the status in China. A blueprint for China's CSP development is elaborated based on China's 13th 5-year program, but also on China's previous success factors in PV and wind power. The results of this study suggest that China could play a more prominent global role in CSP, but this would require stronger efforts in several areas ranging from innovation to policies.

Introduction

During the last years, renewable energy industries have significantly grown, in particular in China, because of favorable domestic and overseas business conditions [1, 2]. Most of the growth in solar energy has originated from photovoltaics which has exceeded a total capacity of 200 GW_p, most of which has been constructed in <10 years [3]. This rapid market growth can be linked to down spiraling costs of PV. The phenomenological growth of PV has overshadowed the other solar technologies. For instance, Concentrating Solar Power technology (CSP), which was earlier identified as a very promising future clean energy option [4], has slowly progressed both in terms of technology development and cost reduction compared to PV.

Albeit the still higher price of CSP, it has several advantages over PV such as easy coupling to other sources of energy and the capability for dispatched use through thermal energy storage. However, to become a worthy option, major technological and economical progress will be necessary. The possible role of China in such a development, which is the subject of this study, is interesting,

as China is the world leader in several clean energy areas such PV and wind power [5]. China has centrally contributed to the price reduction in these in the past [6]. This success can be traced back to the innovation and manufacturing system of China, but also to successful economies of scale efforts. Based on these past experiences, one may ask if Chinese clean energy strategies could still make a significant impact to the worldwide development of other new energy technologies such as CSP.

In case of CSP, most of the development has taken place outside China, in particular in the United States, Spain, and recently also in North Africa. By the end of 2015, the global installed capacity of CSP was 4940 MW, of which just around one percent was found in China [7]. Bibliometric analysis of CSP research similarly shows only modest contributions from China [6, 8–11]. These findings confirm that China is not yet in the forefront of CSP development. However, considering China's strong foothold in the other fields of clean energy, a logical follow-up question would be to find the reasons for such a situation. For example, are the technology and market prospects of CSP not good enough for Chinese industries, or are there may be other reasons such as national prioritization of technologies, missing know-how, lack of innovations, lacking conditions for utilization, etc. that may hamper China to enter the CSP field. Concerning the future of CSP as a clean energy source, one could also ask what would be needed that China became a major player in CSP and could China trig a similar positive cost development as in case of photovoltaics which led to a technology breakthrough. The purpose of this study is to investigate the development and strategies for CSP in China along the above lines.

State-of-the-Art of CSP

The three main technology options for CSP are listed in Table 1. The most common type of CSP, or over 90% of all installed systems [12], is a parabolic trough (PT) system in which a solar-heated high-temperature steam drives a steam turbine to generate electricity in large scale of hundreds of megawatts [13]. The highest conversion efficiency (>30%[14]) is reached with a parabolic dish-type point receiver using a Stirling engine, but it is limited in scale to some megawatts at present. Solar tower systems employing large heliostat mirror fields concentrating solar irradiation to a receiver on the top of a tower can reach high concentration ratios and can be built from medium to very large scale. The technical performance and viability of all three technologies have been demonstrated, but in terms of commercial deployment, the PT systems have proceeded furthest [12]. Most of the PT plants have been built in the United States and Spain, but recently a major system was also erected in Morocco [15].

Each of the three CSP technologies has its specific development needs. In PT systems, steam and salt are often employed as heat transfer medium; for heat storage steam, salt, and oil, respectively. One important component for the success of PT has been the vacuum tube receiver [20], including Chinese involvement, for example, by the Beijing Day Rising Vacuum Technology Development Company, which successfully developed a glass–metal transition seal structure of vacuum tubes. In case of Central Towers, the heliostats and the receiver structure are key components subject to development needs, for example, durability, costs, and efficiency [21–24]. For the Parabolic Dish technology, the receiver structure and the heat engine are the main concerns [25, 26].

Other R&D areas of interest for CSP include high-capacity thermal storage systems and distributed combined cycle systems [27, 28]. Research on thermal storage comprises storage materials and structures, which also affect the efficiency and investment cost of a CSP. Due to temperature limitations, flammability, and costs, steam or thermal oil is not an attractive option as a high-temperature TES material. Whereas molten salts represent an alternative already utilized in the latest CSP plants [29]. An inevitable drawback of molten salt is solidification at low temperature and decomposition at high temperature, which limits the operating temperature range. Recent research shows that a salt mixture could operate between 80°C and 560°C [30]. At present, a two-tank molten salt storage is the only commercially available concept with a large thermal capacity for CSP plants [30]. Hydrated sodium and nitrite are applicable as storage materials, but their efficiency often drops after repeated cycles. Synthetic oils are used as heat transfer fluid (HTF) in CSP, but they have limitations for high-temperature storage use.

Current CSP systems are mostly limited by the temperature limits of the materials used, which impose limitations on the efficiency of the power plant through the turbine inlet temperature. For example, the storage materials used limit the maximum temperature of the HTF. Corrosion and deterioration of molten salts increase with temperature, which requires using more expensive storage tank materials. All these factors restrain maximum temperature levels and hence also the turbine efficiency.

Table 1. Typical characteristics of solar-thermal power technologies.

Property	Parabolic trough (PT)	Parabolic dish (PD)	Central tower (CT)
Typical power range, MW	30–320	3–25	10–200
Concentration ratio	10–100	500–1000	>1000
Conversion efficiency, %	~14 [16]	~30 [19]	>15
Advantages	Commercially available with long-term experience; Modular and suitable for hybrid operation [17]; Can be coupled to heat storage	High conversion efficiency; Modular, suitable for hybrid use	High conversion efficiency; Suitable for hybrid use
Disadvantages	HTF working fluid limits operating temperature to 400°C; Spills/leaks [18]	Commercial viability need to be verified; Cost targets in mass production need to be verified	In experimental phase; Commercial investment and operating costs need to be confirmed
Costs	Potentially low investment costs	Structure of receiver is complex and costly	Still high investment costs

One new development is the use of supercritical carbon dioxide (sCO_2) as heat transfer fluid in a Brayton cycle. While no CSP plant uses sCO_2 yet, it has gained a lot of attention for next-generation power plants [31]. For example, sCO_2 in a closed-loop recompression Brayton cycle has been proposed for CSP. A main benefit from sCO_2 could be a higher cycle efficiency compared to supercritical or superheated steam cycles [32].

Combining solar-thermal power with fossil fuel generation can increase the capacity factor of the solar applications [33]. Rankine, Brayton, and combined cycle power generation schemes have been proposed in this context. The U.S. CSP plants have operated as hybrids employing gas as secondary fuel.

As to future prospects of CSP, the International Energy Agency, European Solar Thermal Energy Association, and Greenpeace forecast that CSP could account for 3–3.6% of the global energy supply in 2030 and 8–11.8% by 2050, which would require two-digit capacity growth in the coming years, which has not yet been demonstrated [34]. Other studies estimate that the price of CSP could decrease to $0.05/kWh by 2025 [35], which would be highly competitive.

Chinese development and deployment of CSP is modest as stated earlier [36]. It is unlikely that the CSP technology could change the structure of electricity supply in China, but it could have potential for large-scale development in the future to be discussed in the next chapter.

Status of CSP Technology Development in China

Resource potential for CSP

China has potential to develop CSP [37] in terms of the solar resource. Figure 1 illustrates the direct normal solar radiation resource available in China [http://swera.unep.net/ (accessed 10 September 2016)]. The best regions are found in the western part of the country with highest daily mean values of direct normal radiation around 9 kWh/m^2 in the Qinghai-Tibet Plateau and Sichuan Basin. A minimum value of 5 kWh/m^2,day is the limit of CSP for economical reasons [38], which is met in most parts of the northern and western China.

Concerning land-use requirements, a CSP plant typically requires 20,000 m^2 of land area per MW [34]. It is estimated that China has 2.63 million km^2 of land area not conflicting with other uses such as food production, most of which is located in the northern and the western China with high solar insolation values [39]. Although the DNI and gross land area available is high, two important factors limit the gross potential of CSP in China: (1) CSP requires very flat land (<2% grade over the entire array field) which eliminates the use of mountainous and hilly sites; and (2) most of the power demand is in the eastern part of China, whereas the potential is in the west, meaning that upgrading of the transmission grid

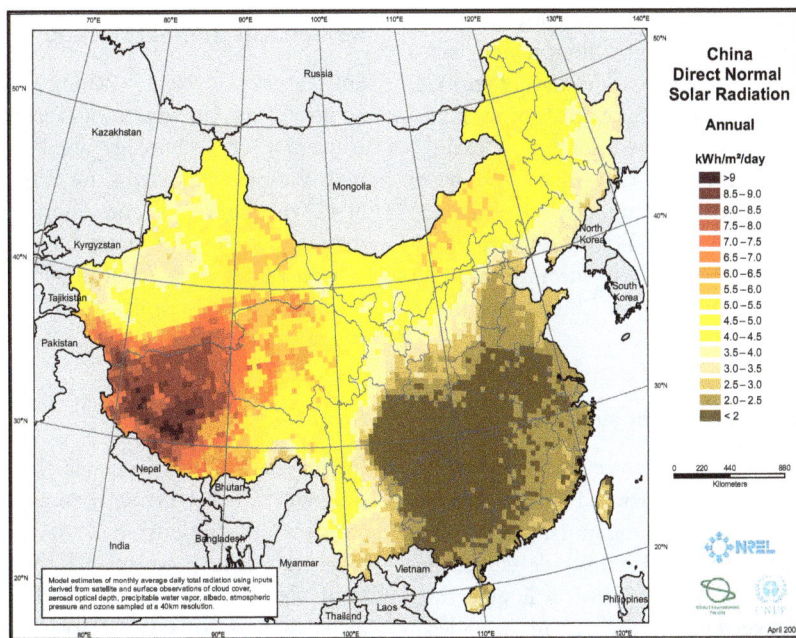

Figure 1. Direct normal solar radiation in China. (Note: This map was created by the National Renewable Energy Laboratory for the U.S. Department of Energy with data provided by UNEP and the Global Environment Facility.)

or employing storage may be necessary for large-scale CSP schemes.

Status of R&D

Since the 1970s, China has carried out relevant basic research on the development of solar-thermal power generation under the different Programmes of the Ministry of Science and Technology [40]. Key players in the CSP R&D in addition to enterprises have included several institutes of the Chinese Academy of Sciences (CAS) (Institute of Electrical Engineering, Changchun Institute of Optics, Fine Mechanics and Physics, Institute of Engineering Thermophysics), Himin solar energy group, and Southeast University [41], among others. Key R&D fields include the design and manufacture of condensers, design of collector field and system control, preparation of key materials, heat exchanger and energy storage systems for molten salts, high-precision heliostat technology, high-temperature tower absorber technology, high-temperature thermal storage technology, manufacturing process of trough evacuated tube, etc.

Key R&D facilities and know-how are presented in detail in the next.

Parabolic trough concentrator and evacuated absorber tubes

The Broad Air Conditioning Company, Institute of Electrical Engineering (CAS), and Hehai University have developed different types of parabolic trough concentrators. In one of the successful designs, honeycomb technology was employed for a 2.5-m-wide and 12-m-long ultra-light reflective structure [42]. China's first high-temperature vacuum receiver, Sanle-3 HCE, was mainly developed by the Southeast University in collaboration with Chinese companies such as Himin, Linuo Paradigma, and IVO (Kunshan) in 2007 [43, 44]. Asia's first parabolic trough power plant (ISCC) was successfully built employing this technology in Ningxia China in October 2011.

Heliostats for solar power tower system

China's first CSP demonstration project, a 70 kW solar tower plant (Fig. 2) [45], was constructed by the Chinese Academy of Engineering near Jiangning in Jiangsu in 2006. The heliostats for this project were jointly developed by Nanjing Chunhui Ltd., Institute of Electrical Engineering (CAS), and Himin. Larger heliostats of 100 m² were also developed and used in a demonstration project of 1 MW size [46, 47]. A 50 MW photothermal project in Delingha was started in 2011, with Supcon Ltd as the main

Figure 2. Nanjing's 70kW solar power tower system [46].

developer. Its first stage included 10 MW and two solar towers, which were completed in 2014 [48].

Solar dish

The Institute of Electrical Engineering (CAS) has developed different types of solar dish condensers. In the best design, the focal temperature of the condenser can reach up to 1600°C. The dish tracks the sun's position with a precision of ±0.2°. The reflectivity of the parabolic mirror reaches 94%.

Molten salt thermal storage

Sun Yat-sen University, Beijing University of Technology, South China University of Technology, and Dongguan University of Technology are key research institutes on multicomponent systems for thermal storage consisting of nitrate, carbonate, and chlorate. In addition, work on thermal storage in nitrates and containers for materials at high temperature (about 550°C) have been carried out at Changzhou Pressure Vessel Inspection Institute for several years.

Pilot plants

In July 2009, China launched the so-called 'Golden Sun' program to boost the solar sector [49]. The central government will support half of the investment costs of large-scale solar power plants. With a nationwide feed-in tariff plan for solar power development, the government plans to have 10 GW of solar power by 2020. Several pilot-plants to test and demonstrate different CSP technologies have been planned, all listed in Table 2. So far three plants have been finished. Considering the pace of

Table 2. Concentrating Solar Power projects in China.

Status	Start	Completion	Location	Size (MW)	Technology
Finished	2004	2005	Nanjing (Jiangsu)	0.07	Solar tower
	2011	2013	Jia yu guan (Gansu)	10	Solar trough
	2010	2013	Yan qing (Beijing)	1	Solar tower
Not finished	2016		Dunhuang (Gansu)	10	Solar tower
	2011		Golmud (Qinghai)	200	Solar tower
	2009		Mon ding (Yunnan)	10	PV& thermal
	2011		Ordos (Inner Mongolia)	50	Solar trough
	2010		De zhou (Shandong)	2.5	Fresnel
	2010		Ruan lin (Hunan)	50	Solar trough
	2009		Golmud (Qinghai)	1000	Solar tower
	2010		Yulin (Shanxi)	2000	Solar tower
	2011		Delingha (Qinghai)	50	Solar trough
Planned			Aba (Sichuan)	100	Tower & trough
			Sanya (Hainan)	100	Solar tower
			Wuwei (Gansu)	100	Solar trough
			Turpan (Xinjiang)	300	Solar trough
			Lhasa (Tibet)	50	Solar tower
			Turpan (Xinjiang)	50	Solar trough
			Jinta (Gansu)	50	Solar trough
			Ningxia	100	Solar trough
			Hangzhou (Zhejiang)	100	Tower & heliostats
			Gansu	100	Multidish

development, it seems unlikely that the goal for year 2020 will be reached.

Chinese Plans for CSP

Chinese policy framework

The present top priorities of Chinese development are stated in the 13th 5-year plan for national economic and social development [50]. According to the plan, the Chinese government will set up green development funds to support green and clean production, promote the green transformation of traditional manufacturing and establishment of low-carbon production and recycling in industry, and advocate enterprises to upgrade technology and renovate equipment [51]. All in all, China seems to strive to build a modern energy system, which is cleaner, less carbon emitting, safer, and more efficient than the present one. This sets the basic conditions for promoting the development of solar-thermal power generation in China.

The economy of China is expected to grow by 6.6% a year on average till year 2020, which also implies increasing demand for electricity. To meet the growing power demand, China would have to install as much as 635–860 GW of new-generation capacity between 2005 and 2020, an amount comparable to EU's total installed capacity in 2003 [52]. China's energy policy target is to reach a 15.4% renewable energy share by the year of 2020, and 27.5% in 2050,

respectively [53]. The Ministry of Science and Technology has listed CSP as an important research issue in its document 'Summary of the national mid & long-term science and technology development plan (2006–2020)' [54]. The official targets for solar energy utilization in China as stated in the 13th 5-year plan by 2020 are shown in Table 3. The CSP target is 10,000 MW [55]. We notice that the capacity of CSP needs to grow by a factor of 700 from 2014 to reach the 2020 goals, which is very unlikely given the other constraints mentioned in this article.

The geographical split down of the national CSP targets is shown in Table 4 indicating that most of the planned CSP will be in the northern and western parts of the country. However, some capacity is perceived in other parts of the country as well.

According to the 5-year plan, the total installed capacity of CSP should be 10 GW [55] by the end of 2020. To reach this goal, investments into CSP will annually be supported by ca 100 Billion yuan (1 Chinese yuan = 0.15 US \$). The target is to drive the investment costs below 20 yuan per W and the generating cost close to 1 yuan/kWh by the end of 2020. By 2030, solar power generation as a whole is envisioned to reach a total installed capacity of 400 GW, which would put Chinese industry into international lead [57]. The first batch of CSP demonstration projects was issued by National Energy Administration in September 2016 consisting of 20 plants (9 tower, 7 trough, and 4 Fresnel projects).

Table 3. Solar energy targets in China by 2020 [56].

Indicator	Solar	2014	2020
Installed capacity (MW)	Centralized PV	23,380	80,000
	Distributed PV	4672	70,000
	CSP	14	10,000
Electricity (TWh)	Gross generation (PV + CSP)	250	2000
Heat utilization indicator (billion m^2)	Solar collector area	–	8

In addition, to facilitate the solar-thermal industry needed to accomplish the targets, efforts on technology progress and demonstration promotion are planned as follows [58]: improving the quality of planning of CSP, establishment of a technical standards system, monitor experiences from demonstration projects, improve the economy and management of CSP projects, and finally, to develop a relevant electricity pricing policy to support CSP. The FiT level for CSP is still under debate and the National Development and Reform Commission's Price Department has not yet settled the level of support [59].

During the demonstration stage of CSP in 2016–2017, China will focus its CSP efforts to the western region mainly with the best solar conditions. This will also include coordination of the access conditions to land, water, and power grid. A number of solar-thermal power-generation demonstration projects with a total installed capacity of at least 50 MW will be constructed, either as standalone or part of hybrid plants. Based on the experiences from the demonstration projects, a gradual move to large-scale CSP is planned during 2018–2020. For this purpose, China plans to construct four MW-class solar-thermal power generation demonstration bases in Qinghai, Gansu, Inner Mongolia, and Xinjiang with a total capacity of hundreds of megawatts.

In order to promote the vigorous development of solar energy utilization, the 13th 5-year plan also proposes that domestic solar product standards should be made compatible to international standards. Meanwhile, stronger international cooperation is also needed to developed advanced energy technology and equipment manufacturing, to improve the industrial technology research and development abilities, and increase the core competitiveness progress [59].

Core technology readiness for CSP in China

Technology and costs are the two major barriers to CSP development in China. Until now, although there are not yet any commercial CSP plants in operation, several research and demonstration projects have been accomplished (see Table 2).

The technical development of CSP in China is foreseen to comprise four stages or four generations of CSP technologies [60, 61] between 2006 and 2025. Through this technology evolution, CSP should become more effective over time reaching higher temperatures, up to 800–1100°C by 2025. The first-generation CSP employing steam and oil as heat transfer medium is already in industrial scale and the second-generation, which also includes molten salt technology, is entering large-scale demonstration, but the third and fourth stages of CSP are still under research. The current first-generation CSP technology could reach an electricity generation cost of $100–120 per MWh, which still is higher than that of traditional thermal power plants [38]. To resolve the critical technological problems blocking cost reductions, the Ministry of Science and Technology is funding more research to improve CSP's market position [http://www.most.gov.cn/tztg/index.htm (accessed 10 September 2016)].

Most of the CSP components originate from traditional industries, for example, employing steel, glass, and cement, among others. For example, a 50 MW CSP with 4–8 h of heat storage needs 100,000–150,000 t of steel, 6000 t of glass, and 10,000 t of cement [62]. China has for the

Table 4. The national layout of solar energy utilization during the 13th 5-Year Plan period [56]. Numbers shown are shares of the total solar electricity target in 2020 (1 = 344 TWh).

Region	Solar PV			Solar CSP	Total
	Centralized	Distributed	Sum		
North China	0.055	0.050	0.105	0.002	0.107
Northwest	0.141	0.012	0.153	0.022	0.174
Northeast	0.010	0.010	0.021	0.001	0.021
East China	0.020	0.071	0.091	0.000	0.091
Central China	0.018	0.032	0.051	0.000	0.051
South	0.014	0.033	0.047	0.000	0.047
Other (Tibet)	0.004	0.001	0.005	0.002	0.008
West	0.146	0.012	0.158	0.024	0.182
Central & East	0.117	0.197	0.314	0.006	0.319

time being a clear oversupply situation of these materials, meaning that CSP will not just offer a clean energy supply scheme [63], but it could also stimulate economic growth in the traditional industry.

A key component of CSP is the concentrator or heliostat, which is composed of an ultra-thin super-white glass silver mirror. Ultra-thin glass technique in China is mature and technologies for self-cleaning glass and long-life silver mirrors already exist. With the advantage of low cost and high performance, heliostats made in China could push domestic CSP development forward. But in spite of such know-how, in practice the mirror products in China lack superior quality, for example, have poor reflectivity and self-cleaning properties [64]. This calls for further technology improvements in the near future.

Lessons learned from wind power and PV sectors

China has made huge progress in successfully scaling up PV and wind power during the last decade. The experiences from this process could be useful for CSP as well.

One important Chinese lesson was that the government needs to introduce a suitable policy framework to open up the market, but also to scale it up at a correct speed based on the market feedback and industrial progress. For CSP, a so-called benchmark price policy could be introduced in the same way as for wind power and PV. The price can be settled, for example, after one or two rounds of bids. In addition, a clear signal of guaranteed financial support and preferential policies to solar-thermal power generation investment enterprises and manufacturing industries is important [62]. In recent years, the Chinese PV industries have encountered major challenges due to the global economic situation, antidumping, and countervailing investigations, but also because the industries outsourced much of the production to overseas markets, which lead to losing core technologies and intellectual property rights. Therefore, from a Chinese perspective, the development of CSP could be accomplished in a more innovative way, so that core technology and the market power would remain in local possession [65]. In addition, building strong industry chains should be enhanced to establish an independent CSP industry [62]. Another problem appearing in both wind power and PV was the difficulty of grid connections, which resulted in forced partial outage, even >30% in some areas. Therefore, thermal energy storage (TES) could be more strongly used in CSP systems to improve solar and load matching. In this way CSP could be used as a flexible peak-shaving option [62].

Roadmap for CSP in China

China's CSP industry faces both great opportunities and challenges. In the long run, three stages are suggested for the development of CSP in China:

1. More small- and medium-size commercial CSP projects should be constructed before up-scaling to gain necessary experience. This requires a supported electricity price (FiT) for CSP to share the risks with the new technologies involved;
2. Distributed solar energy systems could provide an important niche market for CSP, for example, remote power and heat supply, desalination on islands, and industrial and agricultural applications with less financial support;
3. For large-scale CSP, combining with TES and hybrid power plants is recommended. Traditional power plants under 200 MW could in this way be hybridized with CSP leading to major CO_2 reductions [62].

In addition to above general lines, we propose below more specific measures for innovation, costs, and policy measures for CSP in China.

Innovation and collaboration

Chinese CSP industry needs stronger innovation efforts to speed up the technology upgrading. The Ministry of Science and Technology (MOST) has supported technology research and demonstration power plants in the past [46, 62, 64, 66], which lead to mastering some of the CSP core technologies [http://www.most.gov.cn/tztg/index. htm (accessed 10 September 2016)]. Much of this development work has been pursued domestically only, without a clear international context [67], which would need to be reversed now through international collaboration to strengthen own technology base [46, 68]. Several international companies already actively follow the CSP development in China [63, 69], which could offer an opportunity for stronger cooperation and communication with companies or organizations from abroad. We also notice the need of work in thermal storage for CSP considering the previously mentioned challenges in China.

Cost reduction

An important prerequisite for the success with CSP will be reducing the investment costs. For example, the success of parabolic trough systems in the United States can be attributed to cutting down the investments from $4500 to less than $3000 per kW in late 1980s and early 1990s. The CSP electricity cost dropped from $440/MWh for the first systems built to $170/MWh with the ninth system and the system efficiency increased from 9.3% to 13.6% [46]. The same effects need now be reached in the Chinese case by making better use of the scale effects. The industry estimates that in the 1000 MW scale, a power generation cost of 0.7–0.8 yuan per kWh should be possible. However, the required 20 billion yuan investment per 1000 MW is too high for many enterprises to finance [70] as the CSP industry in China is still in a start-up stage. If a continuous flow of smaller projects was realized, the investment cost of CSP could be possibly halved from the present level to <10,000 yuan/kW. The corresponding electricity price would then be close to that of wind power [67].

Policy support

To help achieving its energy and environmental goals and promote green investment, China has promulgated laws and regulations and put forward a series of policies facilitating green development and attracting and steering investments toward clean development [71]. Due to the fact that CSP is not yet a mature technology, policy support would be essential. The Government is expected to provide a reasonable support scheme and related measures soon. If the government can create favorable conditions for CSP, industry development will be greatly stimulated [67]. In addition, there are two fields of concerns, namely, patents and standards, which need to be addressed. To strengthen the creativity dimension, the government needs to improve incentives and mechanisms for intellectual property rights in near future. For example, a leading enterprise in CSP TES technology, Jiangsu Sunhome New Energy Co. Ltd., has independently developed and mastered the design and manufacture of TES systems for CSP applying for more than 30 patents [65]. In 2014, the China National Solar Thermal Energy Alliance convened an examination meeting of standards passing two standards on CSP technologies, which is a good start, but still inadequate for CSP as a whole. Recently, the National Energy Administration issued the first relevant standards for CSP power generation (5 national and 6 international). A framework for China's CSP standard system remains to be built.

Conclusions

In this study we have discussed the status and prospect of China's CSP development. Although the potential for CSP in China is good and there exists experience and know-how on CSP, scaling up CSP to a major energy vector would require large efforts both in terms of technology and cost development, but also stronger policy support. The key limitations for CSP in China need also

more attention. Internationally viewed, China lags behind the international development in this field.

The Chinese government has recognized the promising outlook for CSP, which is demonstrated by the positive plans expressed in the 13th 5-year plan. China has now clear targets for CSP development till year 2020, although these may be difficult to reach considering the short time span. To approach these targets, main development work is needed on a broad front, described and elaborated in this study. Notably, China needs to pay more attention to domestic innovations and cost reductions in CSP, which in turn would require a clear financial support system for CSP investments. Also, the geographical mismatch between the high solar resource in the west and the high power demand in the east will need more attention when moving to large-scale CSP deployment.

In terms of international development, China has several positive factors which could speak in favor of global leadership: China has large areas with excellent solar conditions for CSP (although limited by the geographical mismatch), strong basic capabilities in traditional manufacturing important to CSP, and also to some extent special know-how in CSP technologies. China would also profit from stronger international collaboration in the field, standardization, and IPR legislation and management. We propose here a three-step plan for scaling up CSP, with emphasis on smaller-scale CSP plants in short term to gain more experience and then scaling up to 100 MW scale leaning more to thermal storage and hybridization which give clear system benefits.

Acknowledgments

This work was supported by the National Science Foundation of China (No. 51476099).

Conflict of Interest

None declared.

References

1. Zhu, X., and G. Zhuang. 2014. A review of China's approaches toward a sustainable energy future: the period since 1990. WIREs Energy Environ. 3:409–423.

2. Govinda, R., and K. Lado. 2012. Solar energy: markets, economics and policies. Renew. Sustain. Energy Rev. 16:449–465.

3. Hoogwijk, M. 2004. On the global and regional potential of renewable energy sources. Faculteit Scheikunde, Universiteit Utrecht.

4. Gereffi, G., and K. Dubay. 2008a. Concentrating Solar Power Clean Energy for the Electric Grid. Duke Center on Globalization, Governance & Competitiveness (Duke CGGC). Technical Report.

5. Meneguzzo, F., R. Ciriminna, L. Albanese, and M. Pagliaro. 2015. The great solar boom: a global perspective into the far reaching impact of an unexpected energy revolution. Energy Sci. Eng. 3:499–509.

6. Zhang, S., and Y. He. 2013. Analysis on the development and policy of solar PV power in China. Renew. Sustain. Energy Rev. 21:393–401.

7. Major Growth Market of Future CSP Installed Capacity. Available at http://en.cspplaza.com/major-growth-market-of-future-csp-installed-capacity.html/ (accessed 23 February 2016).

8. Li, J. 2009. Scaling up concentrating solar thermal technology in China. Renew. Sustain. Energy Rev. 13:2051–2060.

9. Qu, H., and J. Zhao. 2008. Prospect of concentrating solar power in China—the sustainable future. Renew. Sustain. Energy Rev. 12:2505–2514.

10. Zhang, Z., and Y. Xia. 2003. Development status quo and prospect analysis of solar thermal power. J. Changjiang Eng. Voc. College 30:1.

11. Kaushika, N. D., K. S. Reddy, and K. Kshitij. 2016. Sustainable energy and the environment: a clean technology approach (7 solar thermal energy and power systems). Springer International Publishing, New York, USA.

12. Giorgio, S. 2013. Concentrating Solar Power Technology Brief. IEA-ETSAP and IRENA Technology Brief E10 – January.

13. Yang, M., X. Yang, R. Lin, and J. Yuan. 2008. Solar energy- based thermal power generation technologies and their dystems. J. Eng. Therm. Energy Power 23:221–228.

14. Stine, W. B., and R. P. Diver. 1994. A Compendium of Solar Dish/Stirling Technology. Sandia National Laboratories, Albuquerque, NM, Report SAND93-7026 UC-236.

15. Meriem, C., and V. Sébastien. 2016. Benchmark of Concentrating Solar Power plants: historical, current and future technical and economic development. Proc. Comp. Sci. 83:782–789.

16. Pilkington Solar International. 1996. Status Report on Solar Thermal Power Plants. Report ISBN 3-9804901-0-6.

17. Williams, T., and M. Bohn. 1995. Solar Thermal Electric Hybridization Issues. Proceedings of the ASME/JSME/JSES International Solar Energy Conference, Maui, HI, March 19-24

18. 1995. Fugitive Emissions Testing – Final Report, AeroVironment Inc. for KJC Operating Company, Monrovia, CA.

19. Washom, B. 1984. Parabolic Dish Stirling Module Development and Test Results. Proceedings of the IECEC, San Francisco, Paper No. 849516.

20. Fan, B., and B. Chen. 2010. The overview of parabolic trough solar power technologies. Power Supp. Technol. Appl. 13:31–36.

21. Zhong, S. 2013. Brief description for tower solar thermal power generation. J. Shenyang Inst. Eng. 9:7–12.

22. Tian, F., and X. Zhu. 2015. The latest development of solar power tower technology. J. Nanjing Norm. Univ. 15:5–10.

23. Ding, T., and X. Zhu. 2012. Study on ground coverage of heliostats field in central receiver solar power plant. Renew. Energy Resour. 30:11–14 (in Chinese).

24. Gong, B., Z. Li, and Z. Wang. 2009. Deflection for reflector plate of heliostat based on thin plate flexure theory. Acta Energiae Solaris Sinica 30:900–903 (in Chinese).

25. Xu, H., H. Zhang, T. Bai, L. Ding, and J. Zhuang. 2009. An Overview of Dish Solar Thermal Power Technology. Thermal Power Generation 6:6–9.

26. West, C. D.. 1986. Principles and applications of stirling engines. Van Nostrand Reinhol, New York, NY.

27. Forrester, J. 2014. The value of CSP with thermal energy storage in providing grid stability. Energy Procedia 49:1632–1641.

28. Okoroigwe, E., and A. Madhlopa. 2016. An integrated combined cycle system driven by a solar tower: a review. Renew. Sustain. Energy Rev. 57:337–350.

29. Steinmann, W. D. 2012. Thermal energy storage systems for concentrating solar power (CSP) plants. Pp. 362–394 in Concentrating solar power technology eds K. Lovegrove and W. Stein. Woodhead Publishing. Cambridge, UK.

30. Bauer, T., N. Breidenbach, N. Pfleger, D. Laing, and M. Eck. 2012. Overview of molten salt storage systems and material development for solar thermal power plants. World Renewable Energy Forum 1-8.

31. Ahn, Y., S. J. Bae, M. Kim, S. K. Cho, S. Baik, J. I. Lee et al. 2015. Review of supercritical CO_2, power cycle technology and current status of research and development. Nucl. Eng. Technol. 47:647–661.

32. Dostal, V., P. Hejzlar, and M. J. Driscoll. 2006. The supercritical carbon dioxide power cycle: comparison to other advanced power cycles. Nucl. Technol. 154:283–301.

33. Lin, R., W. Han, H. Jin, and Y. Zhao. 2013. The integrated solar combined cycle power generation systems. Gas Turbine Technol. 26:9–23.

34. Chen, Y. 2010. Application and prospect of concentrating solar energy power plant (CSP) technology. Power Syst. Clean Energy 26:1–10.

35. Gereffi, G., and K. Dubay. 2008b. Concentrating solar power. Clean energy for the grid. Center on Globalization Governance & Competitiveness, USA.

36. Jing, Y. The Current Situation of Solar Thermal Power Generation and Its Future Trend Analysis. Proceedings of 3rd International Conference on Social Science and Education, (ICSSE 2015), June 27-28, 2015, London, UK.

37. International Energy Agency (IEA). Energy technology perspectives 2008. Available at http://www.iea.org/ (accessed 10 September 2016).

38. Müller-Steinhagen, H., and F. Trieb. 2004. Concentrating solar power – a review of the technology. Ingenia 18:43–50.

39. Hang, Q., J. Zhao, and Y. Xiao. 2008. Prospect of concentrating solar power in China—the sustainable future. Renew. Sustain. Energy Rev. 12:2505–2514.

40. Guo, M., Z. Wang, and W. Liang. 2010. Tracking formulas and strategies for a receiver oriented dual-axis tracking toroidal heliostat. Sol. Energy 84:939–947.

41. Zang, C., Z. Wang, and Y. Wang. 2010. Structural design and analysis of the toroidal heliostat. ASME J. Sol. Energy 132:041007.

42. Yuan, J.. 2007. Research on system integration of a novel solar tower thermal power plant. Chinese Academy of Sciences, Beijing.

43. Wang, J., and Y. M. Zhang. 2007a. Development and study on vacuum absorber tubes. Proceedings of ISES Solar World Congress, Beijing, Pp. 1813–1817.

44. Wang, J., and Y. M. Zhang, 2007b. Development and study on heat-pipe type vacuum absorber tube. Proceedings of ISES Solar World Congress, Beijing, Pp. 1818–1822.

45. Wang, J., Y. Zhang, and D. Liu. 2010. The first solar power tower system in China. Hohai University, Nanjing, Jiangsu, China.

46. Chen, C., Z. Nie, and X. Na. 2009. On the development of parabolic trough concentrating solar power station. J. Eng. Stud. 1:314–318.

47. Zhang, M. 2008. Solar thermal power generation technology in high temperature. High Technol. Indust. 7:22–24.

48. Yi, C., and X. Sun. 2015. Assumptions about CSP and photo-coal complementation. China Electric Power.

49. Zhang, Z. 2013. An analysis of China's energy demand and supply policy framework. WIREs Energy Environ. 2:422–440.

50. The advice of the central committee of the communist party of China to develop the 13th five-year plan for national economic and social development. Xinhua News Agency 18 March 2016. Available at http://politics.chinaso.com/detail/20160318/1000200032851721458258630946197048_2.html/ (accessed 10 September 2016).

51. Li, Z., S. Chen, and X. Hu. 2015. The reform media publisher, editorial director, depth discussion with six famous scholars about 13th five program's execution. Reform 12:5–25.

52. Kahrl, F., and D. Roland-Holst. 2006. China's carbon challenge: insights from the electric power sector. University of California, Berkley, USA. Available at http://are.berkeley.edu/~dwrh/Docs/CCC_110106.pdf (accessed 10 September 2016).

53. He, J. K., and X. L. Zhang. Strategies and policies on promoting massive renewable energy development. Proceeding of China renewable energy development strategy workshop, Beijing, China, 26 October 2005.

54. http://www.most.gov.cn/tztg/index.htm (accessed 10 September 2016).

55. Guo, C. Major Growth Market of Future CSP Installed Cap. Available at http://are.berkeley.edu/~dwrh/Docs/CCC_110106.pdfacity; http://en.cspplaza.com/major-growth-market-of-future-csp-installed-capacity.html (accessed 23 February 2016).

56. National Energy Administration of China. The exposure draft about development of solar energy utilization in 13th five-year plan. Available at http://www.21spv.com/news/show.php?itemid=16598 (accessed 10 September 2016).

57. Extract of exposure draft of 13th five year plan in solar-thermal utilization. Popular Utilization of Electricity 2016;1:23.

58. Wang, Y.. 2016. Advance to the first hundred-goal knowledge and awareness about. Advice of the central committee of the communist party of China to develop the 13th five-year plan for national economic and social development. J. Chin. Acad. Govern. 1:4–12.

59. China CSP FIT is officially taken into the agenda by Price Department of NDRC. Available at http://en.cspplaza.com/china-csp-fit-will-soon-be-issued-pricedepartment-of-ndrc-said.html (accessed 5 May 2015).

60. Wang, Z., and F. Du. 2010. Concentrating solar power tower strategies in China. J. Jap. Inst. Energy 89:331–336.

61. Du, F. 2011. Methods to reduce the cost of concentrating solar power. Sol. Energy 7:11–13.

62. Li, S. 2013. Solar thermal generation: next investment hot spot of new energy. China Invest. 3:30–32.

63. Wen, B. J. 2015. Seize China's opportunity and follow world's trend. Sol. Energy 4:75–76.

64. Yang, G., and Y. Geng. 2010. New application area of CSP and ultra-thin glass. Building Materials in the 21st Century 2:33-36.

65. 2012. How far is CSP from us? Architectural & Functional Glass 7:45-46.

66. Li, W. 2014. Generation type analysis and development trend discussion of CSP. Sci. Technol. Innovat. Appl. 35:80–85.

67. 2010. Three CSP issues demanding solution. Software 9:13-15.

68. Yu, P., and Y. Mu. 2012. Development of concentrated solar power. Glass 39:36–38.

69. 2014. International electronics giants accelerate into China's CSP market. Architectural & Functional Glass 8:45-47.

70. 2011. CSP may become new energy leading investment. Architectural & Functional Glass 1:32-33.

71. Shen, B., and J. Wang. 2013. China's approaches to financing sustainable development: policies, practices, and issues. WIREs Energy Environ. 2:178–198.

12

Energy production advantage of independent subcell connection for multijunction photovoltaics

Emily C. Warmann[1] & Harry A. Atwater[2]

[1]California Institute of Technology, 1200 E California Blvd, Pasadena, California 91125-0002
[2]Kavli Nanosciences Institute, California Institute of Technology, Pasadena, California

Keywords
Energy production, multijunction solar cells, spectral variation, spectrum splitting

Correspondence
Emily C. Warmann, California Institute of Technology, MC 132-801200 E California Blvd, Pasadena, CA 91125-0002. E-mail: warmann@caltech.edu

Funding Information
This project was supported QESST via the National Science Foundation (NSF) and the Department of Energy (DOE) under NSF CA No. EEC-1041895, and also the Advanced Research Projects Agency-Energy (ARPA-E), U.S. Department of Energy, under Award Number DE-AR0000333. ECW acknowledges support from QESST and ARPA-E, and HAA was supported as part of the DOE "Light-Material Interactions in Energy Conversion" Energy Frontier Research Center under grant DE-SC0001293.

Abstract

Increasing the number of subcells in a multijunction or "spectrum splitting" photovoltaic improves efficiency under the standard AM1.5D design spectrum, but it can lower efficiency under spectra that differ from the standard if the subcells are connected electrically in series. Using atmospheric data and the SMARTS multiple scattering and absorption model, we simulated sunny day spectra over 1 year for five locations in the United States and determined the annual energy production of spectrum splitting ensembles with 2–20 subcells connected electrically in series or independently. While electrically independent subcells have a small efficiency advantage over series-connected ensembles under the AM1.5D design spectrum, they have a pronounced energy production advantage under realistic spectra over 1 year. Simulated energy production increased with subcell number for the electrically independent ensembles, but it peaked at 8–10 subcells for those connected in series. Electrically independent ensembles with 20 subcells produce up to 27% more energy annually than the series-connected 20-subcell ensemble. This energy production advantage persists when clouds are accounted for.

Introduction

The most efficient photovoltaic devices under the AM1.5 spectrum use the design concept of dividing the solar spectrum among multiple photovoltaic subcells, or "spectrum splitting" to improve performance beyond the capabilities of single junction designs [1–3]. However, tailoring the selection of subcell band gaps to the specific design spectrum raises the possibility of increased sensitivity of efficiency to spectral variation as experienced by systems operating under natural sunlight, which in turn has the potential to compromise integrated energy production [4, 5]. While efficiency measurements under the AM1.5 spectrum provide an essential means of standardized comparison between different photovoltaic technologies, the true product of photovoltaics is cumulative energy production over days and months with conditions that can vary widely from the standard test. The photovoltaics community has an increasing interest in predicting both the standard test efficiency and the energy production behavior of different photovoltaic technologies, in particular identifying the effects of design decisions on energy production [6–8]. Here, we present an examination of the relative effect on energy production for spectrum splitting designs with 2–20 subcells that are either connected electrically in series or are electrically independent of one another.

The most common form of spectrum splitting photovoltaics at present consists of cells with two to four subcells that are monolithically integrated and connected in series electrically [1, 9, 10]. The incident sunlight is divided among the subcells by sequential absorption, with high band gap junctions filtering out the high-energy photons and transmitting the lower energy light for absorption by the subcells below. The electrical series connection forces all subcells to generate the same current, which in turn determines the optimal combination of band gaps for a given spectrum. An alternate configuration for spectrum splitting uses a separate optical element to direct photons in different energy ranges onto physically isolated subcells [11–13]. In this approach, the subcells can be electrically independent from one another, removing the current matching requirement and constraint on band gap selection. Most of the efficiency difference between spectrum splitting photovoltaic cells will be determined by the number of subcells and how close their band gap energies are to the optimum values for the AM1.5 spectrum [2, 14]. Cells with electrically independent subcells have a small efficiency advantage over series-connected cells with the same number of junctions because they can achieve a better match between band gap energies and the spectrum when relieved of the need to match currents in all subcells [15].

While the monolithic, series-connected spectrum splitting configuration sacrifices only a small amount of efficiency under the design spectrum in exchange for simplified optical and electrical integration requirements, the series connection makes the device sensitive to illumination conditions that differ from the standard [4, 16]. All subcells will absorb an equal number of photons under the AM1.5 spectrum, but under the varying spectrum of natural sunlight, some subcells may be over illuminated relative to the others, and this excess energy cannot be collected. More damaging to cell performance, a subcell that is under illuminated will constrain the current through the entire device and can substantially reduce total device efficiency [17]. The degree to which field performance resembles performance under the standard will be determined by how much the photon density in the spectral bands for each subcell of natural sunlight tends to vary from the AM1.5 spectrum. Currently deployed spectrum splitting photovoltaics use only 2–4 spectral bands, which may reduce sensitivity to fine variations [4]. However, designs with increasing numbers of subcells are under consideration for higher cell efficiencies and may suffer more strongly from this effect in the series-connected configuration [10]. Previous efforts to predict the energy

production of spectrum splitting photovoltaics have examined multijunction solar cells with up to six series-connected subcells and a variety of three subcell, electrically independent band gap combinations [18–21]. This paper extends the analysis to large numbers of subcells and locations that exhibit wider variation in atmospheric conditions, which are important areas of inquiry as spectrum splitting photovoltaics become increasingly ambitious in terms of subcell number and deployment scope.

Determining the impact of electrical configuration and number of subcells on energy production requires analyzing the performance of different designs under a set of spectra that have a degree and type of variation comparable to that experienced in the field. The spectrum of natural sunlight at a particular location on earth and a point in time varies depending on the emission of the sun, the composition-dependent transmission properties of the atmosphere, and the path length through the atmosphere [22]. The sun's spectrum is comparatively stable relative to the other factors, and accounts for only 0.3% of the spectral variation [23]; consequently, this analysis omits this source of variation. The atmospheric path length is determined by location, elevation, and time of day and year, making 1 year a logical length of time over which to simulate energy production [24]. The absorptivity of the atmosphere depends on the concentration of water, CO_2, O_3, and other gaseous pollutants as well as the absorption and scattering of suspended aerosols [25]. While spectrally resolved irradiance data are not typically available at most locations, the spectrum can be simulated using extensive existing databases of typical atmospheric conditions and aerosol optical depth [26]. Although the actual spectrum at a particular location and point in time will likely differ from the simulation based on average conditions, the simulated spectra can reproduce the typical degree of spectral variation for the location and times of year considered. In turn, these spectra can be used to predict the typical performance of photovoltaic systems under that location's varying illumination conditions. This analysis compares the energy production of spectrum splitting ensembles with 2–20 subcells in both electrically independent and in series configurations under one year's worth of simulated spectra for five locations in the United States.

Methods

Spectrum splitting ensembles with 2–20 subcells connected electrically in series and independently and optimized for the AM1.5D spectrum were analyzed by detailed balance calculation under spectra simulated to represent the typical annual range of variation for multiple locations in the United States [27]. The detailed balance calculation assumed ideal cells operating at 300 K and under 500 suns concentration, and the series-connected cells were constrained to pass the same current through all subcells, using Henry's method [28]. The electrically independent cells allowed each subcell to operate at its individual max power point subject to the illumination conditions. The independently connected ensembles were optimized through a simulated annealing process. The band gap combinations for series and independent ensembles are tabulated in the supplemental information. The calculation assumed perfect spectral splitting among the subcells. The assumption of ideal cells will result in higher predictions of energy production than is realistic, however, many common cell nonidealities such as nonradiative recombination are not spectrally sensitive, and consequently, the relative behavior of nonideal cells will follow the same patterns as ideal cells.

Sunny day spectra

Simulated spectra were generated at hourly intervals from 8 am to 5 pm for 365 days to account for the most productive photovoltaic energy production hours of the day. These hours will neglect some period of illumination in the morning and evening of the summer months, however, the analysis will show that these hours cumulatively account for a small portion of the annual irradiance. In addition, neglecting these low-light hours will reduce the effect of the electrical connection difference and understate the advantage of the electrically independent ensembles. Consequently, this limitation of the input spectra is conservative in terms of estimating the effect size. For each spectrum, the air mass was determined by geographic location, date and time, and the temperature, pressure, and dew point were taken from hourly normal values published by NOAA [29]. Daily values for aerosol optical depth and precipitable water were generated using monthly normal and standard deviation values for each location as published from the AERONET data set [30]. CO_2 was set to 370 ppm, the annual average value for the year 2000, and O_3 was set to 0.3438 cm, the value in the U.S. Standard Atmosphere [31]. These atmospheric conditions were used as inputs to the Simple Model of the Atmospheric Radiative Transfer of Sunshine (SMARTS) multiple scattering and absorption model, which then calculated the direct spectral irradiance over wavelengths from 280 to 4000 nm [32]. Spectra were simulated for Phoenix, AZ, Albuquerque, NM, Houston, TX, Tulsa, OK and Knoxville, TN. Phoenix was selected as a location with high direct irradiance (DNI) where concentration photovoltaics (CPV) is already of interest, and because high-quality data on atmospheric conditions were available for input to the SMARTS program. The other locations

were selected for the availability of atmospheric data and to explore the effects of elevation, latitude, and climate. Together these locations span a wide range of climate conditions and latitude in the portion of the United States with a high solar resource. The power production under each spectrum was then multiplied by 1 h to integrate energy production per square meter of aperture area over the course of 1 year.

The simulated spectra will not exactly reproduce the irradiance conditions at any location and time because they use average values for the atmospheric conditions. In addition, the method of generating aerosol optical depth and precipitable water values will not capture the full degree of correlation between those values. However, the range of spectral variation and overall distribution of spectral irradiances generated should resemble the variability experienced at these locations, as with previous analysis efforts [18, 20, 21, 33]. Finally, the simulated spectra all assume an absence of cloud cover at the time and date for which they are generated. Consequently, the simulated spectra will overestimate the total irradiance and energy production for these locations.

Cloud correction

While the sunny day spectra are a good way to investigate the sensitivity of spectrum splitting photovoltaic efficiency to changing atmospheric conditions, all locations experience some amount of cloud cover that will reduce the direct normal irradiance. The prevalence of clouds at a given location can vary greatly with time due to the local climate, and therefore they may interact with the spectral variation to have an unpredictable effect on energy production. Unfortunately, it is difficult to obtain spectrally resolved data that account for cloud cover and identify the direct and diffuse components of the total irradiance. To explore the effect of clouds on energy production for these specific locations, we used data from the National Solar Radiation Database from NREL to determine the average percentage of expected DNI transmission for every hour over the course of the year. The data consist of simulated total DNI, global horizontal, and diffuse total irradiance hourly for every day for years from 1991 to 2010 as obtained from the METSTAT model based on local weather station data, local and satellite-based measurements of aerosol optical depth, ozone and precipitable water, and automated local and satellite-based observations of cloud cover and ground albedo [34]. In addition, the data include expected DNI, global horizontal, and diffuse total irradiance as determined by a sunny day model incorporating the same measurements of AOD, ozone, CO_2, precipitable water, local pressure, and humidity for the location but with the assumption of no cloud cover. For each hour considered in the energy production simulation, we calculated the ratio of DNI modeled including cloud observations to the sunny day model prediction of DNI over each year for all locations. We adopted this percentage as the hourly average DNI transmission for that particular hour. We then averaged the hourly DNI transmission over the 20 years of data for each location and multiplied the expected energy production for each hour by that percentage.

It is important to note two assumptions here. First, this method of derating the DNI transmission assumes that clouds have a uniform spectral impact on transmission. This assumption may be reasonable for circumstances with intermittent clouds that eliminate direct transmission for some percentage of each hour, but it is not likely to be valid for thin, persistent clouds that partially screen the sun while allowing some fraction of direct irradiance. Unfortunately, we were unable to identify a way to distinguish these circumstances in the data or to determine the spectral transmission of clouds in general. The decision to treat the cloud transmission as spectrally uniform may reduce the degree of variation in the simulated spectra relative to a model for cloud transmission with spectral resolution. However, to the best of our knowledge, there are no comprehensive data on the spectral transmission of clouds that would be useful to this analysis. Therefore, our assumption of spectrally uniform transmission is conservative in that it will most likely underestimate the true sensitivity of energy production to atmospheric conditions. Second, the decision to scale the energy production by the DNI transmission ignores any decrease in efficiency due to deconcentration of the light on the solar cells. Because the cells are simulated to operate at 500 suns, the effect on efficiency will be negligible except at very low transmission values, where the effect on cumulative energy production difference will be small.

Results and Discussion

The efficiency of spectrum splitting ensembles with electrically independent subcells is only slightly greater than the efficiency of series-connected ensembles under the design spectrum of AM1.5D as shown in Figure 1(A). The two-cell series ensemble exhibits 99% of the power efficiency of the independent ensemble under AM1.5D, while the 20-subcell series ensemble exhibits 98.4% of the independent power. The efficiency of both sets of ensembles increases with the number of subcells, with some deviation for the 15-subcell series-connected ensemble, which suffers from the current matching constraint aligning poorly with the absorption bands in the AM1.5D spectrum. However, the electrically independent ensembles have a significant energy production advantage that

increases with the number of subcells. Taking Phoenix, AZ as the first case, Figure 1 panel (B) shows the two subcell series ensemble produces 3.4% less energy than the electrically independent two subcell ensemble. For 20 subcells, the series connection reduces energy production by 14.8% relative to the electrically independent design. In addition, the simulated annual energy production of independent ensembles increases as the number of subcells increases from 2 to 20 while the series connected energy production peaks with 12 subcells and declines as more subcells are added, despite the higher design efficiency of the ensembles with more subcells. The discrepancy between the energy production and design efficiency trends indicates a greater difference in efficiency between the series and independent ensembles over some portion of the year.

Performance under varying irradiance levels

A closer examination of the simulated spectra and efficiency trends for Phoenix is useful for identifying the

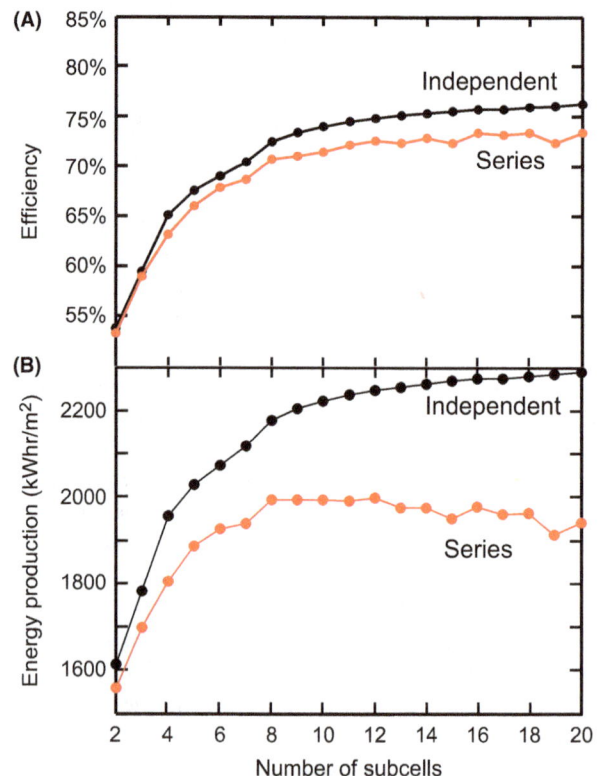

Figure 1. (A) Efficiency versus number of subcells for ideal spectrum splitting ensembles with electrically in series and independent subcells at 500 suns concentration under the AM1.5D spectrum. (B) Simulated annual energy production for Phoenix, AZ for electrically independent and in series spectrum splitting ensembles operating at 500 suns concentration. The energy production units are kWh/m² of aperture area.

cause of this efficiency deviation. The simulated spectra exhibit a large amount of variation over the course of a day and year. This variation includes both spectral composition and total irradiance level. Figure 2 panel (D) shows the number of spectra that have integrated direct irradiance levels falling in 10 different ranges. While many of the simulated spectra have a total irradiance similar to the 900 W/m^2 set by the AM1.5D standard, 485 spectra (13% of the total) have irradiance greater than 950 W/m^2 and 1126 (31%) have irradiance less than 800 W/m^2. The spectra that fall in these 10 irradiance ranges are plotted in the supplemental information. Figure 2 panels (A) and (B) show the efficiency of a subset of independent and series ensembles averaged over all the spectra in the power ranges shown in panel (D). The small inset panel at the right shows the efficiency of the ensembles under the AM1.5D design spectrum.

Panel (A) shows the efficiency of the independent ensembles is on average very consistent over a wide range of irradiance levels. The average efficiency at the lowest irradiance range is 5.7% lower than the efficiency over the 900–950 W/m^2 range for two cells and 4.7% lower for 20 cells. At every power range, the average efficiency increases with increasing subcell number, and overall, the average efficiencies are close to the efficiency under AM1.5D.

By contrast, panel (B) shows the efficiency of the series-connected ensembles averaged over the spectra in the same power bins. Again, the inset panel at right shows the efficiency of the ensembles under the AM1.5D spectrum. The series-connected ensembles show a much larger sensitivity to the irradiance level in their average efficiency. The two-cell ensemble efficiency in the lowest power range is only 72% of the average efficiency in the 900–950 W/m^2 range. Ensembles with larger numbers of subcells exhibit a larger decrease in efficiency as the irradiance level deviates from the 900 W/m^2 level, including a strong decrease in efficiency at the highest irradiance levels.

The decrease in efficiency with declining irradiance becomes steeper as more subcells are incorporated into the ensembles, and at irradiance levels less than 500 W/m^2, the average efficiency decreases as the number of subcells increases. In addition, the peak average efficiency for the series-connected ensembles does not correspond to the 20-subcell ensemble. Instead, the 16-subcell ensemble has the highest average efficiency in the 900–950 W/m^2 range. Finally, at all irradiance levels, all series-connected ensembles have an average efficiency that is substantially lower than the efficiency under the AM1.5D spectrum, indicating significant differences between the standard spectrum and the realistic spectra regardless of irradiance level. For these ensembles, the design efficiency is not a

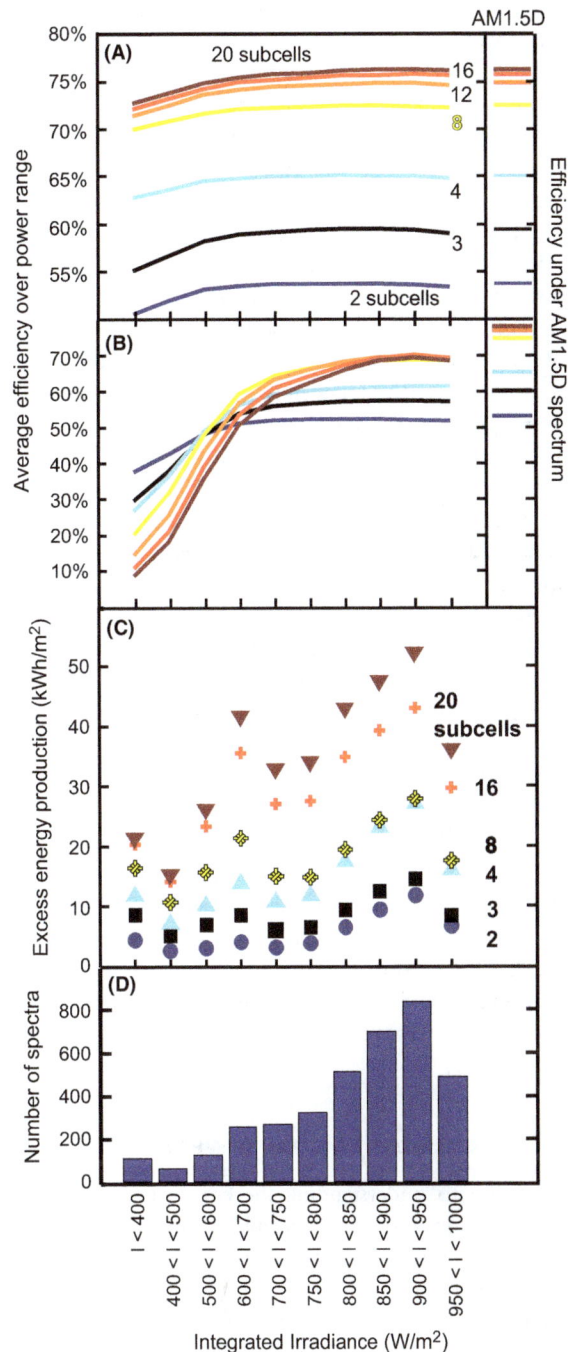

Figure 2. (A) Efficiency of electrically independent spectrum splitting ensembles averaged over all spectra in 10 different irradiance level ranges. The far right panel shows efficiency of the independent ensembles under the AM1.5D spectrum. (B) Efficiency of series-connected spectrum splitting ensembles averaged over all spectra in the irradiance level ranges. The far right panel shows the series-connected efficiency under AM1.5D. Note the difference in efficiency range for panels (A) and (B). (C) The cumulative energy produced by the electrically independent ensembles in excess of that produced by the series-connected ensembles over the then different irradiance level ranges. (D) Histogram of spectra showing the relative prevalence of the different irradiance levels.

good predictor of efficiency under the changing conditions in deployment.

Figure 2 panel (C) shows the combined energy production advantage of the independent ensembles relative to the series-connected ensembles for the different irradiance ranges. The energy production advantage increases with number of subcells at all irradiance levels, and the relative magnitude of the advantage compared to the combined irradiance at the different ranges suggests that the independent ensembles enjoy two slightly different advantages over the series ensembles. First, note that averaged over all numbers of subcells, 60% of the excess energy of the independent ensembles is generated under spectra with more than 750 W/m^2 irradiance. The spectra in this range constitute 84% of the total irradiance over the year. This additional energy production at high irradiance levels can be considered an efficiency advantage for independent ensembles due to improved spectral utilization. At the low-power level, spectra with less than 750 W/m^2 constitute 16% of the cumulative annual irradiance. Series-connected ensembles exhibit low average efficiency levels under these low irradiance spectra and produce little energy. Compared to the low energy production of the series-connected ensembles, the energy production of the independent ensembles under these low-power spectra can be considered an extended capacity factor allowing the independent ensembles to generate power under illumination conditions that are unfavorable to series-connected ensembles. Averaged over all numbers of subcells, this extended capacity factor accounts for 40% of the excess energy generated by the independent ensembles (on average 7% of the total annual energy production).

Performance over course of year

In addition to understanding the energy production advantage of independent ensembles relative to series-connected ensembles at different irradiance levels, it is also interesting to examine their relative performance over the course of a year. Figure 3 panel (A) shows the monthly average relative efficiency of series-connected ensembles normalized to the independent ensemble efficiency, again for Phoenix, AZ. Panel (B) shows the monthly mean, maximum, and minimum irradiance level. While the relative performance generally tracks the monthly average irradiance levels, the seasonal variation is not simply connected to day length. The series ensembles have a strong relative performance dip in the months of July, August, and September, most likely due to higher aerosol optical depth and precipitable water presence during the monsoon season.

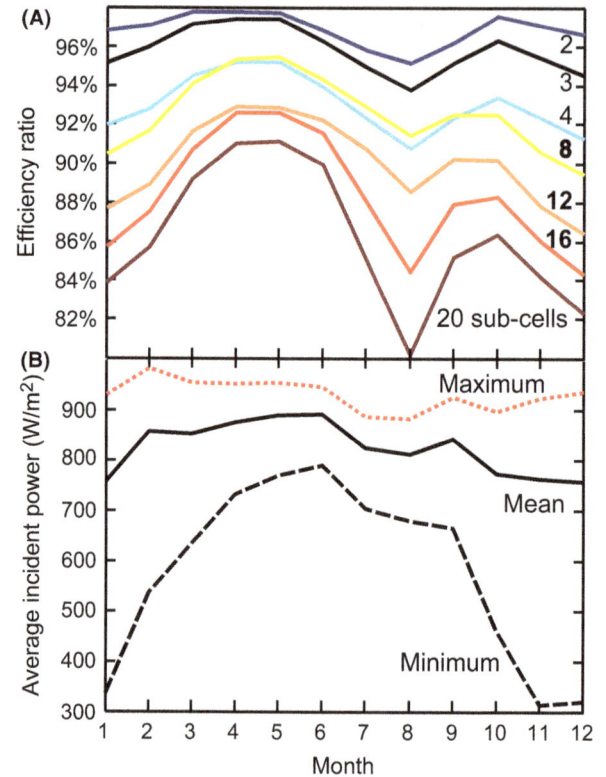

Figure 3. (A) Efficiency ratio of series-connected to electrically independent spectrum splitting ensembles averaged over all spectra in each month of the year in Phoenix, AZ. (B) Minimum, mean, and maximum irradiance level for each month of the year in Phoenix, AZ.

Performance at different locations

The particular performance penalty of series relative to independent ensembles in the summer months for Phoenix appears to be specific to the climate of that location. To explore the effect of geographic location on the relative performance of spectrum splitting ensembles with electrically in series and independent subcells, the annual energy production simulation was duplicated for a variety of different locations: Houston, TX, Albuquerque, NM, Tulsa, OK, and Knoxville, TN. Figure 4 panel (A) shows the annual energy production for the independent ensembles at all five locations and panel (B) shows the annual energy production for the series ensembles. The broad trend of energy production as a function of subcell number is the same at all locations, though the expected energy production varies based on differences in the total sunny day irradiance simulated for the various cities. Importantly, the series-connected ensembles all exhibit the same maximum energy production with 8–12 subcells and declining energy production with larger ensembles, consistent with the performance for Phoenix.

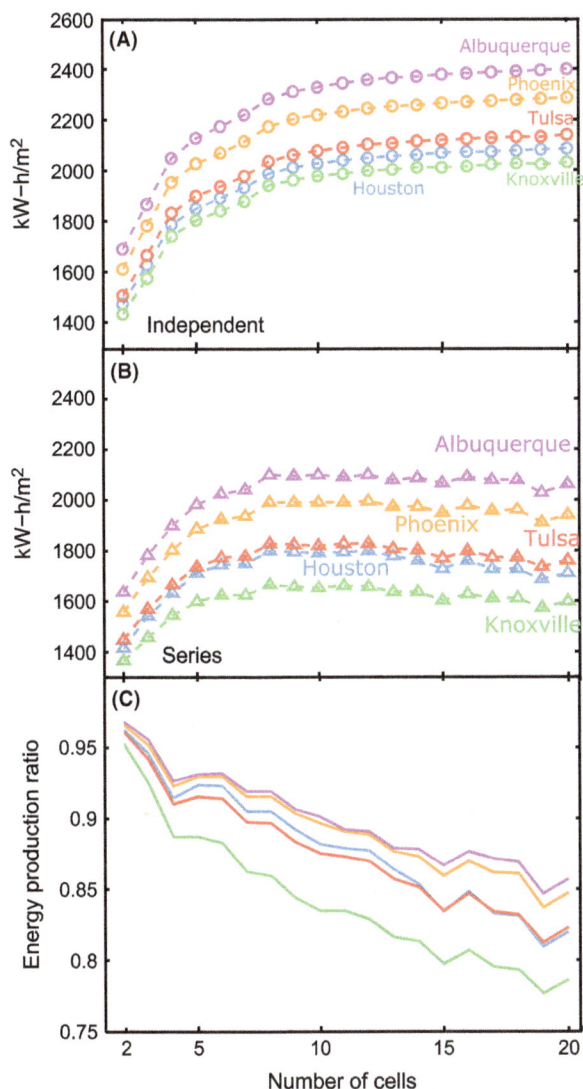

Figure 4. (A) Simulated annual energy production for ideal spectrum splitting ensembles with electrically independent subcells operating at 500 suns in Phoenix, AZ, Albuquerque, NM, Tulsa, OK, Knoxvile, TN, and Houston, TX. All simulated spectra are for sunny days. (B) Simulated annual energy production for spectrum splitting ensembles with series-connected subcells. (C) Ratio of series-connected energy production to electrically independent energy production.

Figure 4 panel (C) shows the ratio of series-connected to independent energy production for all five locations versus the number of subcells per ensemble. The series-connection penalty ranges from 2.5% to 5% of independent energy production for ensembles with two cells. The penalty increases to range from 14% to 22% for the 20-cell ensembles, which suggests that not all locations experience an equal amount of spectral deviation over the course of the year. The unequal series-connection penalty among different locations raises an important point. First, it is

current practice to target concentrating photovoltaic (CPV) installations with series-connected subcells in locations with a high percentage of direct irradiance, which helps ensure most incident spectra resemble the AM1.5D standard and cell performance is close to specification, as indicated in Figure 3. However, as CPV expands into more marginal locations with higher latitude or more atmospheric absorption and scattering, systems with series-connected cells will underperform their specification, with this effect worsening for ensembles with larger numbers of subcells. By contrast, CPV systems with electrically independent subcells will produce power that is a more consistent percentage of the total direct irradiance.

Energy production with cloud correction

Figure 5 shows the expected annual energy production for independent (panel A) and series-connected (panel B) ensembles exposed to cloud-adjusted direct normal irradiance at 500 suns concentration. While the baseline energy production levels for the different cities is significantly different from the levels shown in Figure 4, the relative performance of series and independent ensembles at a given location are preserved. All locations project increasing energy production with increasing number of subcells for independently connected ensembles, and all locations show annual energy production peaking with 8–12 subcells for the series-connected ensembles. In addition, the series-connected energy production as a percentage of independent energy production is not substantially different from the trends shown in Figure 4(C). The consistency of this behavior with and without cloud correction and across multiple locations with differing climates suggests that the energy production advantage of independent ensembles is not tied to local climate effects other than the degree of spectral variability. While the behavior does neglect the impact of clouds on the direct spectrum, the efficiency trends shown in Figure 2(A) and (B) suggest that accounting for spectral transmission through screening clouds will increase the energy production advantage of the independent ensembles, making the projections in Figure 5 conservative.

Conclusions

While the efficiency of spectrum splitting ensembles with electrically independent subcells is only slightly higher than the efficiency of ensembles with the same number of subcells connected in series under the AM1.5D design spectrum, the energy production potential of the two electrical configurations differs widely under varying spectral conditions. The energy production advantage of ensembles with electrically independent cells results from

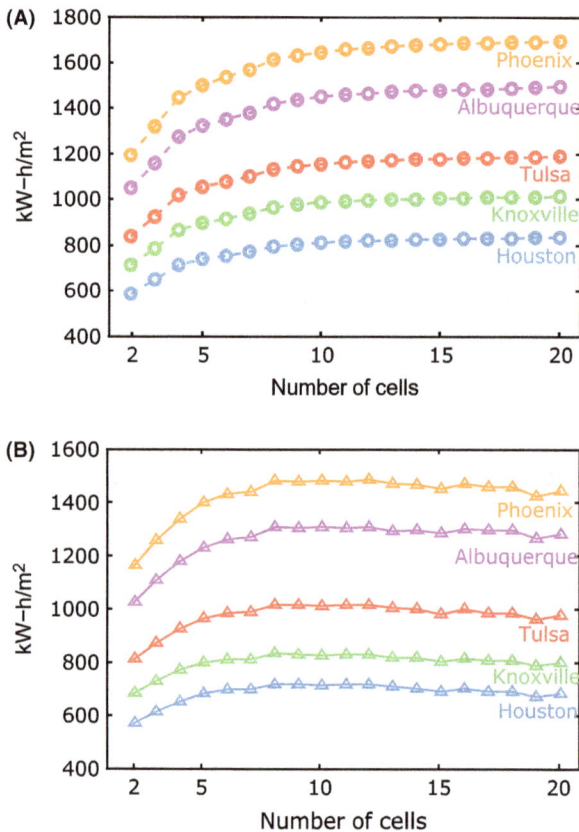

Figure 5. (A) Simulated annual energy production of ideal spectrum splitting ensembles with electrically independent subcells under 500 suns. Simulated spectra have intensity uniformly decreased by the average percent cloud coverage for each hour. (B) Simulated annual energy production of ideal spectrum splitting ensembles with series-connected subcells under cloud-adjusted spectra.

the consistency of the efficiency of these ensembles under spectra that differ from the design spectrum. By contrast, the series-connected ensembles decrease in efficiency rapidly as the incident spectrum deviates from the AM1.5D irradiance level, and the efficiency decreases faster with increasing subcell number.

The energy production advantage of independent ensembles can be roughly broken into two components. A total of 60% of the additional energy comes from better performance under high illumination conditions and will appear as additional peak power production. The other 40% comes from energy production at low irradiance conditions when the series-connected ensembles produce negligible power. While the power production of the independent ensembles under these conditions is low, it does constitute an extended capacity factor that adds up over the course of the year and may have considerable economic value.

While the energy production advantage of independent ensembles is consistent for a wide variety of locations

both with and without accounting for the effects of clouds reducing direct transmission, the relative advantage throughout the year for a particular location will be sensitive to local climate trends. This may result in the series-connected ensembles producing less power than expected at different times of the year, as in the case of Phoenix, AZ. Combining the local power production trends with energy pricing trends for specific installation locations can give a clearer evaluation of the relative economic benefit of independent electrical connection for spectrum splitting ensembles.

Finally, as the cost of concentrating photovoltaic tracking and optical systems decreases, the range of territory where these types of installations can produce cost-competitive power will increase. For example, if a generation plant with series-connected subcells can produce cost-effective power in Albuquerque, NM, Figure 5 suggests that a similar plant with independent subcells will be economically feasible in Tulsa, OK, assuming comparable costs for the system components. The independent electrical connection's spectral insensitivity and improved performance have the potential to increase the scope of the market for CPV into areas previously considered uneconomical.

The energy production advantage of electrically independent spectrum splitting ensembles relative to series-connected ensembles is significant and consistent for all numbers of band gaps and all locations. This performance advantage is a powerful argument for pursuing new designs with electrically independent subcells. Historically, the small efficiency advantage of electrically independent subcells was not considered to justify the additional technical complexity required to achieve those connections or optically distribute the spectrum among the subcells. The comparative simplicity of monolithic series-connected designs has made them more practical for systems with small numbers of subcells. However, as efforts turn to larger numbers of subcells, this analysis shows that the energy production advantage of independently connected cell designs will increase and potentially justify the pursuit of more complex systems.

Conflict of Interest

None declared.

References

1. Dimroth, F., M. Grave, P. Beutel, U. Fiedeler, C. Karcher, T. N. D. Tibbits, et al. 2014. Wafer bonded four-junction GaInP/GaAs//GaInAsP/GaInAs concentrator solar cells with 44. 7% efficiency. Prog. Photovoltaics 22:277–282.

2. Marti, A., and G. L. Araujo. 1996. Limiting efficiencies for photovoltaic energy conversion in multigap systems. Sol. Energy Mater. Sol. Cells 43:203–222.

3. King, R. R., D. C. Law, C. M. Fetzer, R. A. Sherif, K. M. Edmondson, S. Kurtz, et al. 2005. Pathways to 40% -efficient concentrator photovoltaics. *20th European Photovoltaic Solar Energy Conference and Exhibition*, June, 6–10.

4. Kurtz, S. R., J. M. Olson, and P. Faine. 1991. The difference between standard and average efficiencies of multijunction compared with single-junction concentrator cells. Sol. Cells 30:501–513.

5. Philipps, S. P., G. Peharz, R. Hoheisel, T. Hornung, N. M. Al-Abbadi, F. Dimroth, et al. 2010. Energy harvesting efficiency of III-V multi-junction concentrator solar cells under realistic spectral conditions. AIP Conf. Proc. 1277:294–298.

6. Hoang, P., V. Bourdin, Q. Liu, G. Caruso, and V. Archambault. 2014. Coupling optical and thermal models to accurately predict PV panel electricity production. Sol. Energy Mater. Sol. Cells 125:325–338.

7. Chan, N. L. A., H. E. Brindley, and N. J. Ekins-Daukes. 2014. Impact of individual atmospheric parameters on CPV system power, energy yield and cost of energy. Prog. Photovoltaics Res. Appl. 22:1080–1095.

8. Leloux, J., E. Lorenzo, B. García-Domingo, J. Aguilera, and C. A. Gueymard. 2014. A bankable method of assessing the performance of a CPV plant. Appl. Energy 118:1–11.

9. Chiu, P., S. Wojtczuk, X. Zhang, C. Harris, D. Pulver, and M. Timmons. 2011. 42.3% Efficient InGaP/GaAs/InGaAs concentrators using bifacial epigrowth. *2011 37th IEEE Photovolt. Spec. Conf.* 000771–000774.

10. King, R. R., D. Bhusari, D. Larrabee, X. Liu, E. Rehder, K. Edmondson, et al. 2012. Solar cell generations over 40% efficiency. Prog. Photovoltaics:801–815.

11. Barnett, A., D. Kirkpatrick, C. Honsberg, D. Moore, M. Wanlass, K. Emery, et al. 2009. Very high efficiency solar cell modules. Prog. Photovoltaics Res. Appl. 17:75–83.

12. Imenes, A. G., and D. R. Mills. 2004. Spectral beam splitting technology for increased conversion efficiency in solar concentrating systems: a review. Sol. Energy Mater. Sol. Cells 84:19–69.

13. Mitchell, B., G. Peharz, G. Siefer, M. Peters, T. Gandy, J. C. Goldschmidt, et al. 2011. Four-junction spectral beam-splitting photovoltaic receiver with high optical efficiency. Prog. Photovoltaics 19:61–72.

14. Kurtz, S., D. Myers, W. E. McMahon, J. Geisz, and M. Steiner. 2008. A comparison of theoretical efficiencies of multi-junction concentrator solar cells. Prog. Photovoltaics 16:537–546.

15. Tobias, I., and A. Luque. 2002. Ideal efficiency of monolithic, series-connected multijunction solar cells. Prog. Photovoltaics 10:323–329.

16. Faine, P., S. R. Kurtz, C. Riordan, and J. M. Olson. 1991. The influence of spectral solar irradiance variations on the performance of selected single-junction and multijunction solar cells. Sol. Cells 31:259–278.

17. Kinsey, G. S., A. Nayak, M. Liu, V. Garboushian, and S. Beach. 2011. Increasing power and energy in amonix solar power plants. IEEE J. Photovolt. 1:3–4.

18. Mols, Y., L. Zhao, G. Flamand, M. Meuris, and J. Poortmans. 2012. Annual energy yield: A comparison between various monolithic and mechanically stacked multijunction solar cells BT - 38th IEEE Photovoltaic Specialists Conference, PVSC 2012, June 3, 2012 – June 8, 2012. 2092–2095.

19. Dimroth, F., S. P. Philipps, G. Peharz, E. Welser, R. Kellenbenz, T. Roesener, et al. 2010. Promises of advanced multi-junction solar cells for the use in CPV systems. 1231–1236.

20. Philipps, S. P., G. Peharz, R. Hoheisel, T. Hornung, N. M. Al-Abbadi, F. Dimroth, et al. 2010. Energy harvesting efficiency of III-V multi-junction concentrator solar cells under realistic spectral conditions. Sol. Energy Mater. Sol Cells 94:877–897.

21. Wang, X., A. Barnett, and L. Fellow. 2012. The effect of spectrum variation on the energy production of triple-junction solar cells. IEEE J. Photovoltaics 2:417–423.

22. Wild, M., H. Gilgen, A. Roesch, A. Ohmura, C. N. Long, E. G. Dutton, et al. 2005. From dimming to brightening: decadal changes in solar radiation at Earth's surface. Science 308:847–850.

23. Foukal, P., C. Fröhlich, H. Spruit, and T. M. L. Wigley. 2006. Variations in solar luminosity and their effect on the Earth's climate. Nature 443:161–166.

24. Myers, D. R., and K. E. Emery. 2002. Terrestrial solar spectral modeling tools and applications for photovoltaic devices preprint. *29th IEEE PV Specialists Conference Conference.*

25. Dubovik, O., and M. D. King. 2000. A flexible inversion algorithm for retrieval of aerosol optical properties from Sun and sky radiance measurements. J. Geophys. Res. 105:20673.

26. Tossa, A. K., Y. M. Soro, Y. Azoumah, and D. Yamegueu. 2014. A new approach to estimate the performance and energy productivity of photovoltaic modules in real operating conditions. Sol. Energy 110:543–560.

27. Warmann, E. C., C. Eisler, E. Kosten, M. Escarra, and H. A. Atwater. 2013. Spectrum splitting photovoltaics: Materials and device parameters to achieve ultrahigh system efficiency. *2013 IEEE 39th Photovolt. Spec. Conf.* 1922–1925.

28. Henry, C. H. 1980. Limiting efficiencies of ideal single and multiple energy gap terrestrial solar cells. J. Appl. Phys. 51:4494.

29. 1981-2010 U.S. Climate Normals | National Climatic Data Center (NCDC). [Online]. Available at http://www.ncdc.noaa.gov/data-access/land-based-station-data/land-based-datasets/climate-normals/1981-2010-normals-data (accessed 10 March 2015).

30. Buis, J. P., A. Setzer, B. N. Holben, T. F. Eck, I. Slutsker, D. Tanre, et al. 1998. AERONET — a federated instrument network and data archive for aerosol characterization. Remote Sensing of Environment 66:1–16.

31. Gueymard, C. A., D. Myers, and K. Emery. 2002. Proposed reference irradiance spectra for solar energy systems testing. Sol. Energy 73:443–467.

32. Myers, D. R., and C. A. Gueymard. 2004. Description and availability of the SMARTS spectral model for photovoltaic applications Proc. SPIE 5520, Organic Photovoltaics V, 56.

33. Kinsey, G. S. 2015. Spectrum sensitivity, energy yield, and revenue prediction of PV modules. IEEE J. Photovoltaics 5:258–262.

34. NSRDB: 1991–2010 update. [Online]. Available at http://rredc.nrel.gov/solar/old_data/nsrdb/1991-2010/. (accessed 10 March 2015).

13

Design of photovoltaics for modules with 50% efficiency

Emily C. Warmann[1], Cristofer Flowers[1], John Lloyd[1], Carissa N. Eisler[2], Matthew D. Escarra[3] & Harry A. Atwater[1]

[1]California Institute of Technology, Pasadena, California, USA
[2]E O Lawrence Berkeley National Laboratory, Berkeley, California, USA
[3]Tulane University, New Orleans, Louisiana, USA

Keywords
Efficiency, photovoltaic design, photovoltaic modules, spectrum splitting

Correspondence
Emily C. Warmann, California Institute of Technology, 1200 E. California Blvd Pasadena, CA, 91125, USA.
E-mail: warmann@caltech.edu

Funding Information
Dow Chemical Company (Grant/Award Number: 'Full spectrum photovoltaics'); National Science Foundation (Grant/Award Number: 'EEC-1041895'); Advanced Research Projects Agency – Energy (Grant/Award Number: 'DE-SC0001293').

Abstract

We describe a spectrum splitting solar module design approach using ensembles of 2–20 subcells with bandgaps optimized for the AM1.5D spectrum. Device physics calculations and experimental data determine radiative efficiency parameters for III-V compound semiconductor subcells and enable modification of conventional detailed balance calculations to predict module efficiency while retaining computational speed for a wide search of the design space. Accounting for nonideal absorption and recombination rates due to realistic material imperfections allows us to identify the minimum subcell quantity, quality, electrical connection configuration, and concentration required for 50% module efficiency with realistic optical losses and modeled contact resistance losses. We predict a module efficiency of 50% or greater will be possible with 7–10 electrically independent subcells in a spectral splitting optic at 300–500 suns concentration, assuming a 90% optical efficiency and 98% electrical efficiency, provided the subcells can achieve an average external radiative efficiency of 3–5% and a short circuit current that is at least 90% of the ideal. In examining spectrum splitting solar cells with both series-connected and electrically independent subcells, we identify a new design trade-off independent of the challenges of fabricating optimal bandgap combinations. Series-connected ensembles, having a single set of electrical contacts, are less sensitive to lumped series resistance losses than ensembles where each subcells are contacted independently. By contrast, ensembles with electrically independent subcells can achieve lower radiative losses when the subcells are designed for good optical confinement. Distributing electrically independent subcells in a concentrating receiver module allows flexibility in subcell selection and fabrication, and can achieve ultra-high efficiency with conventional III-V cell technology.

Introduction

A wide variety of photovoltaic cell technologies have shown dramatic performance improvements over the past decade, yet the prospect of a practical module capable of 50% efficiency remains remote. Experimentally achieved single-cell devices have achieved a record efficiency of 28.8% [1], which is close to the theoretical limit of 33.8% for such devices [2]. However, the single-cell limit is far below the fundamental efficiency limit for solar energy conversion of 74.0% for global illumination and 92.8% for direct [2] because a single pn junction can only efficiently convert photons with energy close to the value of its energy bandgap. The best single junction cell will lose more than 40% of the energy in the incident light to transmission of subbandgap photons and thermalization of carriers with photon energy in excess of the bandgap [3]. Spectrum splitting, which divides the solar spectrum into spectral bands of different energy and directs the bands onto multiple subcells with bandgap values matched to the energy of their photon allocation, is a necessary feature of any photovoltaic design capable of achieving >33.8% efficiency. The use of multiple subcells to increase conversion efficiency is well known. In these designs, the subcells are grown monolithically in a stacked configuration and are electrically in series. The incident spectrum is divided among the subcells by sequential absorption, with the top subcells absorbing and converting high energy photons while transmitting lower energy photons to the subcells below. State-of-the-art high-efficiency solar cells using this monolithic multijunction stack technology have achieved efficiency as high as 46% [4–8].

While the 46% record efficiency for monolithic multi-junction solar cells (MJSCs) appears close to 50%, the performance of these cells is measured under high-intensity illumination that simulates an ideal optical concentration system. Once practical high-efficiency devices are installed in the necessary concentrating optics, the overall module efficiency drops, with the record module achieving only 38.9% [9, 10] with commercial optics and 43.3% with ultra-high-efficiency optics in a minimodule [11, 12]. This suggests that an integrated module efficiency of 50% or greater will not be achieved by a commercially practical module through continued incremental improvements to the monolithic MJSC architecture, such as material quality improvements or adjustment of the subcell bandgaps. Prospective designs incorporating 4, 5, and 6 subcells into the monolithic MJSC stack are projected to raise the photovoltaic cell efficiency as high as 50.91% under ideal concentrating optics [13], which will result in a module efficiency well below 50% once integrated with realistic optics and electronics.

An alternative approach to the monolithic MJSC uses a separate optical element to split the incident light among subcells that are electrically independent from one another and may be physically isolated [14–17]. While the addition of the spectrum splitting optic increases the complexity of these designs, removing the requirement for monolithi-cally integrated subcells confers benefits that may be overall advantageous. First, the subcells no longer need to be grown on the same substrate, which allows a wider variety of materials and bandgaps to be combined. Secondly, the subcells can be optimized independently to maximize electrical performance. Thirdly, isolated subcells that are not connected in series electrically are not constrained to collect equal numbers of photons for current matching, and consequently can better match the target spectrum. To date, all designs using separate spectrum splitting optics have used six or fewer subcells. The highest demonstrated efficiency is 38.5% for a minimodule with two physically separated dual-junction subcells [18]. A hybrid approach using a monolithic triple junction and an electrically in-dependent, isolated fourth subcell in a submodule achieved 40.4% efficiency [19].

Both types of spectrum splitting modules lack an obvi-ous path of incremental improvements that could enable them to achieve 50% efficiency. If photovoltaics are to reach that efficiency target, they will require designs that are much more ambitious.

Here, we present a systematic investigation of the design requirements for photovoltaic modules capable of achieving

50% efficiency with multijunction architecture based on established single-junction cell technology. We consider spectrum splitting module designs that include many more bandgap combinations than previously demonstrated de-signs and we specifically analyze the effect in series electrical connection as well as concentration. By incorporating performance parameters representing realistic material per-formance for the subcells, and by accounting for a realistic amount of loss in the optical and electrical systems required by a spectrum splitting module, we identify the number of subcells, degree of concentration, cell quality, and elec-trical configuration required to achieve 50% module ef-ficiency for designs with up to 20 subcells.

The efficiencies predicted by this design approach rep-resent a large improvement over current state-of-the-art designs, but do not require disruptive innovation in cell technology. Recent advances in cell material growth, epi-taxial liftoff and transfer of cells, and pick-and-place au-tomation all set the stage for this design approach that features the use of many high-quality independent subcells [20–22]. Combined with optical systems capable of split-ting and concentrating the solar spectrum into desirable subbands with high efficiency [23], a module incorporating many subcells has realistic potential to achieve the ultra-high-efficiency target of 50%.

Methods

Exploring the design space for multijunction photovoltaics requires a mechanism for predicting the realistic perfor-mance of a large number of designs. The detailed balance calculation is the standard tool for evaluating potential solar cell designs because it presents the ideal limiting efficiency for the combination of energy bandgaps and concentration level [24]. The calculation determines an ideal J-V relationship for a cell by balancing the number of absorbed photons with the combined number of col-lected carriers and radiatively emitted photons. Radiative emission is assumed to consist of the temperature-dependent black body spectrum of the cell, plus lumi-nescence caused by recombination of excited carriers, which depends on the degree of quasi-Fermi-level splitting present in the cell. Conventional detailed balance calcula-tions assume no carriers are lost to nonradiative recom-bination. In addition, the calculation typically assumes perfect absorption of photons with energy greater than the cell energy bandgap. The detailed balance calculation allows rapid assessment of a potential cell design, enabling a wide search of the design space.

To explore the benefit of designs with multiple subcells, we start by calculating the conventional detailed balance efficiency of ensembles with 2–20 subcells, both electrically in-series and independent. The set of bandgaps in each ensemble was optimized. In the case of the series-connected cells, this optimization is constrained by the current-matching requirement and the technique is straightforward, see ref [25]. The systems with electrically independent cells are not constrained by the need for each cell to absorb an equal number of photons, and consequently the design space for these ensembles grows exponentially with the number of subcells. For the electrically independent ensembles, a computational technique termed "simulated annealing" identified the optimal sets of bandgaps for each number of subcells [26]. The simulated annealing algorithm used consists of two stages of optimization: first, a randomly seeded search allows all subcell bandgaps to vary over a wide range; secondly, a subsequent search is seeded with the best candidate identified in previous search and allows one subcell bandgap at a time to vary over a narrow range. This process was repeated multiple times for each ensemble in search of repeated optimum values, which suggest a global optimum.

The conventional detailed balance calculation informs us of the limiting efficiency of these spectrum splitting designs, however, the practical efficiency of any device falls far short of this limiting efficiency value [27]. At present computationally intensive device physics and optical simulations are required for a realistic performance estimate, and these techniques require a fully specified design for every device under consideration. Because data on the optical and electrical properties of materials at every possible energy bandgap value are not documented, and because the computational costs and design effort required for each device physics calculation are large, device physics techniques are not practical for a comprehensive search of the photovoltaic system design space. To address the shortcomings of conventional detailed balance calculation while retaining the speed of calculation, we introduce three parameters, the external radiative efficiency (ERE), absorption efficiency, and a lumped series resistance as key parameters for modified detailed balance calculations, which together capture a large portion of the nonideal behavior. We use this modified detailed balance approach to identify systems whose designed efficiency exceeds 50% in the presence of realistic material performance and optical and electrical system losses.

It is straightforward to incorporate these ERE and absorption efficiency parameters into the detailed balance calculation. The modified equation balances the incident flux as modified by the absorption efficiency (abs) with the collected carriers and the radiative and nonradiative recombination. Radiative emission from the cell is assumed to follow the same modified Kirchhoff relation as in the unmodified calculation, but that quantity is now assumed to be a percentage – equal to the ERE – of the total recombination. With these assumptions we calculate a

new J-V relation for each cell and determine its fill factor and maximum power point under illumination by its particular spectral slice.

$$J(V) = q\left[\text{abs} \int_{E_G}^{\infty} \text{AM1.5}(E)\,\mathrm{d}E - \frac{1}{\text{ERE}} \frac{n^2 \sin^2 \theta_c}{4\pi^2 \hbar^3 c^2} \int_{E_G}^{\infty} \frac{E^2}{e^{(E-qV)/k_B T} - 1}\,\mathrm{d}E \right].$$
(1)

Note that we assume unity absorptivity for photons with energy at or above the cell bandgap value, and zero absorptivity for lower energy photons. This allows us to disregard the cell thickness in these calculations and allows us to treat the absorption efficiency and ERE as independent parameters. The J-V curve produced by the modified detailed balance calculation captures the performance of the cell with perfect current collection and will therefore overestimate the fill factor of the cell. Incorporating a lumped series resistance captures the loss in voltage due to contact and other resistances. With these three parameters, the ERE, absorption efficiency, and series resistance (R_s), the modified detailed balance approach can accurately reproduce the performance of realistic cells while retaining computational speed.

To determine realistic values for the absorption and ERE parameters to use in the detailed balance calculations, we selected seven materials that span a large range of bandgap values and performed one-dimensional device physics simulations on candidate designs of cells made from these materials using AFORS-HET [28, 29]. We incorporated realistic doping-dependent mobility and lifetime data for the candidate materials [30–33] to optimize device designs for each cell, including doping and thicknesses for emitter, base, front and back window layers, and contact layers. A schematic of the cell design showing the layers included for each material is shown in Figure 1. The design did not account for top contact shadowing losses and the top contact layer was treated as optically transparent. Table 1 summarizes the layer thicknesses, compositions, and doping levels for the Afors-Het simulations. In all cases, the cells were optimized both for performance on the growth wafer and for performance after epitaxial liftoff (ELO) or substrate removal, which allows the use of back reflectors to improve light trapping and allow thinner absorption layers. The epitaxial liftoff cells were modeled assuming perfect back reflectors. Because ERE and absorption efficiency are effected by the interaction of cell design, material quality, and optical environment, the AFORS-HET simulations serve to bound realistic values of ERE and absorption efficiency for subsequent modified detailed balance optimizations and to illustrate the possible degree of variation in these parameter values between subcells.

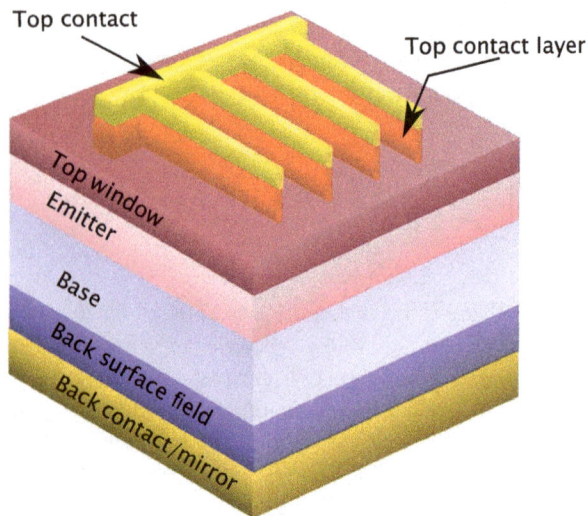

Figure 1. Schematic of basic solar cell design optimized through 1D device simulations using AFORS-Het. Layer thickness, doping, and device polarity were optimized based on material parameters from literature.

The performance results for these simulations are summarized in Table 2. In all cases the simulated power production of each cell was substantially lower than the ideal detailed balance power, as expected. However, the lowest bandgap cells exhibited the largest departure from the ideal efficiency. This trend suggested the need to reoptimize the bandgaps for the ensembles to account for realistic material behavior. Table 2 also includes the ERE values extracted from the simulated open circuit voltage and short circuit current for each cell material [34] and the absorption efficiency (abs), which is the short circuit current as a percentage of the value predicted by ideal detailed balance. The values for the InGaP and GaAs epitaxial liftoff cells are taken from record cell performance reported in the literature [21] and [34], respectively. Comparing the ERE values for the simulated epitaxial liftoff cells to the values for on-wafer cells shows the tremendous value of epitaxial liftoff for improving cell performance. These ELO cells benefit from improved optical confinement with the back reflector, which prevents photons from escaping the cell into the growth substrate, and from thinner absorber layers, which reduces bulk nonradiative recombination. Table 2 shows that epitaxial liftoff is an essential component of high external radiative efficiency for the III–V cells considered for spectrum splitting PV.

The ERE values for the ELO cells in Table 2 vary over a wide range among the cells. Because the cell open circuit voltage decreases linearly with the natural logarithm of ERE [35], a direct average of ERE values will overstate the expected average performance of these cells. Instead,

taking the geometric mean of the ERE values incorporates their logarithmic scaling and results in an average ERE of 4.7%. The performance of an ensemble of cells with these bandgaps and that average ERE value will have the same total efficiency as the ensemble with the cell-specific ERE values. While this value is higher than the simulated value for most of the cells, the highest ERE values included in the average come from experimentally realized cells, suggesting that optimization of growth and other device parameters can produce III–V devices with very high radiative efficiency. The arithmetic mean of the absorption efficiency values is 90.6%. For simulation purposes, the value 90% was chosen as a conservative yet realistic derating value. Table 2 also shows the experimentally verified ERE values for cells on their growth substrates. These values differ from the simulated ERE values, with five of the seven cells having lower ERE compared to the simulation. The geometric mean ERE for the on-wafer experimental cells is 0.22%, whereas the geometric mean for the simulated cells is 0.97%, with the highest bandgap cell contributing most of the difference. The gap between simulated and experimental cells suggests a range of expected realistic performance after additional work on cell development.

With realistic values for the ERE and absorption parameters determined, we reoptimized the bandgaps for the spectrum splitting ensembles with both series-connected and electrically independent subcells using the modified detailed balance calculation to determine whether the nonideal material behavior would change the desired bandgap combinations. We used the appropriate mean values for the ERE and absorption because the optimization allowed bandgap values to range freely between 0.3 and 4 eV, and because the effects of the ERE and absorption on cell performance are independent. Finally, we varied the ERE value and concentration for the system to map out the full design space for ultra-high-efficiency spectrum splitting photovoltaics.

As with the ERE and absorption parameters, the value for the lumped series resistance parameter, R_s, was determined from the geometric mean value of lumped resistance from a set of electrical simulations. For each of the bandgaps in Table 2, a distributed circuit model was built with dark current and short circuit current densities determined by the corresponding ERE and absorption parameters from Table 2 [36]. The circuit model was used to optimize an inverted square contact grid for 1 mm^2 square cells at concentrations ranging from 1 sun to 1000 suns. Because the loss due to series resistance is proportional to the square of the current, higher concentrations require much lower series resistances for optimal performance. The lumped R_s for each concentration for the seven cells is plotted in Figure 2.

Table 1. Layer thickness, composition, and doping level for all devices simulated in Afors-Het. Subcells with the "ELO" designation are optimized for operation on their growth substrate, whereas cells with the "wafer" designation are assumed to be removed from their growth substrate and placed on a reflective back surface.

Subcell	Bottom contact			Bottom window			Base		
	Composition	t (nm)	N	Composition	t (nm)	N	Composition	t (nm)	N
InGaAs ELO	n++ InGaAs	300	1.0E+19	n+ InP.	15	1.0E+17	n InGaAs	2000	1.0E+17
InGaAs wafer	n++ InGaAs	300	1.0E+19	n+ InP	15	1.0E+17	n InGaAs	4000	1.0E+17
InGaAsP 1 ELO	p++ InP	200	1.0E+19	p+ InP	100	3.0E+18	$p\ In_{0.71}Ga_{0.29}As_{0.62}P_{0.38}$	3000	5.0E+17
InGaAsP 1 wafer	p++ InP	200	1.0E+19	p+ InP	100	3.0E+18	$p\ In_{0.71}Ga_{0.29}As_{0.62}P_{0.38}$	5000	5.0E+17
InGaAsP 2 ELO	p++ InP	300	1.0E+19	p+ InAlAs	10	1.0E+18	$p\ In_{0.87}Ga_{0.13}As_{0.28}P_{0.72}$	3000	5.0E+17
InGaAsP 2 wafer	p++ InP	300	1.0E+19	p+ InAlAs	10	1.0E+18	$p\ In_{0.87}Ga_{0.13}As_{0.28}P_{0.72}$	5000	5.0E+17
GaAs ELO	n++ GaAs	300	5.0E+18	$n+\ Ga_{0.73}In_{0.63}P$	30	3.0E+18	n GaAs	2000	2.0E+17
GaAs wafer	n++ GaAs	300	5.0E+18	$n+\ Ga_{0.73}In_{0.63}P$	30	3.0E+18	n GaAs	4000	2.0E+17
AlGaAs ELO	n++ GaAs	300	5.0E+18	$n+\ Ga_{0.37}In_{0.63}P$	30	3.0E+18	$n\ Al_{0.1}Ga_{0.9}As$	1000	2.0E+17
AlGaAs wafer	n++ GaAs	300	5.0E+18	$n+\ Ga_{0.37}In_{0.63}P$	30	3.0E+18	$n\ Al_{0.1}Ga_{0.9}As$	2000	2.0E+17
InGaP ELO	p++ GaAs	250	1.0E+19	$p+\ Al_{0.2}Ga_{0.32}In_{0.48}P$	30	2.0E+17	$p\ Ga_{0.37}In_{0.63}P$	800	5.0E+17
InGaP wafer	$p+\ Ga_{0.37}In_{0.63}P$ p++ GaAs $p+\ Ga_{0.37}In_{0.63}P$	10 250 10	5.0E+18 1.0E+19 5.0E+18	$p+\ Al_{0.2}Ga_{0.32}In_{0.48}P$	30	2.0E+17	$p\ Ga_{0.37}In_{0.63}P$	1400	5.0E+17
AlGaAsP ELO	$p++\ Ga_{0.52}In_{0.48}P$ p++ GaAs	10 300	5.0E+18 1.0E+19	$p+\ Al_{0.5}In_{0.5}P$	20	2.0E+17	$p\ Al_{0.2}Ga_{0.32}In_{0.48}P$	700	5.0E+17
AlGaAsP wafer	$p++\ Ga_{0.52}In_{0.48}P$ p++ GaAs	10 300	5.0E+18 1.0E+19	$p+\ Al_{0.5}In_{0.5}P$	20	2.0E+17	$p\ Al_{0.2}Ga_{0.32}In_{0.48}P$	1200	5.0E+17

Subcell	Emitter			Top window			Top contact		
	Composition	t (nm)	N	Composition	t (nm)	N	Composition	t (nm)	N
InGaAs ELO	p+ InGaAs	350	1.0E+18	p++ InP	15	1.0E+19	p++ InGaAs	300	1.0E+19
InGaAs wafer	p+ InGaAs	350	1.0E+18	p++ InP	15	1.0E+19	p++ InGaAs	300	1.0E+19
InGaAsP 1 ELO	$n+\ In_{0.71}Ga_{0.29}As_{0.62}P_{0.38}$	270	8.0E+17	n+ InP	200	5.0E+18	n++ InGaAs	300	1.0E+19
InGaAsP 1 wafer	$n+\ In_{0.71}Ga_{0.29}As_{0.62}P_{0.38}$	270	8.0E+17	n+ InP	200	5.0E+18	n++ InGaAs	300	1.0E+19
InGaAsP 2 ELO	$n+\ In_{0.87}Ga_{0.13}As_{0.28}P_{0.72}$	200	1.0E+18	n+ InAlAs	10	5.0E+18	n++ InGaAs	300	1.0E+19
InGaAsP 2 wafer	$n+\ In_{0.87}Ga_{0.13}As_{0.28}P_{0.72}$	200	1.0E+18	n+ InAlAs	10	5.0E+18	n++ InGaAs	300	1.0E+19
GaAs ELO	$p+\ A\ Al_{0.2}Ga_{0.8}As$	250	2.0E+18	$p+\ Ga_{0.37}In_{0.63}P$	50	8.0E+18	p++ GaAs	250	1.0E+19
GaAs wafer	$p+\ Al_{0.2}Ga_{0.8}As$	250	2.0E+18	$p+\ Ga_{0.37}In_{0.63}P$	50	8.0E+18	p++ GaAs	250	1.0E+19
AlGaAs ELO	$p+\ Al_{0.1}Ga_{0.9}As$	300	2.0E+18	$p+\ Ga_{0.37}In_{0.63}P$	50	8.0E+18	p++ GaAs	250	1.0E+19
AlGaAs wafer	$p+\ Al_{0.1}Ga_{0.9}As$	300	2.0E+18	$p+\ Ga_{0.37}In_{0.63}P$	50	8.0E+18	p++ GaAs	250	1.0E+19
InGaP ELO	$n+\ Ga_{0.37}In_{0.63}P$	60	3.0E+18	$n+\ Al_{0.2}Ga_{0.32}In_{0.48}P$	10	5.0E+18	n++ GaAs	300	5.0E+18
InGaP wafer	$n+\ Ga_{0.37}In_{0.63}P$	60	3.0E+18	$n+\ Al_{0.2}Ga_{0.32}In_{0.48}P$	10	5.0E+18	n++ GaAs	300	5.0E+18
AlGaAsP ELO	$n+\ Al_{0.2}Ga_{0.32}In_{0.48}P$	50	5.0E+18	$n+\ Al_{0.5}In_{0.5}P$	10	5.0E+18	GaAs n	300	5.0E+18
AlGaAsP wafer	$n+\ Al_{0.2}Ga_{0.32}In_{0.48}P$	50	5.0E+18	$n+\ Al_{0.5}In_{0.5}P$	10	5.0E+18	GaAs n	300	5.0E+18

Table 2. Simulation results for candidate subcells for ideal detailed balance and 1D device physics models. The cell designs for the device physics simulations are detailed in Table 1. The ELO (epitaxial liftoff) cells assume a perfect back reflector. Experimental cell ERE values are for cells that remain on their growth substrates. Note that the performance parameters and ERE and absorption values for the GaAs and InGaP ELO cells are taken from record cells reported in the literature.

Cell	E_g	Detailed balance prediction			Device physics simulation			Extracted parameters		Experiment
		Jsc (mA/cm²)	Voc (mV)	FF (%)	Jsc (mA/cm²)	Voc (mV)	FF (%)	Absorption	ERE	ERE
InGaAs ELO	0.74	6.68	399	77.2	6.24	417	76.0	0.938	0.147	
InGaAs wafer	0.74				5.97	366	75.1	0.897	0.021	0.0095
InGaAsP 1 ELO	0.94	8.57	595	82.7	8.56	566	80.4	0.993	0.024	
InGaAsP 1 wafer	0.94				8.41	540	79.2	0.975	0.009	0.017
InGaAsP 2 ELO	1.15	9.75	798	86.1	9.36	769	84.6	0.958	0.025	
InGaAsP 2 wafer	1.15				9.17	741	83.2	0.940	0.008	0.004
GaAs ELO	1.42	32	1154	89.5	29.43	1107	87.6	0.920	0.225	
GaAs wafer	1.42				11.48	1034	87.4	0.923	0.026	0.018
AlGaAs ELO	1.58	7	1449	91.3	5.82	1220	86.9	0.925	0.123	
AlGaAs wafer	1.58				5.36	1199	86.9	0.898	0.082	0.00175
InGaP ELO	1.84	19.7	1506	91.4	16.00	1458	88.8	0.816	0.080	
InGaP wafer	1.84				5.80	1345	88.6	0.975	0.001	0.0065
InAlGaP ELO	2.13	6.13	1733	92.4	5.81	1642	91.0	0.629	0.0022	
InAlGaP wafer	2.13				5.81	1639	91.0	0.629	0.0019	2.1E-6

Once the lumped series resistance, the ERE, and the absorption efficiency are incorporated into the modified detailed balance calculation it can reproduce the short circuit current and open circuit voltage of the cells modeled by the distributed circuit model to within 1% (relative) and the modified detailed balance can reproduce the fill factor to within 2.5%. Overall the efficiency of the modified detailed balance model of each cell captures the efficiency predicted by the lumped circuit model to within 2%. While this does constitute a loss of accuracy relative to the more physically detailed lumped circuit model, the computational speed of the modified detailed balance approach makes it more practical for searching the large design space.

The series resistance values plotted in Figure 2 are optimized for the specific cells and performance parameters in Table 2. An average lumped series resistance, analog to the average ERE and absorption values is extracted from fitting the geometric mean versus concentration for all data points in Figure 2. The dependence of the average Rs on concentration, C, is described by a power law:

$$R_s(C) = 454C^{0.9352}. \tag{2}$$

The individual lumped series resistances plotted in Figure 2 vary over a wide range for any particular concentration and the performance of individual cells modeled using the average R_s value will differ from the performance using the optimized R_s value by as much as 12%. However, the total performance of the entire ensembles of cells modeled using the average R_s value is within 2% of the ensemble performance modeled with the optimized R_s values for each cell, which validates the utility of the average for modeling ensemble behavior.

Results

Figure 3 shows the conventional detailed balance efficiencies of optimized spectral splitting ensembles with 2–20 subcells at different concentration values. The plot shows the increase in efficiency with increasing number of subcells is primarily due to the increase in spectral efficiency, which we define as

$$S.E. = \frac{\sum_{E_{gi}} \int_{E_{gi}}^{E_{gi+1}} E_{gi} \frac{dn}{dE} dE}{\int_0^\infty \mathrm{AM1.5D}(E)\, dE}, \tag{3}$$

where dn/dE is the spectral flux density of the AM1.5D standard spectrum in photons/cm²-s-eV and E_{gi} is the bandgap value of the ith subcell in the ensemble. The spectral efficiency is percentage of energy in the incident spectrum not lost to thermalization or subbandgap transmission. The increase in spectral efficiency is most dramatic (17 percentage points) up to eight cells, with only four percentage points of efficiency gained by increasing from 8 to 20 cells. Achieving a spectral efficiency of 90%, which corresponds to reducing the combined thermalization and transmission losses to 10% of the incident power, requires an ensemble of at least eight cells. Figure 2 also shows that increasing the concentration on the system from 1 sun to 1000 suns results in approximately 10 percentage points in efficiency improvement. This suggests that a combination of concentration and spectrum splitting will be valuable in exceeding 50% system efficiency. Finally, the dashed lines for the series-connected spectrum splitting ensembles are consistently 1–2 percentage points lower than those for the independently connected designs. This is a consequence of the current-matching constraint forcing the selection of bandgap combinations with a lower spectral efficiency in order to ensure that all subcells absorb an equal number of photons.

The efficiencies in Figure 3 suggest that 50% efficiency can be exceeded by a three-cell ensemble at any concentration. However, these conventional detailed balance

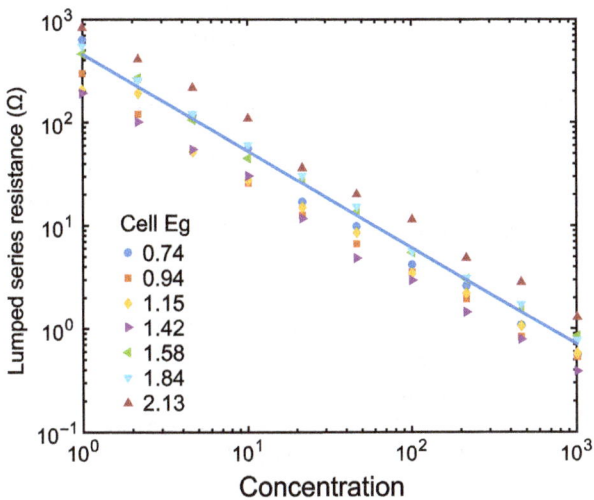

Figure 2. Optimized series resistance for simulated subcells with bandgaps ranging from 0.74 to 2.13 eV at concentrations from 1 to 1000 suns. The blue line shows the trend for the optimized values.

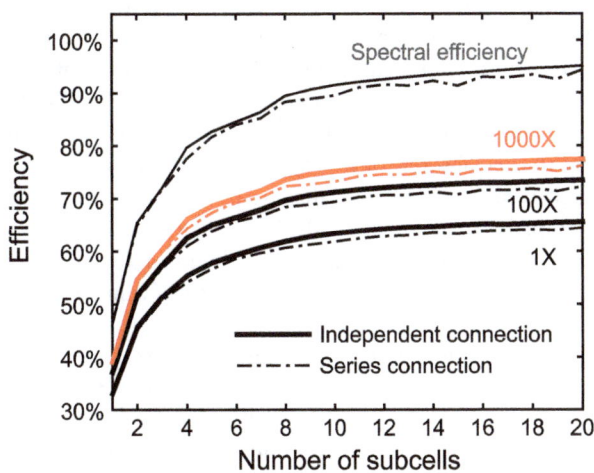

Figure 3. The ideal detailed balance efficiency for spectral splitting ensembles with 2–20 cells, both series connected and electrically independent at concentrations of 1, 100, and 1000 suns. The spectral efficiency of the independently connected and series-connected ensembles is also plotted showing the source of the efficiency improvement with increasing cell number.

calculations assume ideal behavior for photon absorption and radiation and perfect carrier collection. Practical cells have less than ideal absorption due to finite thickness, as well as reflection and transmission losses and will have some nonradiative recombination due to material imperfections. In order to understand the efficiency trends for cells with realistic material quality, we repeated the optimization and efficiency calculations for ensembles with 2–20 subcells (both series connected and with electrically independent subcells) using nonunity values for the ERE and absorption efficiency.

Including nonunity ERE and absorption efficiency did not change the optimal bandgaps for the series-connected ensembles. The constraint of series electrical connection and the need to maximize spectral efficiency dominate the subcell bandgap selection regardless of material quality. However, the bandgaps for the independently connected designs did change upon the inclusion of nonideal material behavior. The designs of ensembles with fewer than 10 cells exhibited the strongest dependence on material quality. Once optimized with nonunity ERE and absorption efficiency, the electrically independent designs uniformly increased the bandgap energy of the lowest energy subcell. This trend is a result of low bandgap cells being particularly sensitive to decreases in ERE. At low bandgap energies the loss of voltage due to nonunity ERE consumes a greater percentage of the open circuit voltage and the fill factor also degrades more significantly, which combine to eliminate the benefit of capturing more low energy photons. Ensembles with more than 10 cells did not have large changes in subcell bandgap values after optimization with nonideal material parameters. The expected efficiencies with 3% ERE and 90% absorption efficiency are ~10 percentage points lower than the ideal detailed balance efficiencies.

While the modified detailed balance approach gives a realistic prediction of the efficiency of an ensemble of cells, the total system has additional losses that must be taken into account. A practical spectrum splitting photovoltaic system will require some optical system to split and concentrate the incident spectrum into the desired spectral range for each subcell in the design. Such a system will inevitably introduce inefficiency through misallocation of photons to the wrong cell and reflections. In addition, an electrical system to combine the power of the subcells at a single output voltage will add electrical losses. In order to accommodate these optical and electrical losses and still produce module efficiency <50%, the ensemble of subcells must have a combined efficiency of much <50%. In our analysis we have assumed an optical system that concentrates the light and divides the spectrum with 90% optical efficiency, including losses from reflection and photon misallocation. The optical efficiency of the spectrum splitting optic is assumed to be equal for the electrically independent and series-connected designs. We assume an electrical system of 98% efficiency for designs with electrically independent subcells to account for losses in a DC–DC voltage combination circuit. The contact resistance losses for both the series-connected subcells and electrically independent subcells are accounted for by the lumped series resistance parameter.

Figure 4 shows the two-dimensional plot of efficiency versus number of cells and concentration for independently connected modules with 2–20 subcells. The two panels of the figure show the cell ensemble efficiency under two different material parameters: (1) 3% ERE and 90% absorption efficiency; and (2) 5% ERE and 90% absorption efficiency. Marked on each plot are contours showing total *module* efficiency with a spectrum splitting optic with 90% optical efficiency and electrical system of 98% efficiency. Also included in Figure 4 are dashed contours showing the required concentration for a 50% efficient module with electrically series-connected subcells and a 90% optical efficiency (series-connected cell efficiency not shown on this 2D plot). The color scale corresponds to the efficiency of the photovoltaic cell under ideal illumination, analog to the flash test efficiency used to evaluate current monolithic MJSCs [37]. As the color scale indicates, both 3% and 5% ERE cells with four or more subcells can achieve 50% cell efficiency even at one sun concentration, in contrast to the current record four-junction cell efficiency of 46% at 508 suns. The higher efficiency predicted for the four subcell ensembles in Figure 4 is due in part to the higher spectral efficiency of the four optimal independent bandgaps (79.6%) compared to record cell spectral efficiency (77.8%).

These plots show the importance of ERE for achieving a high module efficiency. With the lower external radiative efficiency, as shown in panel (A), it requires at least seven subcells at a concentration of 400 suns to achieve a total module power conversion efficiency of 50% with 90% efficient optics and 98% efficient electronics. By contrast, the set of designs with 5% ERE and 90% absorption efficiency can achieve 50% module efficiency with realistic optical and electrical losses using a design with six subcells at 620 suns concentration. The concentration required decreases with larger numbers of subcells, and only 59 suns are required to achieve 50% module efficiency with 10 subcells at 5% ERE. This highlights the trade-off in complexity between the optical design and the cell design in achieving very high module efficiency.

The dashed contour showing the minimum concentration required for a module with series-connected subcells to achieve 50% efficiency shows the advantage conferred by series electrical connection for the subcells. The

series-connected module assumes 90% optical efficiency and no additional electrical losses to account for the easier task of routing and combining power for cells that are already series connected [38]. Despite the disadvantage in spectral efficiency the series-connected ensembles have higher performance than the corresponding independent ensembles once the losses due to contact resistance are accounted for. This efficiency advantage for contacted systems is due to the electrical configuration. Monolithic series-connected cells have one set of contacts for the entire ensemble and consequently the voltage loss due to contact resistance is applied once to the entire ensemble's voltage. By contrast, ensembles with independent subcells require unique contacts for each subcell, which in turn cause a loss of voltage for each subcell. Because we expect the lumped series resistance for series-connected ensembles to be similar to that for independent ensembles, the loss due to lumped series resistance will increase with the number of subcells for the independent ensembles while remaining relatively constant across all series-connected designs, for a given concentration.

However, the contours in Figure 4 assume that monolithic series-connected subcells are capable of achieving the same ERE as physically separated cells. In fact, as Table 2 shows, the ERE of a subcell is highly dependent on its optical environment, as well as dependent on the material quality of the cell [39, 40]. Placing the subcells on high-quality back reflectors, as is possible for the electrically independent subcells, improves the cell ERE by preventing light from escaping through the back of the cell and by allowing thinner absorber layers, which reduces bulk recombination. Monolithic series-connected MJSCs cannot incorporate back reflectors between the subcells, and consequently must use thicker absorber layers to achieve the equivalent light path through the material. Monolithic cells exhibit lower radiative efficiency due to the larger amount of nonradiative bulk recombination in the thicker absorber layers and due to the loss of radiatively recombined photons that escape into the surrounding material. For a monolithic MJSC to achieve the ERE values in Figure 4, the material would require a much lower defect density than an ensemble of electrically independent subcells placed on back reflectors. This is consistent with the fact that practical series-connected MJSCs with four subcells require substantial concentration to achieve efficiencies over 45% [7, 11]. Finally, this analysis has assumed that the optimal bandgap combinations for series-connected ensembles can be grown monolithically, a feat that has not yet been achieved for ensembles with more than three subcells.

Figure 5 presents another view of the interaction among number of cells, concentration, and ERE in determining overall module efficiency. This plot shows the efficiency

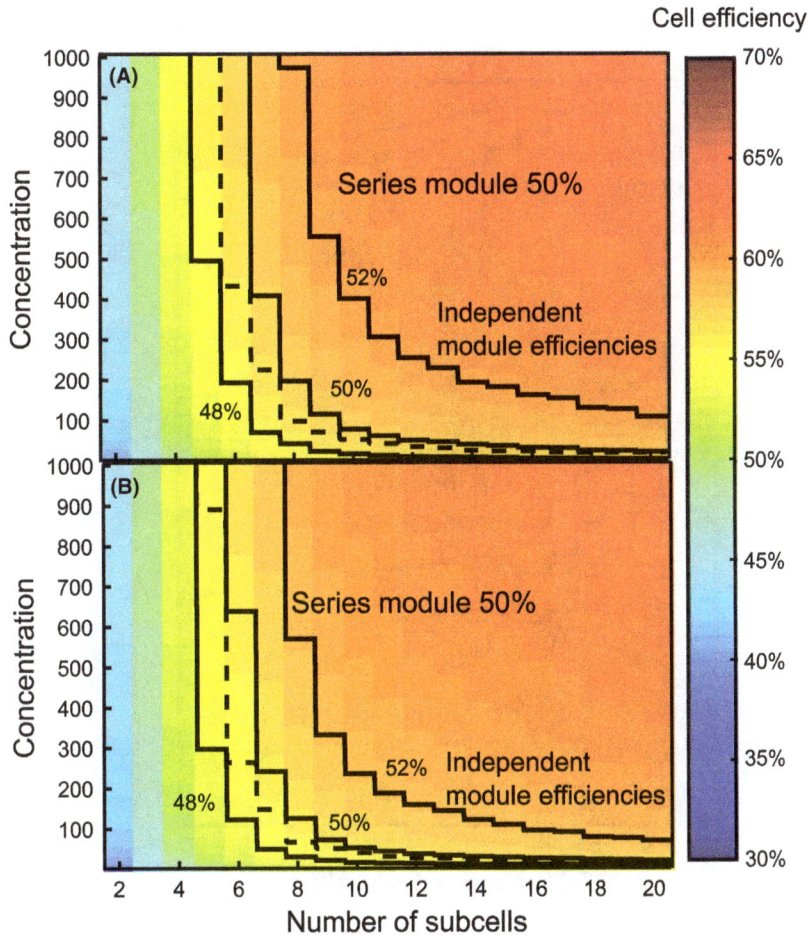

Figure 4. Efficiency maps for independent cell spectrum splitting ensembles with 2–20 cells at different concentrations and in panel (A) 3% external radiative efficiency (ERE), 90% absorption, and panel (B) 5% ERE and 90% absorption. The color maps indicate the efficiency of the independent photovoltaic cell ensemble under ideal illumination and with series resistance losses. The solid contours indicate the minimum concentration required for an overall module efficiency assuming electrically independent subcells, an optical system with 90% efficiency and a 98% efficient electrical system. The dashed contours indicate the minimum concentration required for a 50% efficient module with series-connected subcells and 90% optical efficiency.

of independently connected cell ensembles at 10 suns and 500 suns concentration with different ERE values. The solid contours again show total module efficiency with 90% efficient optics and a 98% efficient electrical system for electrically independent subcells and the dashed contour shows the module efficiency for series-connected subcells with 90% efficient optics. Considering second panel (B), the plot at 500 suns concentration, at 1% ERE, this plot indicates that nine independent subcells will be required to achieve 50% module efficiency. The steepness of the contours in the region from 4 to 10 cells highlights the value of improvements in ERE. An increase from 1% to 2% ERE reduces the number of independent subcells required to achieve 50% from nine to eight, which would constitute a reduction in potential cost and complexity. By contrast, the plot for 10 suns shows that low concentration modules will require 10 or more independent

subcells with average radiative efficiency of roughly 20%, equal to current record performance devices [34].

The contour showing the minimum ERE for series-connected modules to achieve 50% highlights the challenge for designs with monolithic cells to achieve ultra-high efficiency. The reduced loss from series resistance does translate to a lower minimum ERE required for the series-connected ensembles. However, note that 0.2% ERE, the average value for the experimental on-wafer cells in Table 2, corresponds to a need for an ensemble with 13 series-connected subcells with optimal bandgaps at 500 suns to achieve 50% module efficiency. At 1% ERE, a series-connected ensemble with seven subcells at the optimal bandgaps can achieve 50% efficiency at 500 suns. Achieving an ERE of 1% in a monolithic ensemble that lacks optical confinement will require exceptionally high material quality. Identifying materials for the optimal bandgaps of these

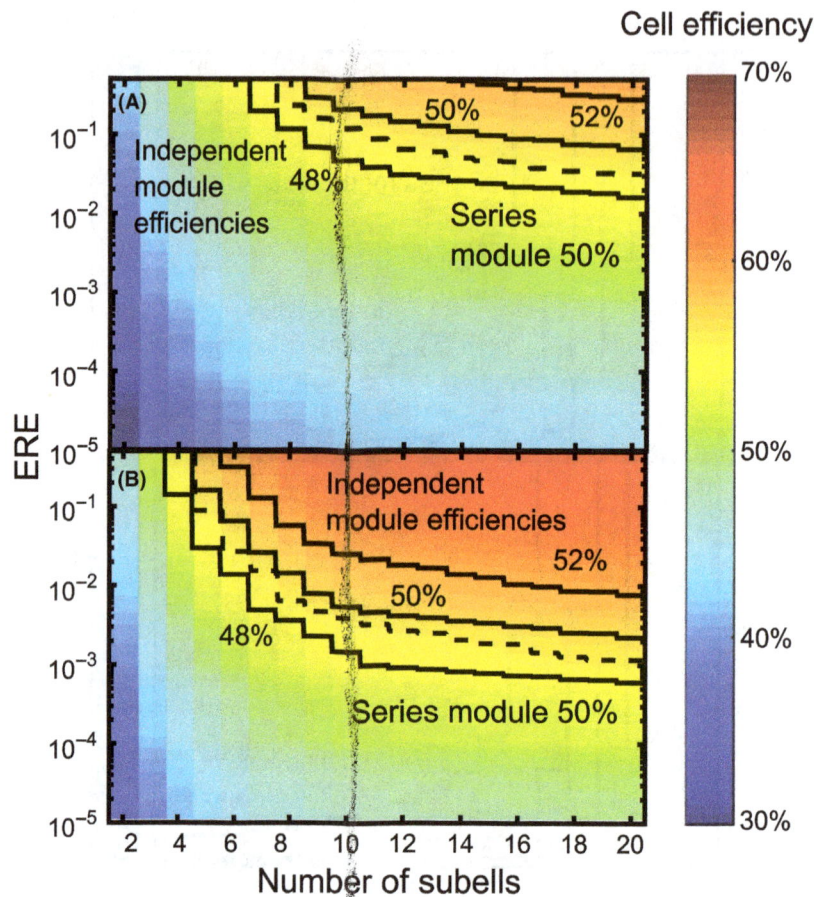

Figure 5. Efficiency map for independently connected spectrum splitting ensembles with 2–20 cells at (A) 10 suns and (B) 500 suns at different external radiative efficiency (ERE) values. The solid contours show the minimum ERE required for overall module efficiency for ensembles with electrically independent subcells with 90% optical efficiency and 98% DC electrical combination efficiency. The dashed contours show the minimum ERE required for modules with series-connected subcells and 90% optical efficiency.

subcells that can be grown or assembled while maintaining that material quality adds to the challenge and mitigates the efficiency advantage of the lower series resistance losses.

Conclusion

Achieving very high module efficiency (>50%) for photovoltaic solar conversion requires a combination of a large number of high-quality cells, an efficient optical system to split the incident spectrum correctly among those cells and a moderate-to-high degree of concentration. The radiative quality of the cells will determine the optimum bandgaps of the cells in an ensemble and the radiative and absorption efficiencies together with a lumped series resistance determine how far the cell performance departs from the ideal detailed balance limit. By including these three parameters into the detailed balance calculation, we have predicted that a module efficiency of 50% or greater will be possible with 7–10 electrically

independent subcells in a spectral splitting optic at 300–500 suns concentration, assuming a 90% optical efficiency and 98% electrical efficiency. Alternatively, a series-connected cell with 6–8 subcells at 100–300 suns could achieve this efficiency if the cells can reach an ERE of 3% or greater and can be manufactured with the optimal bandgaps.

While this analysis has been presented as an exploration of the minimum requirements for the photovoltaic cells, it is important to note that it also contains some aggressive requirements for the quality of the optical and electrical systems. This analysis has not presented any specific optical spectrum splitting concepts. Instead, the 90% optical efficiency used to predict module performance serves as a baseline performance requirement for any spectrum splitting concept under consideration. This value is consistent with reported spectrum splitting optical efficiencies [15, 41]. Similarly, the 98% efficiency of the DC electrical system serves as a target for power management systems [38]. This approach was used to develop specific implementations of spectrum splitting with practically achievable cells and optics. These designs are being developed as prototypes and are detailed in other publications [42–44].

Conflict of Interest

None declared.

References

1. Green, M. A., K. Emery, Y. Hishikawa, W. Warta, and E. D. Dunlop. 2012. Solar cell efficiency tables (version 40). Prog. Photovoltaics Res. Appl. 20:606–614.

2. Green, M. A. 2011. Limiting photovoltaic efficiency under new ASTM International G173-based reference spectra. Prog. Photovoltaics Res. Appl. 19: Pp 954–959.

3. Polman, A., and H. A. Atwater. 2012. Photonic design principles for ultrahigh-efficiency photovoltaics. Nat. Mater. 11:174–177.

4. King, R. R., D. C. Law, C. M. Fetzer, R. A. Sherif, K. M. Edmondson, S. Kurtz et al. 2005. Pathways to 40% efficient concentrator photovoltaics. Pp. 6–10 in 20th European Photovoltaic Solar Energy Conference and Exhibition.

5. Chiu, P., S. Wojtczuk, X. Zhang, C. Harris, D. Pulver, and M. Timmons. 2011. 42.3% Efficient InGaP/GaAs/InGaAs concentrators using bifacial epigrowth. Pp. 000771–000774 in 37th IEEE Photovoltaic Specialists Conference.

6. Wesoff, E. 2013. Sharp hits record 44.4% efficiency for triple-junction solar cell: greentech media. *Greentech Media*. Available at http://www.greentechmedia.com/articles/read/Sharp-Hits-Record-44.4-Efficiency-For-Triple-Junction-Solar-Cell (accessed 27 June 2013).

7. Green, M. A., K. Emery, Y. Hishikawa, W. Warta, and E. D. Dunlop. 2015. Solar cell efficiency tables (Version 45). Prog. Photovoltaics Res. Appl. 23:1–9.

8. Colthorpe, A. Soitec-Fraunhofer ISE multi-junction CPV cell hits world record 46% conversion efficiency | PV-Tech. Available at http://www.pv-tech.org/news/soitec_fraunhofer_ise_multi_junction_cpv_cell_hits_world_record_46_conversi (accessed 15 July 2015).

9. Gifford, J. 2015. Soitec hits 38.9% with four-junction CPV cell: pv-magazine. *PV Magazine*. Available at http://www.pv-magazine.com/news/details/beitrag/soitec-hits-389-with-four-junction-cpv-cell_100019952/#axzz3lGi8yNcg (accessed 9 September 2015).

10. Green, M. A., K. Emery, Y. Hishikawa, W. Warta, and E. D. Dunlop. 2015. Solar cell efficiency tables (version 46). Prog. Photovoltaics Res. Appl. 23:805–812.

11. Steiner, M., G. Siefer, T. Schmidt, M. Wiesenfarth, F. Dimroth, and A. W. Bett. 2016. 43% sunlight to electricity conversion efficiency using CPV. IEEE J. Photovolt. 6:1020–1024.

12. Green, M. A., K. Emery, Y. Hishikawa, W. Warta, and E. D. Dunlop. 2016. Solar cell efficiency tables (version 47). Prog. Photovoltaics Res. Appl. 24:3–11.

13. King, R. R., D. Bhusari, D. Larrabee, X. Liu, E. Rehder, K. Edmondson et al. 2012. Solar cell generations over 40% efficiency. Prog. Photovoltaics Res. Appl. 20:801–815.

14. Moon, R. L., L. W. James, H. A. VanderPlas, and N. J. Nelson. 1978. Performance of an $Al_{0.92}Ga_{0.08}As/Al_{0.14}Ga_{0.86}As$ solar cell in concentrated sunlight. Appl. Phys. Lett. 33:196.

15. Barnett, A., D. Kirkpatrick, C. Honsberg, D. Moore, M. Wanlass, K. Emery et al. 2009. Very high efficiency solar cell modules. Prog. Photovoltaics Res. Appl. 17:75–83.

16. Imenes, A. G., and D. R. Mills. 2004. Spectral beam splitting technology for increased conversion efficiency in solar concentrating systems: a review. Sol. Energy Mater. Sol. Cells 84:19–69.

17. Gross, B., G. Peharz, G. Seifer, M. Peters, J. C. Goldschmidt, M. Steiner et al. 2009. Highly efficient light splittig photovoltaic receiver. Pp. 21–25 in 24th European Photovoltaic Solar Energy Conference.

18. Wang, X., N. Waite, P. Murcia, K. Emery, M. Steiner, F. Kiamilev et al. 2012. Lateral spectrum splitting concentrator photovoltaics: direct measurement of component and submodule efficiency. Prog. Photovoltaics Res. Appl. 20:149–165.

19. Green, M. A., M. J. Keevers, I. Thomas, J. B. Lasich, K. Emery, and R. R. King. 2015. 40% efficient sunlight to electricity conversion. Prog. Photovoltaics Res. Appl. 23:685–691.

20. Kayes, B. M., H. Nie, R. Twist, S. G. Spruytte, F. Reinhardt, I. C. Kizilyalli et al. 2011. 27.6% Conversion efficiency, a new record for single-junction solar cells under 1 sun illumination. Pp. 000004–000008 in Conference Record of the IEEE Photovoltaic Specialists Conference.

21. Geisz, J. F., M. A. Steiner, I. García, S. R. Kurtz, and D. J. Friedman. 2013. Enhanced external radiative efficiency for 20.8% efficient single-junction GaInP solar cells. Appl. Phys. Lett. 103:41118.

22. Yoon, J., S. Jo, I. S. Chun, I. Jung, H.-S. Kim, M. Meitl et al. 2010. GaAs photovoltaics and optoelectronics using releasable multilayer epitaxial assemblies. Nature 465:329–333.

23. Eisler, C. N., E. D. Kosten, E. C. Warmann, and H. A. Atwater. 2013. Spectrum splitting photovoltaics: polyhedral specular reflector design for ultra-high efficiency modules. Pp. 1848–1851 in IEEE 39th Photovoltaic Specialists Conference.

24. Shockley, W., and H. J. Queisser. 1961. Detailed balance limit of efficiency of p-n junction solar cells. J. Appl. Phys. 32:510.

25. Henry, C. H. 1980. Limiting efficiencies of ideal single and multiple energy-gap terrestrial solar cells. J. Appl. Phys. 51:4494–4500.

26. Kirkpatrick, S., C. D. Gelatt, and M. P. Vecchi. 1983. Optimization by simulated annealing. Science 220:671–680.

27. Kurtz, S., D. Myers, W. E. McMahon, J. Geisz, and M. Steiner. 2008. A comparison of theoretical efficiencies of multi-junction concentrator solar cells. Prog. Photovolt. 16:537–546.

28. Froitzheim, A., R. Stangl, L. Elstner, M. Kriegel, and W. Fuhs. 2003. AFORS-HET: a computer-program for the simulation of heterojunction solar cells to be distributed for public use. Pp. 279–282 *in* Proceedings of the 3rd world conference on photovoltaic energy conversion.

29. Varache, R., C. Leendertz, M. E. Gueunier-Farret, J. Haschke, D. Muñoz, and L. Korte. 2015. Investigation of selective junctions using a newly developed tunnel current model for solar cell applications. Sol. Energy Mater. Sol. Cells 141:14–23.

30. Ahrenkiel, R. K., R. Ellingson, S. Johnston, J. Webb, J. Carapella, and M. Wanlass. 1999. Recombination lifetime of In[sub x]Ga[sub 1−x]As alloys used in thermophotovoltaic converters. AIP Conference Proceedings 460:282–288.

31. Vurgaftman, I., J. R. Meyer, and L. R. Ram-Mohan. 2001. Band parameters for III–V compound semiconductors and their alloys. J. Appl. Phys. 89:5815.

32. Sermage, B., J. L. Benchimol, and G. M. Cohen. 1998. Carrier lifetime in p-doped InGaAs and InGaAsP. Pp. 758–760 *in* 1998 International Conference on Indium Phosphide and Related Materials (Cat. No. 98CH36129), 2.

33. King, R. R., C. M. Fetzer, K. M. Edmondson, D. C. Law, P. C. Colter, H. L. Cotal et al. 2004. Metamorphic III–V materials, sublattice disorder, and multijunction solar cell approaches with over 37% efficiency. *in* 19th European Photovoltaic Solar Energy Conference and Exhibition.

34. Green, M. A. 2012. Radiative efficiency of state-of-the-art photovoltaic cells. Prog. Photovoltaics Res. Appl. 20:472–476.

35. Ross, R. T. 1967. Some thermodynamics of photochemical systems. J. Chem. Phys. 46:4590–4593.

36. Steiner, M., S. P. Philipps, M. Hermle, A. W. Bett, and F. Dimroth. 2011. Validated front contact grid simulation for GaAs solar cells under concentrated sunlight. Prog. Photovoltaics Res. Appl. 19:73–83.

37. Fanetti, E. 1981. Flash technique for GaAs concentrator solar cell measurement. Electron. Lett. 00:469–470.

38. Flowers, C. A., C. N. Eisler, and H. A. Atwater. 2014. Electrically independent subcircuits for a seven-junction spectrum splitting photovoltaic module. Pp. 1339–1343 *in* IEEE 40th Photovoltaic Specialist Conference, PVSC 2014.

39. Braun, A., E. A. Katz, D. Feuermann, B. M. Kayes, and J. M. Gordon. 2013. Photovoltaic performance enhancement by external recycling of photon emission. Energy Environ. Sci. 6:1499.

40. Kosten, E. D., J. H. Atwater, J. Parsons, A. Polman, and H. A. Atwater. A path to a 40% efficient single junction solar cell by limiting light emission angle.

41. Eisler, C. N., E. C. Warmann, C. A. Flowers, M. Dee, E. D. Kosten, and H. A. Atwater. 2014. Design improvements for the polyhedral specular reflector spectrum-splitting module for ultra-high efficiency (>50%). Pp. 2224–2229 *in* IEEE 40th Photovoltaic Specialist Conference (PVSC).

42. Eisler, C. N., C. A. Flowers, P. Espinet, S. Darbe, E. C. Warmann, J. Lloyd et al. 2015. Designing and prototyping the polyhedral specular reflector, a spectrum-splitting module with projected >50% efficiency. *in* IEEE 42nd Photovoltaic Specialist Conference (PVSC).

43. Xu, Q., Y. Ji, B. Riggs, A. Ollanik, N. Farrar-foley, H. Jim et al. 2016. A transmissive, spectrum-splitting concentrating photovoltaic module for hybrid photovoltaic-solar thermal energy conversion. Sol. Energy 137:585–593.

44. Escarra, M. D., S. Darbe, E. C. Warmann, and H. A. Atwater. 2013. Spectrum-splitting photovoltaics: holographic spectrum splitting in eight-junction, ultra-high efficiency module. Pp. 1852–1855 *in* IEEE 39th Photovoltaic Specialist Conference.

14

Potential induced degradation of n-type crystalline silicon solar cells with p⁺ front junction

Soohyun Bae[1], Wonwook Oh[2], Kyung Dong Lee[1], Seongtak Kim[1], Hyunho Kim[1], Nochang Park[2], Sung-Il Chan[2], Sungeun Park[1], Yoonmook Kang[3]*, Hae-Seok Lee[1]* & Donghwan Kim[1]*

[1]Department of Materials Science and Engineering, Korea University, 145, Anam-ro, Seongbuk-gu, Seoul, Korea
[2]Electronic Convergence Material and Device Research Center, Korea Electronic Technology Institute, Seongnam, Korea
[3]KU•KIST Green School, Graduate School of Energy and Environment, Korea University, 145, Anam-ro, Seongbuk-gu, Seoul, Korea

Keywords
Boron-doped emitter, n-type silicon solar cell, photovoltaic module, potential induced degradation, quantum efficiency

Correspondence
Donghwan Kim, Department of Materials Science and Engineering, Korea University, 145, Anam ro, Seonbuk gu, Seoul, Korea.
E-mail: solar@korea.ac.kr

Funding Information
Korea Institute of Energy Technology Evaluation and Planning (20143030011960).

*Hae-Seok Lee, Yoonmook Kang and Donghwan Kim have equally been contributed as corresponding authors.

Abstract

N-type silicon-based solar cells are currently being used for achieving high efficiency. However, most of the photovoltaic modules already constructed are based on p-type silicon solar cells, and there are few studies on potential induced degradation (PID) in n-type solar cells. In this study, we investigated PID in n-type silicon solar cells with a front p+ emitter. Further, the PID characteristics of n-type solar cells are compared with those of p-type solar cells. The electrical properties of PID in solar cells are observed with the light I-V, quantum efficiency (QE), and electroluminescence (EL). The possible causes for the change in the external quantum efficiency (EQE) after PID are interpreted using PC1D and are discussed by comparing the experimental results with the simulation results.

Introduction

With the increase in solar energy generation, photovoltaic (PV) modules are connected in series for generating high voltages and power. Solar panels, however, can be exposed to high-voltage stress up to several hundreds of volts between the grounded module frame and the solar cell. Frequent high-voltage stress causes a power drop in the modules, and this type of degradation is called potential induced degradation (PID) [1–4]. Most of the PV modules already constructed are based on p-type crystalline silicon solar cells, and the various studies on PID focus on p-type silicon solar cells and modules. Studies of the PID mechanism in p-type solar cells showed that the Na ion

decoration of the stacking fault in the silicon is responsible for the decrease in the shunt resistance after the PID test [5–9]. To reduce or minimize PID, various efforts have been developed from the cell level to the system level. SiNx films used for antireflection coating on solar cells are modified to reduce its resistivity. [10]. An encapsulant having high volume resistivity or different types of polymers such as polyolefin or ionomer is adopted instead of conventional ethylene vinyl acetate (EVA) [11, 12]. Chemically modified cover glass is used for the module structure [13]. Choosing proper grounding poles in a solar power system is also one of the possible solutions for PID.

Of late, to overcome the limitations in the efficiency of the silicon solar cells, studies on n-type crystalline silicon solar cells are being conducted. Power conversion efficiencies of over 25% are reported with an n-type silicon-based structure [14–16]. Compared to p-type silicon solar cells, there are fewer studies on PID in n-type silicon solar cells. PID in an n-type interdigitated back contact (IBC) solar cell was first reported by SUNPOWER in 2005 [17]; the PID test results of the IBC solar cell with different poles and voltage magnitudes were reported [18]. A p-n junction is formed at the rear side of the silicon wafer in the IBC solar cells; however, the junction is located at the front side of the silicon wafer in most high-efficiency n-type solar cells such as the HIT, TOPCON, bifacial solar cell, etc. Halm et al. [19] have reported PID results for an n-type IBC solar cell with a front floating emitter, and Hara et al. [20] have reported PID results for an n-type solar cell with a front p-n junction. In addition, Yamaguchi et al. [21] showed the PID results of silicon solar modules having a rear-side emitter fabricated by turning bifacial solar cells upside down. The overall results of PID in solar cell-based n-type wafers showed similar trends. The fill factor remains similar; on the other hand, current and voltage decrease with continuous tests. PID in n-type solar cells is attributed to the surface polarization effect [17] and the increase in the surface recombination velocity. The cause of PID in n-type solar cells, however, is still unclear and needs to be verified. Therefore, a study of PID in n-type solar cells is important owing to the increasing market trend for n-type solar cells.

In this study, the PID characteristics of n-type silicon solar cells with front p-n junctions having p+ emitters are investigated and compared with those of p-type silicon solar cells. From electrical analysis such as the light I-V and QE, the PID phenomenon in n-type solar cells is observed. Further, the possible reasons for the change in the EQE after PID are interpreted using PC1D, a solar cell simulation tool. Finally, the experimental results are compared with the simulated results, and the reasons for PID in n-type solar cells are discussed.

Experimental

Conventional structures of bifacial silicon solar cells with a front emitter and rear back surface field (BSF) are prepared for the PID test. The base wafers used in this study are phosphorous-doped n-type Czochralski (CZ), 6-inch monocrystalline silicon. The emitter and BSF within the solar cells are fabricated by BBr_3 and $POCl_3$ furnace doping, respectively. Both sides of the doped layers are passivated with stacked layers composed of silicon oxide and silicon nitride, and the electrodes are fabricated using a conventional screen printing of Ag paste and a firing process.

To analyze the PID characteristics of n-type solar cells, tests are conducted. The PID test for the solar cell is

carried out without a modulated structure because the breakdown analysis of a modulated solar cell is difficult, and the quantum efficiency (QE) results of the solar cells measured at the structure of the module are inaccurate. Hence, the PID test scheme reported in the PID mechanism study without modulation is introduced [7]. The test setup is shown in Figure 1. The EVA and cover glass are stacked on the solar cells similar to the PV modules. The temperature of each solar cell is controlled by a hot plate at 60°C. The (+) electrode is connected to the front surface of the glass, and the (−) electrode is connected to the rear side of the solar cell. The area of the (+) electrode used in the PID test is 100 cm². A 1000 V bias is applied between the front and rear electrodes for 48 h using a high-voltage source measure unit (Keithley 2410). To evaluate the electrical properties, the light I-V (Wacom Electric Co., Ltd., Tanaka, Fukaya-shi, Saitama, Japan solar simulator), QE (PV Measurements, solar cell IPCE/QE measurement system), and electroluminescence (EL) were measured.

Next, for further investigation, the transfer length method (TLM) is employed, and a capacitance–voltage (C–V) measurement is conducted. To measure the TLM, the solar cells in which PID occurred are cut using a mechanical dicing machine into sizes of 1 cm × 2 cm with only fingers in the front metal design. The resistances between two fingers with different electrode distances are measured by a prove station system and a source measure unit (Keithley 2400). To measure the C–V, a metal–insulator–silicon structure is prepared. The capacitance and conductance with the voltage are measured by an LCR meter (HP 4284A).

Results and Discussions

Electrical characteristics of PID

Figure 2 shows the light I-V curves of the n-type silicon solar cell before and after the PID test. The short circuit

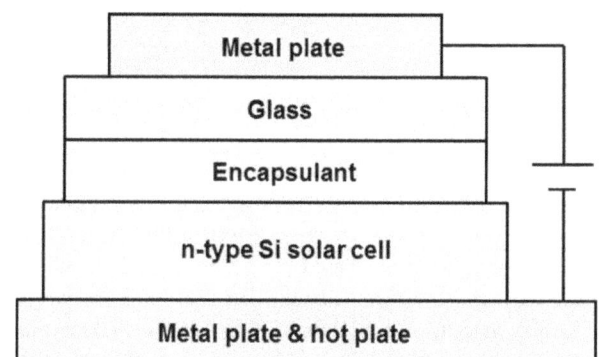

Figure 1. Schematic structure of potential induced degradation test setup.

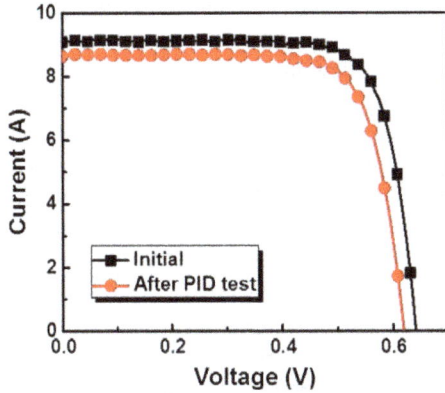

Figure 2. Light I-V curves of the n-type silicon solar cell before and after the potential induced degradation test.

current density (J_{sc}) and the open circuit voltage (V_{oc}) are mainly reduced, but the fill factor (FF) is maintained after the PID test. A similar result for PID in an n-type solar cell was reported in the literature [19–21]. Based on the result that there is no decrease in the FF for the n-type solar cell PID test and considering that the FF is mostly degraded after PID in the p-type solar cell owing to a decrease in the shunt resistance, it can be considered that the behavior of PID differs according to the type of base and emitter doping. It is reported that Na ions are introduced into the stacking faults within the emitter, n^+ doped, in p-type silicon solar cells and form shunt paths from the base silicon to the surface of the emitter after PID [5–7]. Stacking faults can also be formed inside boron-doped emitter [22], and the possibility of forming a shunt path still existed in the n-type solar cell. Shunt resistance and FF, however, are not decreased after PID in the n-type solar cell, and it can be argued that the stacking faults in the emitter of the n-type solar cells are

not active after PID. From a study of the interaction energy between the silicon doping type and the stacking faults, this phenomenon can be interpreted by the relationship between the dopant in the silicon and the stacking faults [23, 24]. P-type and n-type dopants show (−) charges and (+) charges after they are activated in the silicon, respectively. From research, it is known that p-type dopants have a weaker interaction than n-type dopants with stacking faults, and the stacking fault is broadened only when n-type dopants agglomerate and reduce the energy of the stacking fault. The passage of the Na ions through the path of the stacking fault will then be difficult or easy depending on whether the p-type or n-type dopant is near the stacking fault, respectively.

Figure 3 shows the EQE of the n-type and p-type solar cells before and after the PID test. Although the reflectance data of the solar cells are not shown here, there is no difference in the reflectance before and after the test. This result suggests that the antireflection coating (ARC) of the solar cell is not changed after the PID test in both the solar cells, and the decrease in Isc is not because of the additional reflection of the sunlight. From the EQE curve of the n-type solar cell (Fig. 3A), it can be seen that the EQE decreases only at short wavelengths below 600 nm. Hence, it is considered that the bulk and the rear side of the solar cell are not degraded, but the front side including the emitter and the surface of the emitter are affected after the PID test. This may be owing to the increase in the front surface recombination after PID [20]. The details will be discussed in the next section. On the other hand, the EQE of the p-type solar cell decreases in parallel for the entire wavelength after the PID test (Fig. 3B). It is reported that the EQE decreases when the shunt resistance of the solar cell is extremely low [25]. The normalized EQE curve derived from the ratio between

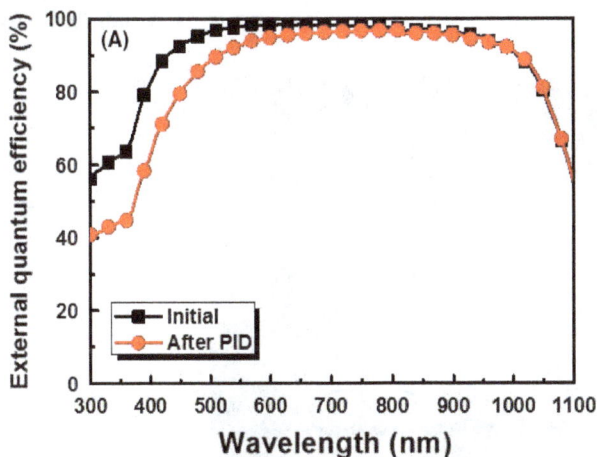

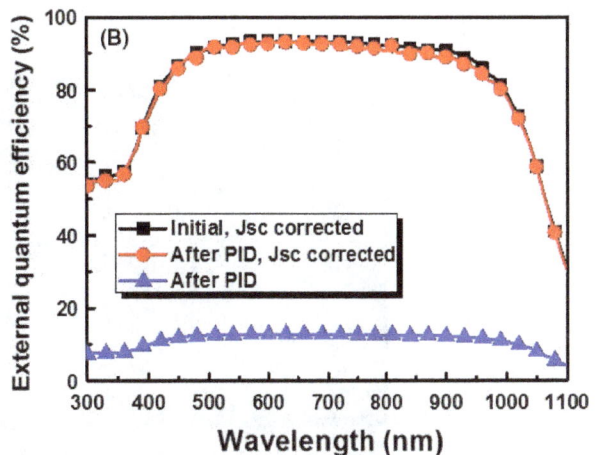

Figure 3. External quantum efficiency of the (A) n-type silicon solar cell (B) p-type silicon solar cell before and after the potential induced degradation test.

the J_{sc} measured by the light I-V and by the QE, however, shows no change compared to the initial curve. It is demonstrated that there is no degradation within the solar cell in the case of the p-type solar cell, and only the electrical shunt path affects the final performance after PID. It is concluded that there are different causes for PID in n-type solar cells, excluding the shunt path.

To observe the shape of the degradation, the electroluminescence (EL) is measured on the samples in which PID occurred (Fig. 4). As shown in the EL images, the entire affected area is degraded uniformly in the n-type solar cell, but local degradation spots are observed in the p-type solar cell after PID. These two differences support the opinion that only the stacking faults of the p-type solar cell are affected, but different types of degradation such as surface recombination, inversion of the emitter, and changes in the doping density occur in the n-type solar cell.

Causes of PID in n-type silicon solar cells with p+ front junctions

Interpretation of EQE changes using PC1D simulation

It is experimentally observed that the EQE of n-type solar cells at short wavelengths decreases after PID. There are three possible causes for the decrease in the EQE only at short wavelengths: (1) decrease in the emitter doping density; (2) increase in the positive fixed charge of the front passivation layer; and (3) increase in the surface recombination velocity (SRV) owing to the interface defect density. Figure 5 shows the simulated EQE results obtained by PC1D. The other material and electrical parameters are fixed, and one of the three parameters – the emitter sheet resistance, the fixed charge, or the front SRV – is varied. The initial parameters used in Figure 5A are as follows: emitter sheet resistance = 63 $\Omega/\square$, fixed charge

density = 1.1E12 cm^{-2}, and the front SRV = 4000 cm/sec. First, by decreasing the emitter sheet resistance to 23 $\Omega/\square$, or by increasing the doping density, the simulated EQE is fitted to the experimental curve after PID (Fig. 5B). The EQE at short wavelengths can be decreased if the emitter doping density is decreased. Next, by increasing the fixed charge density in the front passivation layer to 4.95E12 cm^{-2}, the simulated EQE is fitted to the experimental curve (Fig. 5C). Finally, by increasing the front SRV to 45,000 cm/sec, the simulated EQE is also matched with the experimental curve (Fig. 5D). If the other parameters are fixed, the SRV indicates the interface defect density between the passivation layer and the emitter surface. In summary, it is found by simulation that three possible scenarios – increasing the doping density, increasing the fixed charge, and increasing the interface defect density – can exist for the decrease in the EQE at short wavelengths.

Experimental proof of causes of PID in n-type solar cells

Figure 6 shows the sheet resistance calculated by the TLM before and after the PID test. The TLM is a well-known method for measuring the contact resistance between the contact electrode and the semiconductor [26]. By comparing the resistance between two different electrodes with the distances between the electrodes, the sheet resistance is calculated. From the result, it is observed that the sheet resistance is not changed after the PID test. Therefore, the change in the doping density is not the reason for PID in n-type solar cells as seen from the simulated EQE. Further, even if considerable amounts of Na ions drift into the emitter, the Na atoms act as the n-type dopant in silicon [27]. Next, to verify the changes in the fixed charge density and the interface defect density, a capacitance–voltage (C–V) measurement is conducted. By

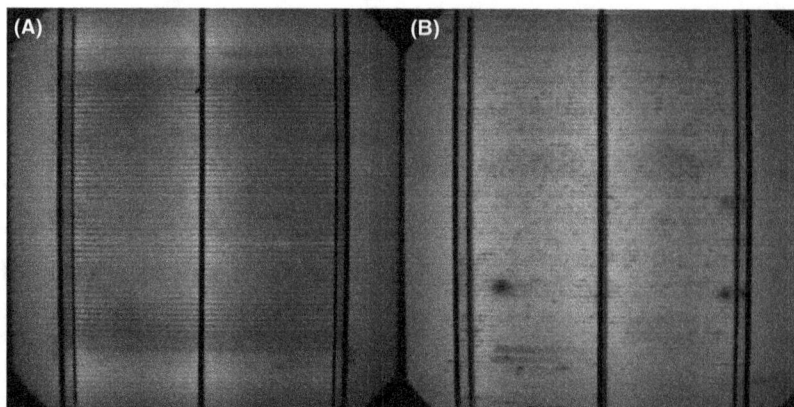

Figure 4. Electroluminescence image of the (A) n-type silicon solar cell (B) p-type silicon solar cell after the potential induced degradation test.

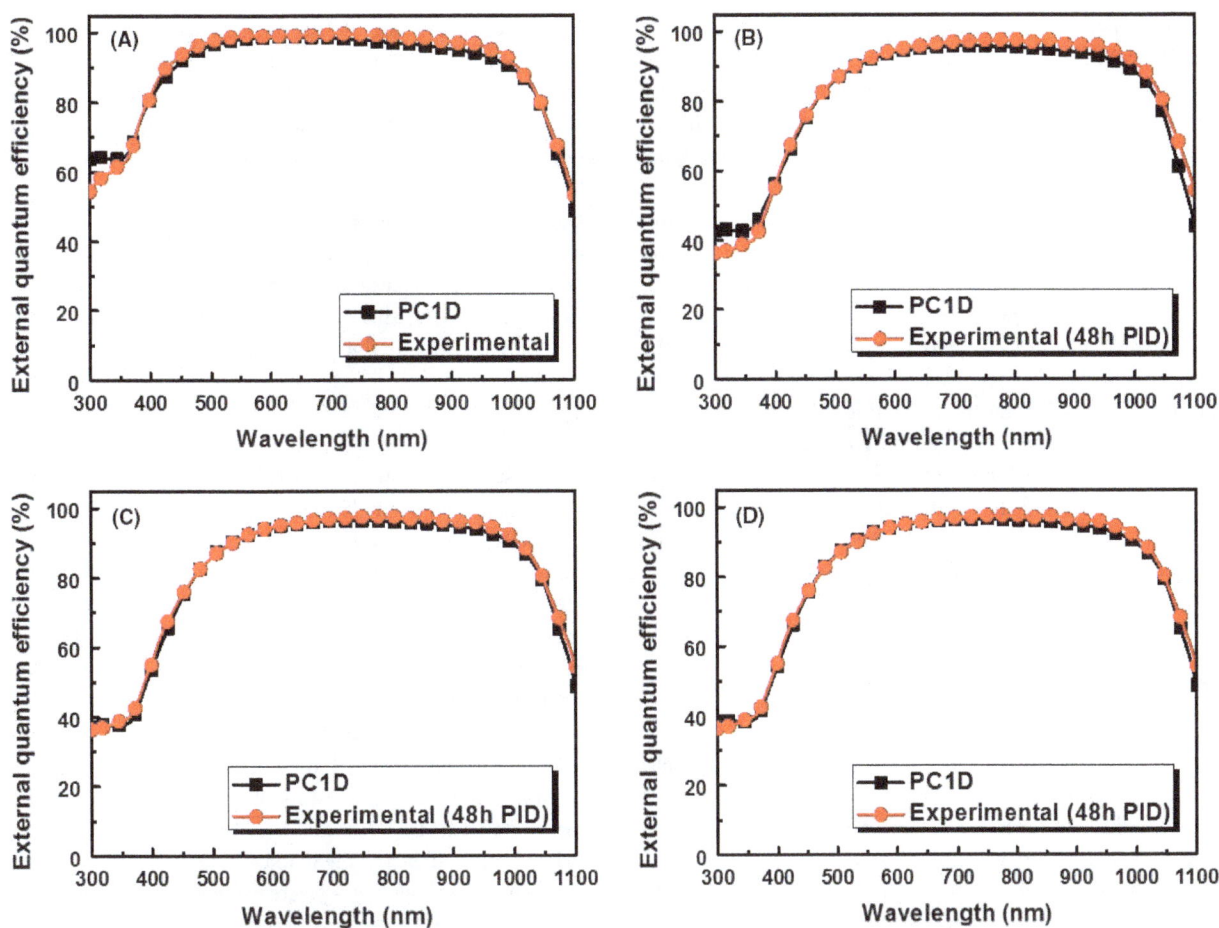

Figure 5. PC1D simulated external quantum efficiency curve compared with the experimental result when (A) initial, (B) the sheet resistance of the emitter is changed to 23 Ω/□, (C) the fixed charge in the passivation layer is changed to 4.9E12 cm^{-2}, and (D) the front surface recombination velocity is changed to 45,000 cm/sec.

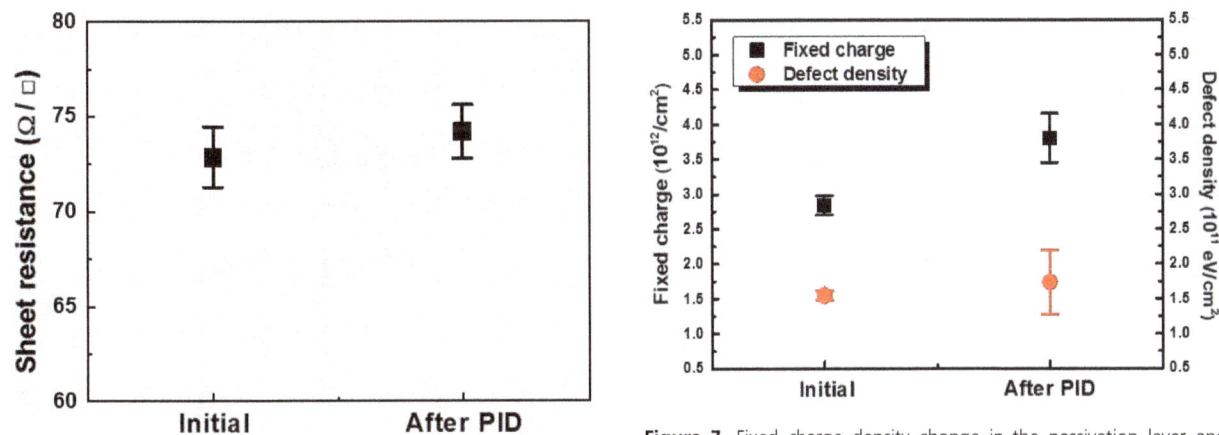

Figure 6. Doping density (sheet resistance) changes after the potential induced degradation test calculated by the transfer length method.

Figure 7. Fixed charge density change in the passivation layer and interface defect density change between the passivation layer and the silicon.

comparing the capacitance–voltage curves, the fixed charge density in the dielectric layer and the interface defect density can be calculated [26, 28]. The calculated results

for the fixed charge density and the interface defect density before and after the PID test are shown in Figure 7. The fixed charge density increases, but the interface defect density remains the same. From the literature, it is known that the SRV and the emitter saturation current are proportional to the interface defect density [29]. The PC1D simulation results show that the SRV increases by an order of magnitude. Although it is impossible to compare the absolute values, the fixed charge density increases within one order of magnitude. The simulation result shows that a slight increase in the fixed charge density can affect PID. When the positive fixed charge is increased, the p-type-doped emitter in the n-type silicon solar cell is depleted or inverted owing to the repulsion of the majority carrier, the (+) charged holes. The diffusion of the minority carrier toward the surface will then be easier, and the emitter saturation current can be increased. In conclusion, when a high voltage is induced between the n-type solar cell and the module frame, it is demonstrated that PID occurs owing to an increase in the fixed charge in the passivation on the surface of the front p^+ junction.

Cause of increase in positive fixed charges in passivation layer

It is reported that the cause of PID in p-type solar cells is the drift of Na ions to the solar cell under an external electric field. To confirm the role of the Na ions in PID in n-type solar cells, a simple PID test is conducted. Na is directly deposited on the surface of the solar cell by thermal evaporation, and the PID test is performed with the EVA only, without the glass. For comparison, the PID test is conducted on a solar cell without a Na source. To exclude the effects of the Na ions remaining within the EVA, the EVA is immersed in deionized (DI) water for 1 day before the PID test. Although not depicted here, degradation did not occur when the same EVA was used for the PID test in a p-type solar cell. The test results are shown in Figure 8. PID occurs regardless of the Na source. Interestingly, a power drop is observed in the sample using only the EVA without the Na source.

Therefore, the origin of the positive fixed charge in the passivation layer over the emitter of the n-type solar cell is not the migrated ions but the polarization owing to the dielectric property. It is speculated that PID in an n-type solar cell is inevitable. In conclusion, to reduce PID in n-type solar cells, an appropriate choice of the properties of the passivation layer is needed, minimizing the dielectric constant and enhancing the conductivity. Further, using an encapsulation material having a higher volume resistivity, the PID can be reduced by decreasing the voltage drop in the passivation layer.

Conclusion

In this study, PID in n-type solar cells with front junctions was investigated. The characteristics of PID in n-type solar cells were observed by measuring the light I-V, EQE, and EL, and by comparison with PID in p-type solar cells. The parameters that were mainly reduced are J_{sc} and V_{oc}; there was no drop in the FF after the PID test. Further, PID occurred uniformly over the affected area, and it is understood that PID in n-type solar cells cannot be explained by known mechanisms, including stacking faults in the emitter and the formation of shunt paths. By comparing the EQE results obtained from the experiments with the PC1D simulation, three possible changes – the doping density, the fixed charge density, and the interface defect density – were suggested. It was found that an increase in the positive fixed charge in the front passivation layer was the most likely cause of PID in n-type silicon solar cells, based on the experiments. Finally, to investigate the reason for the increase in the positive fixed charge, a PID test was conducted on an incomplete module structure, and it was observed that it was not caused by migrated ions such as Na, but because of the dielectric polarization of the passivation layer. PID in n-type solar cells can be prevented by introducing a passivation layer having a higher conductivity and a lower dielectric constant.

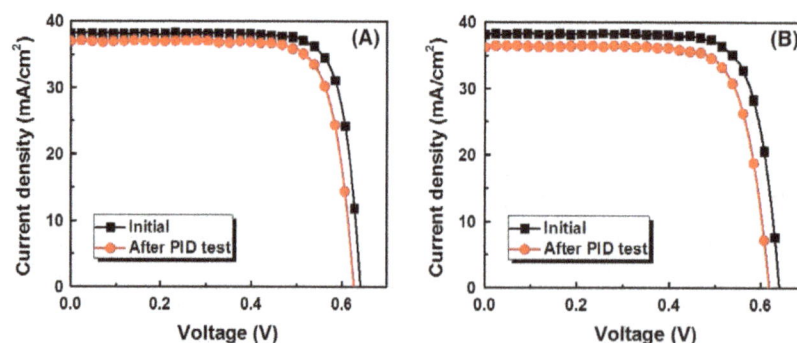

Figure 8. Potential induced degradation results of the n-type solar cell (A) with an Na source (glass) and (B) without an Na source using only ethylene vinyl acetate.

Acknowledgments

This work was supported by the New & Renewable Energy Core Technology Program of the Korea Institute of Energy Technology Evaluation and Planning (KETEP), granted financial resource from the Ministry of Trade, Industry & Energy, Republic of Korea. (No. 20143030011960).

Conflict of Interest

None declared.

References

1. Berghold, J., O. Frank, H. Hoehne, S. Pingel, B. Richardson, and M. Winkler. 2010. Potential induced degradation of solar cells and panels. *25th EUPVSEC.* Pp. 3753–3759.
2. Pingel, S., O. Frank, M. Winkler, S. Daryan, T. Geipel, H. Hoehne et al., eds. 2010. Potential induced degradation of solar cells and panels. *Photovoltaic Specialists Conference (PVSC), 2010 35th IEEE.* IEEE.
3. Hacke, P., K. Terwilliger, R. Smith, S. Glick, J. Pankow, M. Kempe et al., eds. 2011. System voltage potential-induced degradation mechanisms in PV modules and methods for test. *IEEE – 2011 37th IEEE Photovoltaic Specialists Conference (PVSC).*
4. Schütze, M., M. Junghänel, M. B. Koentopp, S. Cwikla, S. Friedrich, J. W. Müller et al., eds. 2011. Laboratory study of potential induced degradation of silicon photovoltaic modules. *IEEE – 2011 37th IEEE Photovoltaic Specialists Conference (PVSC).*
5. Naumann, V., D. Lausch, A. Graff, M. Werner, S. Swatek, J. Bauer et al. 2013. The role of stacking faults for the formation of shunts during potential-induced degradation of crystalline Si solar cells. Phys. Status Solidi. Rapid Res. Lett. 7:315–318.
6. Naumann, V., D. Lausch, A. Hähnel, J. Bauer, O. Breitenstein, A. Graff et al. 2014a. Explanation of potential-induced degradation of the shunting type by Na decoration of stacking faults in Si solar cells. Sol. Energy Mater. Sol. Cells 120:383–389.
7. Lausch, D., V. Naumann, O. Breitenstein, J. Bauer, A. Graff, J. Bagdahn et al. 2014. Potential-induced degradation (PID): introduction of a novel test approach and explanation of increased depletion region recombination. IEEE J. Photovolt. 4:834–840.
8. Ziebarth, B., M. Mrovec, C. Elsässer, and P. Gumbsch. 2014. Potential-induced degradation in solar cells: electronic structure and diffusion mechanism of sodium in stacking faults of silicon. J. Appl. Phys. 116:093510.
9. Naumann, V., D. Lausch, A. Hähnel, O. Breitenstein, and C. Hagendorf. 2015. Nanoscopic studies of 2D-extended defects in silicon that cause shunting of Si-solar cells. Phys. Status Solidi C 12:1103–1107.
10. Mishina, K., A. Ogishi, K. Ueno, T. Doi, K. Hara, N. Ikeno et al. 2014. Investigation on antireflection coating for high resistance to potential-induced degradation. Jpn. J. Appl. Phys. 53:03CE01.
11. Reid, C. G., S. A. Ferrigan, J. I. F. Martinez, and J. T. Woods, eds. 2013. Contribution of PV encapsulant composition to reduction of potential induced degradation (PID) of crystalline silicon PV cells. *28th EUPVSEC.* Pp. 3340–3346.
12. Kapur, J., K. M. Stika, C. S. Westphal, J. L. Norwood, and B. Hamzavytehrany. 2015. Prevention of potential-induced degradation with thin ionomer film. IEEE J. Photovolt. 5:219–223.
13. Kambe, M., K. Hara, K. Mitarai, S. Takeda, M. Fukawa, N. Ishimaru et al., eds. 2014. Chemically strengthened cover glass for preventing potential induced degradation of crystalline silicon solar cells. *IEEE – 2013 IEEE 39th Photovoltaic Specialists Conference (PVSC).*
14. Masuko, K., M. Shigematsu, T. Hashiguchi, D. Fujishima, M. Kai, N. Yoshimura et al. 2014. Achievement of more than 25% conversion efficiency with crystalline silicon heterojunction solar cell. IEEE J. Photovolt. 4:1433–1435.
15. Feldmann, F., M. Bivour, C. Reichel, H. Steinkemper, M. Hermle, and S. W. Glunz. 2014. Tunnel oxide passivated contacts as an alternative to partial rear contacts. Sol. Energy Mater. Sol. Cells 131:46–50.
16. Moldovan, A., F. Feldmann, M. Zimmer, J. Rentsch, J. Benick, and M. Hermle. 2015. Tunnel oxide passivated carrier-selective contacts based on ultra-thin SiO_2 layers. Sol. Energy Mater. Sol. Cells 142:23–27.
17. Swanson, R., M. Cudzinovic, D. De Ceuster, V. Desai, J. Jürgens, N. Kaminar et al., eds. 2005. The surface polarization effect in high-efficiency silicon solar cells. *Proceedings of 15th International Photovoltaic Science and Engineering Conference (PVSEC-15), Shanghai, China.*
18. Naumann, V., T. Geppert, S. Großer, D. Wichmann, H.-J. Krokoszinski, M. Werner et al. 2014b. Potential-induced degradation at interdigitated back contact solar cells. Energy Procedia. 55:498–503.
19. Halm, A., A. Schneider, V. D. Mihailetchi, L. J. Koduvelikulathu, L. M. Popescu, G. Galbiati et al. 2015. Potential-induced degradation for encapsulated n-type IBC solar cells with front floating emitter. Energy Procedia. 77:356–363.
20. Hara, K., S. Jonai, and A. Masuda. 2015. Potential-induced degradation in photovoltaic modules based on n-type single crystalline Si solar cells. Sol. Energy Mater. Sol. Cells 140:361–365.
21. Yamaguchi, S., A. Masuda, and K. Ohdaira. 2016. Changes in the current density-voltage and external quantum efficiency characteristics of n-type

single-crystalline silicon photovoltaic modules with a rear-side emitter undergoing potential-induced degradation. Sol. Energy Mater. Sol. Cells 151:113–119.

22. Sopori, B., H.-C. Yuan, S. Devayajanam, P. Basnyat, V. LaSalvia, A. Norman et al., eds. 2014. Bulk defect generation during B-diffusion and oxidation of CZ wafers: mechanism for degrading solar cell performance. *IEEE – 2014 IEEE 40th Photovoltaic Specialists Conference (PVSC).*

23. Ohno, Y., T. Taishi, Y. Tokumoto, and I. Yonenaga. 2010. Interaction of dopant atoms with stacking faults in silicon crystals. J. Appl. Phys. 108:073514.

24. Ohno, Y., Y. Tokumoto, H. Taneichi, I. Yonenaga, K. Togase, and S. R. Nishitani. 2012. Interaction of dopant atoms with stacking faults in silicon. Phys. B 407:3006–3008.

25. Oh, J., S. Bowden, and G. Tamizh Mani. 2015. Potential-induced degradation (PID): incomplete recovery of shunt resistance and quantum efficiency losses. IEEE J. Photovolt. 5:1540–1548.

26. Schroder, D. K. 2006. Semiconductor material and device characterization. John Wiley & Sons, New Jersey 541 Tanaka, Fukaya-Shi, Saitama 369–1108, Japan.

27. Korol, V. 1988. Sodium-ion implantation into silicon. Phys. Status Solidi A 110:9–34.

28. Nicollian, E. H., and J. R. Brews. 1982. MOS (metal oxide semiconductor) physics and technology. Wiley, New York, NY.

29. Jin, H., and K. Weber. 2008. Relationship between interface defect density and surface recombination velocity in (111) and (100) silicon/silicon oxide structure. *23rd European Photovoltaic Solar Energy Conference.*

15

Performance analysis of a tower solar collector-aided coal-fired power generation system

Liqiang Duan[1], Xiaohui Yu[1], Shilun Jia[1], Buyun Wang[1] & Jinsheng Zhang[2]

[1]School of Energy, Power and Mechanical Engineering, National Thermal Power Engineering and Technology Research Center, Key Laboratory of Condition Monitoring and Control for Power Plant Equipment of Ministry of Education, Beijing Key Laboratory of Emission Surveillance and Control for Thermal Power Generation, North China Electric Power University, Beijing 102206, China
[2]Shenhua Guohua (Beijing) Electric Power Research Institute Co., Ltd., Beijing 100025, China

Keywords

Annual performance, molten salt tower, performance analysis, solar energy, TRNSYS

Correspondence

Liqiang Duana, School of Energy, Power and Mechanical Engineering, National Thermal Power Engineering and Technology Research Center, Key Laboratory of Condition Monitoring and Control for Power Plant Equipment of Ministry of Education, Beijing Key Laboratory of Emission Surveillance and Control for Thermal Power Generation, North China Electric Power University, Beijing 102206, China. E-mail: dlq@ncepu.edu.cn

Funding Information

This study has been supported by the National Nature Science Foundation Project (No. 51576062 and No. 51276063) and the National Basic Research Program of China (No. 2015CB251505).

Abstract

In this paper, a tower solar collector-aided coal-fired power generation (TSCACPG) system is proposed and studied in order to save the fossil energy and protect the environment. The integration scheme of tower solar collector and conventional coal-fired power plant is proposed. Based on the simulation platform TRNSYS, the TSCACPG system model is established and the dynamic performance of the TSCACPG system with the operating mode of coal saving is studied. The TSCACPG system performances of 1 day and 1 year are all discussed by using the DNI data of typical year in Chinese typical city of Dunhuang. Then, the sensitivity analysis of the TSCACPG system is carried out by changing the heliostat field scale (the size of the molten salt tower also changes accordingly). The annual performance of the TSCACPG system is also acquired. In consideration of the economic costs, the heliostat field area with the maximum annual solar-to-electric efficiency is selected as the optimal value. The results show that, for the case studied, the optimal heliostat field area is 101,400 m^2, and the maximum annual solar-to-electric efficiency is 16.74%. And under the optimal situation, the standard coal consumption rate of the original coal-fired power plant is reduced from 301.5 g/kWh to 294.5 g/kWh.

Introduction

Over a long period, the fossil fuel as the primary energy has played a key role in the human survival and social development, but at the same time, it is accompanied by the serious environment pollution problem and great energy consumptions [1, 2]. In order to realize the sustainable development of the living environment, new energy resources should be explored and the clean renewable energy should be utilized on a larger scale in order to gradually reduce the destruction of the environment and

the dependence on the fossil energy [3]. The complementary utilization of the renewable energy and the fossil energy has gradually become an important development trend of the future energy system [4–7]; especially in China, about 70% of the electricity comes from coal-fired power plants. Currently, restricted by the resource shortage and environment pollution problems, many medium and small thermal power plants that consume a large amount of coal will face the fate of shutting down. The utilization of the solar energy with its unique advantage has been paid more and more attentions. The solar energy is inexhaustible energy without greenhouse gases (mainly CO_2, NO_X) or toxic gases (SO_2 and particulate) emissions. As a kind of special renewable energy, the solar energy is not only abundant but also cheap [8].

Presently, the solar thermal power generation has been widely applied and developed at home and abroad, such as the power generation of solar parabolic trough and solar energy tower [9]. But for the independent solar thermal power generation system, both the high initial investment and lower thermal performance are major obstacles to its development [10]. However, the solar energy-aided power generation system can integrate the solar energy into a fossil fuel (coal or gas)-fired power plant at a relatively low cost [11]. For instance, the United States Colorado project that started in July 2010 is the first solar-aided coal-fired power plant with the grid-connected power generation mode. The heat capacity is 4 MW (total capacity of power station is 44 MW), and the collector area is 25,899 m^2. The power plant efficiency is improved by 3–5%, and 2000 tons per year of greenhouse gas emissions is reduced. In addition, the huge power generation project of Kogan Creek Solar Boost, combination of 750-MW coal-fired power station and 44-MW solar heat collection system, is located in Queensland, Australia, adopting the CLFR type (linear Fresnel) technology. The collector area is about 300,000 m^2, and the annual power generation capacity is 44 GWh. The reduced carbon dioxide emission is 35,600 tons per year, and the investment cost is $105 million.

In recent years, many researches have been carried out on the integration of the solar thermal energy with the traditional fossil fuel-fired power system. A typical manner to employ the solar-aided coal-fired power generation technique is to heat the boiler feedwater using the solar thermal energy instead of the regenerative extraction steam in the coal-fired power unit. In the studies of Y. Yang team and H. Jin team, the solar heat with about 300°C medium-or-low temperature is added into the regenerative system of the conventional power plant by using the trough solar collectors [12, 13]. Hou et al. and Peng et al. also studied the solar-aided coal-fired power generation system on the basis of the second law of thermodynamics [14, 15]. Gupta and Kaushik concluded that using the solar energy heating the boiler feedwater of a thermal power plant is more efficient than using the same solar energy in a stand-alone solar thermal power plant [16]. Eck and Zarza proved the feasibility of the direct steam generation process in a horizontal parabolic trough concentrator [17]. Popov Dimityr studied the three replacements of low-pressure heaters, high-pressure heaters, high-pressure heaters and economizer by using the solar energy, respectively [18]. Jian et al. [19] analyzed the performance of a trough

solar energy collector-aided 330-MW plant with part of extraction steam in high-pressure (HP) heaters replaced in the fuel-saving operation mode. Larrain et al. [20] established the 100-MW fossil energy-aided feedwater heating of solar thermal power plant, and using the climatic parameters of Chile desert to find the best efficiency in the hybrid system with the minimum amount of fossil energy. Zhao et al. [21] analyzed the process of energy transfer of the solar-aided coal-fired power generation system, and the reason why the thermodynamic performance of the solar-aided system is better than that of the original unit is revealed.

Although many researches on the integration of the parabolic trough solar energy collector system with the traditional coal-fired power plant, there is few study on the tower solar collector-aided coal-fired power generation system. And in the previous researches, the impact of DNI is usually neglected. In this paper, the high-temperature tower solar energy collector is integrated into the superheating and reheating system of the boiler, and the impact of variation in DNI data is considered. The conventional coal-fired power plant system integrated with the tower solar energy collector with the molten salt as the working medium is discussed. In consideration of the variation in DNI and heliostat field area, the dynamic performance of the overall system is analyzed in order to seek the best annual performance at the design point. In addition, the standard coal consumption rate of the overall system is also discussed in this paper.

The novelty of this paper lies in the following several aspects:

1. The coal-fired power plant integrated with tower solar collector with the operating mode of coal saving is proposed and deeply investigated.
2. The solar radiation intensity is adequately considered in the annual performance analysis.
3. On the basis of the annual performance, the heliostat field area is optimized and discussed.

System Descriptions

Coal-fired power generation system (base system)

A 660-MW supercritical coal-fired power unit is selected as the base system. Its feedwater regenerative part includes three high-pressure feedwater heaters, one deaerator, and four low-pressure feedwater heaters. The simplified scheme of the coal-fired power generation system and its main parameters is shown in Figure 1 and Table 1, respectively.

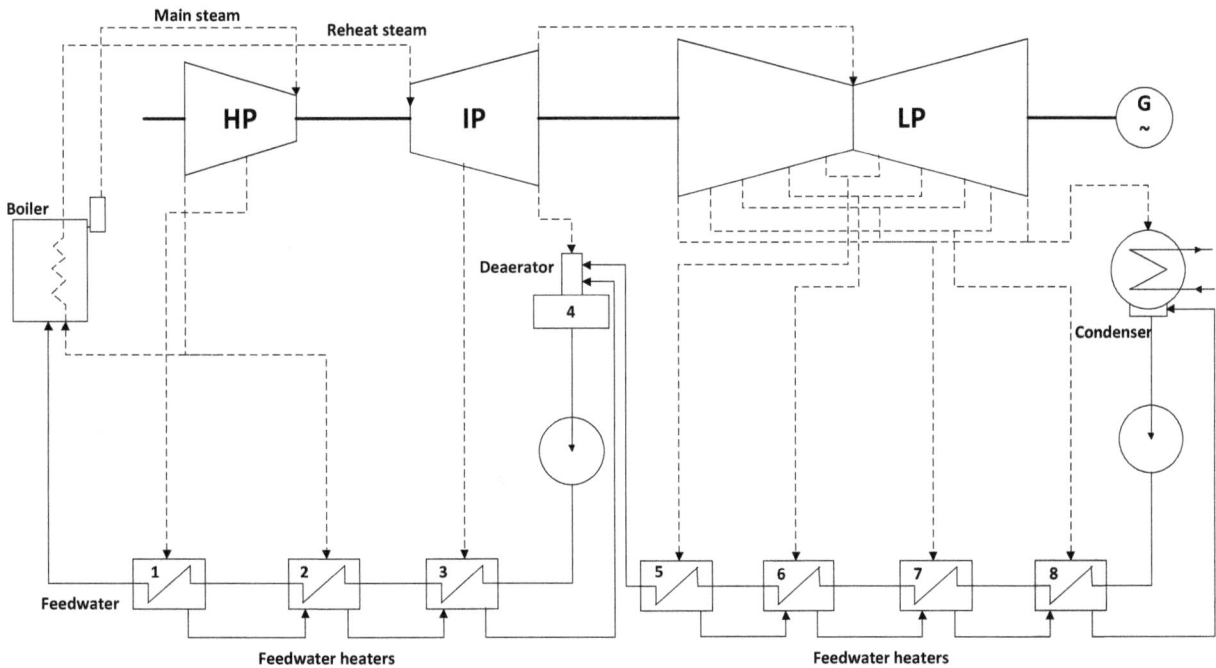

Figure 1. Schematic diagram of the coal-fired power generation system.

Table 1. Main parameters of coal-fired power plant at design point (rated load).

Item	Unit	Value
Power	MW	660
Main steam pressure	MPa	24.2
Main steam temperature	°C	566
Reheat steam temperature	°C	566
Boiler feedwater temperature	°C	275.3
Mass flow rate of the boiler feedwater	t/h	2150.6
Turbine exhausted steam pressure	kPa	5.3
Turbine exhausted steam enthalpy	kJ/kg	2351.5

The integration of tower solar collector system with the boiler system of the base system

In this paper, the tower solar collector system uses the molten salt as the working medium to absorb the solar energy and release the heat energy to the steam from the base coal-fired power generation system. And two solar energy heaters are adopted. Figure 2 shows the local schematic diagram of the integration of the tower solar collector system with the boiler system of the base system.

Part of the extraction steam from the boiler steam-water separator is heated by the #1 solar heater and then mixes with the outlet steam of the boiler high-temperature superheater with the reduced steam temperature because of the operation mode of coal saving in order to maintain the temperature of the main steam (high-pressure cylinder

inlet steam) at the original level (566°C). The #2 solar heater heats the outlet steam of high-temperature reheater with the reduced steam temperature because of the operation mode of coal saving in order to maintain the reheat steam temperature at the original level (566°C).

TSCACPG system

In this paper, a tower solar collector-aided coal-fired power generation (TSCACPG) system is proposed and studied with the operating mode of coal saving (the boiler feedwater flow rate is constant, and the coal consumption rate is reduced with the introduction of solar energy). Figure 3 shows the schematic diagram of the TSCACPG system.

The main advantages of the TSCACPG system lie in the following three aspects.

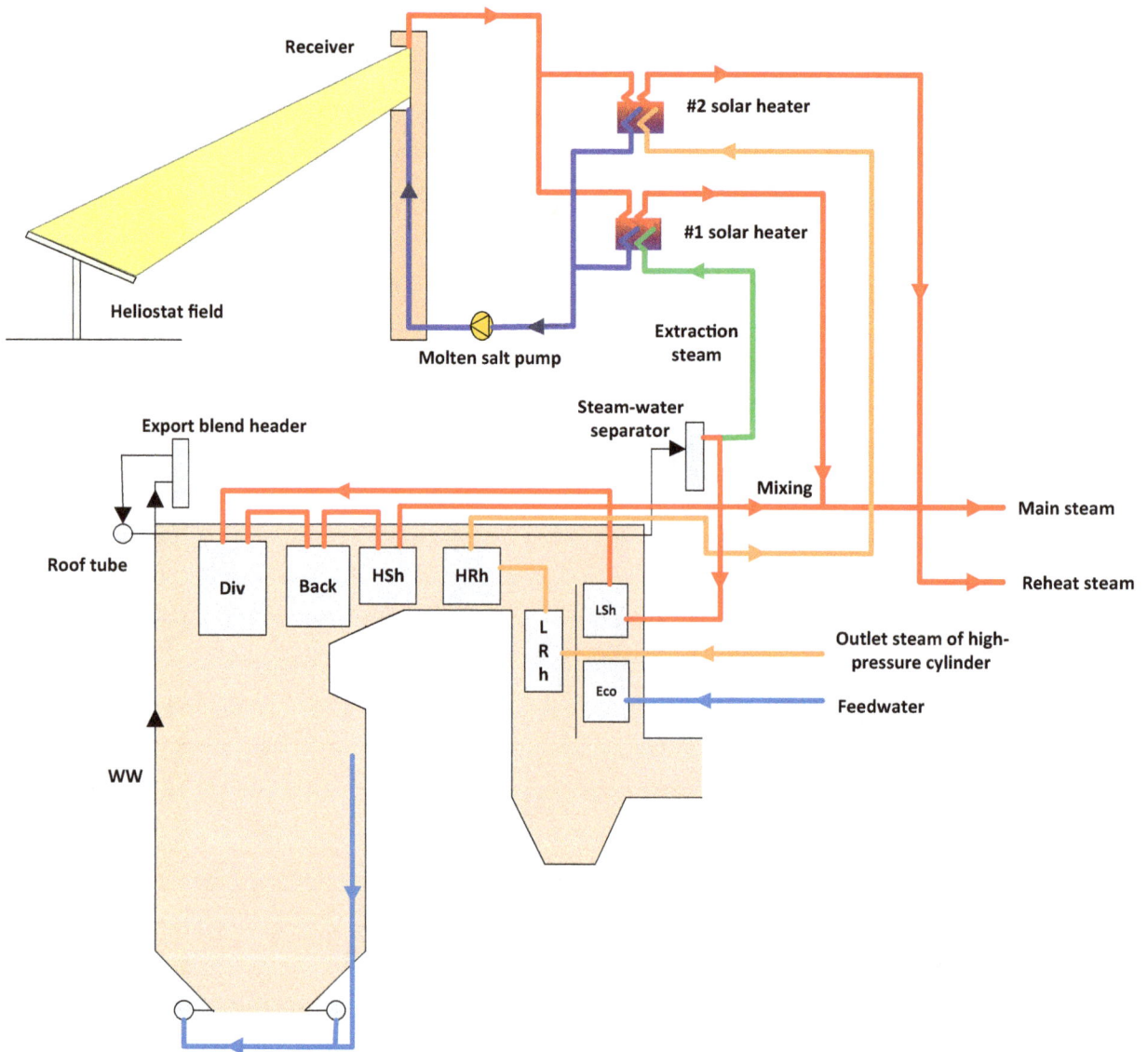

Figure 2. Local schematic diagram of integration of the solar tower system with the boiler system of the base system.

1. Firstly, compared with the reheat steam temperature adjustment mode by using the flue gas damper, the TSCACPG system avoids the disadvantages of a large time lag and the limited adjustment range.

As we know, the reheat steam temperature adjustment method with the flue gas damper has the advantages of simple structure and convenient operation, which is used by many large-scale power plant boilers. But both the opening degree of the flue gas damper and the change of steam temperature are nonlinear, and the effective range is narrow. Generally, the flue gas damper has a good performance in the 40~60% opening degree. But, if the opening degree is below 25% or above 75%, the flue gas damper will be useless. This kind of the reheat steam temperature adjustment method will be ineffective and

cannot guarantee the rated reheat steam temperature. As in the actual operation, there will be some fluctuations in the coal and the load. After that, both the composition of flue gas and the opening degree will be changed. Coupled with the instability of solar energy, a higher requirement for the sensitivity and the damper adjustment speed is put forward. In this paper, the #2 solar heater is used to adjust the reheat steam temperature; at the same time, the flue gas damper is only used as an auxiliary temperature adjustment method. So it avoids the above-mentioned problem.

2. Secondly, compared with heating the feedwater or condensation water mode by using the solar energy, because of without the change in the regenerative extraction steam flow, the TSCACPG system will not affect the

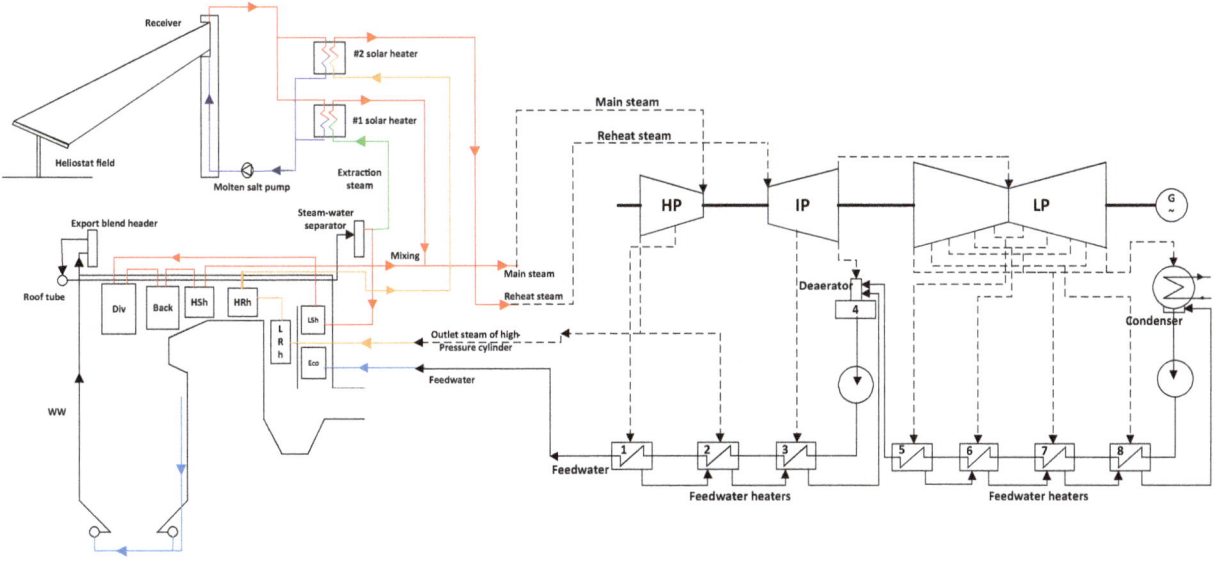

Figure 3. Schematic diagram of the TSCACPG system.

changes in steam flow, pressure, temperature and the output power of the steam turbine. And it is not necessary to calculate and evaluate the steam turbine performance at all levels once again to ensure the safe operation of the unit.

3. Thirdly, as the supercritical once-through boiler has no steam drum, the heat storage capacity is small. Therefore, in the same degree of disturbance, the steam temperature changes more abruptly. And the TSCACPG system can better control the temperatures of both the main steam and reheat steam.

Simulation of the System

In this paper, the TSCACPG system is simulated with the software platform TRNSYS that has been a perfect transient simulation software with the extended function developed by the University of Wisconsin. And Figure 4 is the TSCACPG simulation system built by the TRNSYS.

Heliostat field

The heliostat field model requires a user-supplied field efficiency matrix to evaluate the field efficiency, η_{field}, as a function of the date and time. The field efficiency matrix will be read in and interpolated by the TRNSYS routine data. It comes from the actual operating parameters of a solar power station. Its specific values are shown in Table 2. And its three-dimensional graph is shown in Figure 5. The power to the receiver is calculated by the following equation:

$$\dot{Q}_{rec} = A_{field} \cdot \rho_{field} \cdot I \cdot \eta_{field} \cdot \Gamma. \tag{1}$$

The model also considers the parasitic electrical energy consumption for tracking, start-up and shutdown. The shutdown is performed automatically at the high wind speed [22].

Receiver

The receiver is a photo-thermal conversion device in the solar power tower system. The molten salt, which consists of LiCl and KCl (the component ratio of LiCl and KCl is 0.595:0.405, and the working temperature ranges from 355°C to 1400°C), is used as the working medium of tower solar collector. The outlet temperature of the molten salt receiver is designed to a constant value of 620°C.

The receiver outlet parameters are calculated depending on the inlet condition of the molten salt and the radiation input. And the receiver body and piping heat losses can be calculated [23].

The model of the receiver in the paper is described as follows:

$$\overline{T}_{abs} = \frac{t_{in} + t_{out}}{2} + 273.15 \tag{2}$$

$$\dot{Q}_{loss\_rad} = A_p \cdot eps_{abs} \cdot 3600 \cdot \sigma \cdot (\overline{T}_{abs})^4 \tag{3}$$

$$\dot{Q}_{abs} = eta_{opt} \cdot \dot{Q}_{rec} - \dot{Q}_{loss\_rad} \tag{4}$$

$$\dot{Q}_{loss\_pipe} = A_{pipe} \cdot \left(eps_{pipe} \cdot 3600 \cdot \sigma \cdot (t_{out} + 273.15)^4 + k_{pipe} \cdot (t_{out} - t_{amb}) \right) \tag{5}$$

$$\dot{Q}_{loss\_cool} = f_{cooling\_loss} \cdot \dot{Q}_{abs} \tag{6}$$

$$\dot{Q}_{net} = \dot{Q}_{abs} - \dot{Q}_{loss\_pipe} - \dot{Q}_{loss\_cool} \tag{7}$$

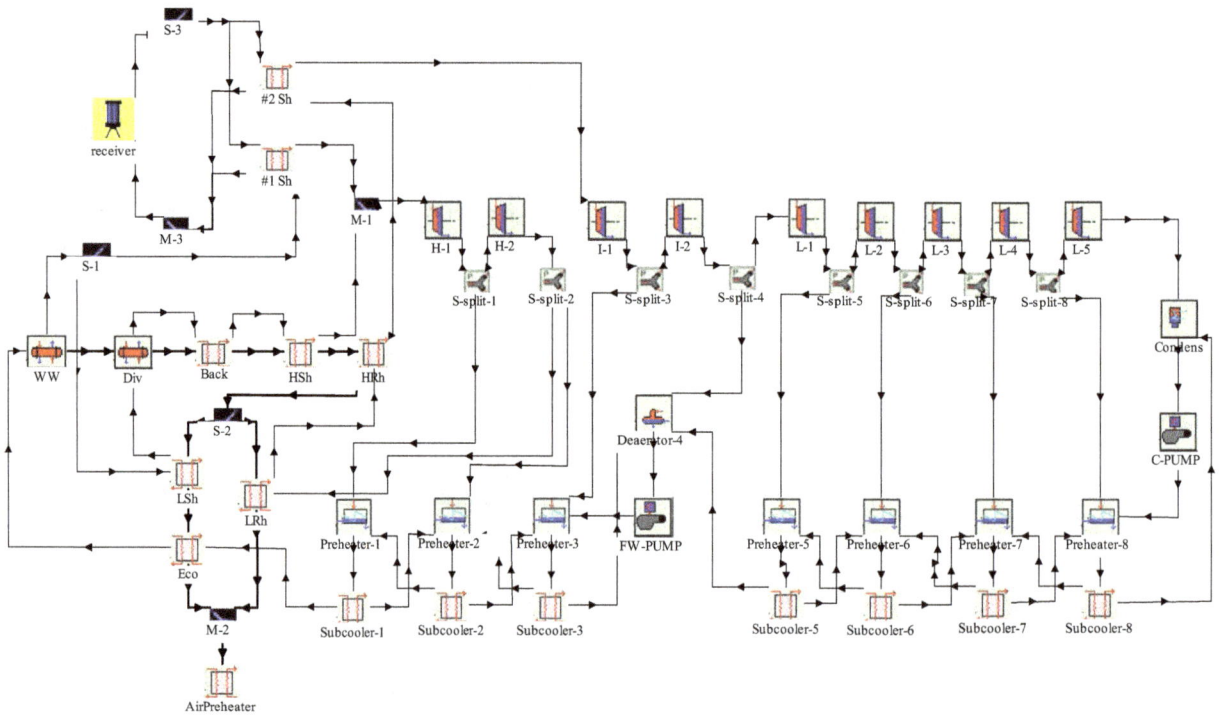

Figure 4. The TSCACPG simulation system built by TRNSYS.

Table 2. The field efficiency 7 × 9 matrix including 7 zenith angles and 9 azimuth angles.

	−120	−90	−60	−15	0	15	60	90	120
0	0.1710	0.1430	0.1523	0.1782	0.3187	0.3658	0.1803	0.1523	0.1482
25	0.5513	0.6363	0.6839	0.7140	0.7280	0.7264	0.6373	0.5710	0.4819
45	0.6352	0.6974	0.7409	0.7751	0.7808	0.7793	0.7130	0.6528	0.5855
65	0.6839	0.7171	0.7440	0.7658	0.7705	0.7699	0.7316	0.6964	0.6611
75	0.7016	0.7212	0.7378	0.7513	0.7554	0.7565	0.7337	0.7119	0.6912
85	0.7140	0.7202	0.7264	0.7306	0.7342	0.7368	0.7285	0.7223	0.7150
90	0.7223	0.7223	0.7223	0.7223	0.7223	0.7223	0.7223	0.7223	0.7223

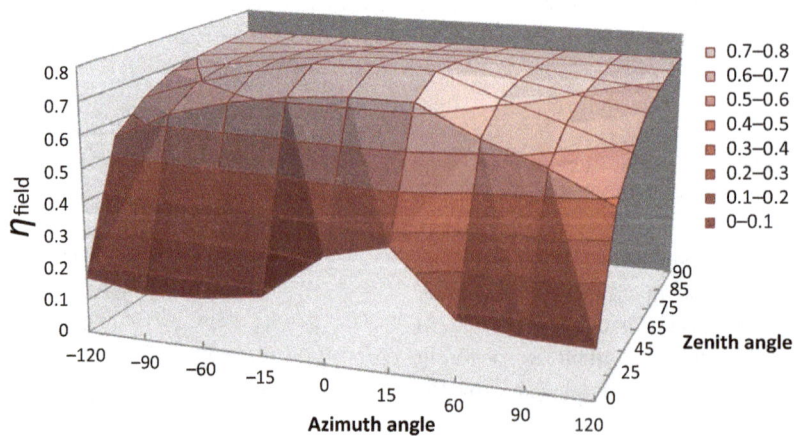

Figure 5. The three-dimensional graph of the field efficiency matrix.

$$\dot{Q}_{th} = c_{salt} \cdot \dot{m}_{salt} \cdot (t_{out} - t_{in}) \tag{8}$$

$$\eta_{rec} = \frac{\dot{Q}_{net}}{\dot{Q}_{rec}} \tag{9}$$

#1 solar heater

The #1 solar heater heats the extraction steam from the boiler steam-water separator. The model adopted for the #1 solar heater in the paper is described as follows:

$$\dot{m}_{ms1} \cdot (h_{i,ms1} - h_{o,ms1}) = \dot{m}_{es} \cdot (h_{0,es} - h_{1,es}) \tag{10}$$

#2 solar heater

The #2 solar heater heats the outlet steam of high-temperature reheater. The model adopted for the #2 solar heater in the paper is described as follows:

$$\dot{m}_{ms2} \cdot (h_{i,ms2} - h_{o,ms2}) = \dot{m}_{rs} \cdot (h_{o,rs} - h_{i,rs}) \tag{11}$$

System performance evaluation

To evaluate the benefits or the efficiency of the solar heat utilization in the TSCACPG, the solar-to-electricity efficiency (η_{se}) is defined as:

$$\eta_{se} = \frac{3.6 \cdot P_S}{\dot{Q}_S} = \frac{3.6 \cdot \left(P_Z - \frac{Q_b}{3.6} \cdot \eta_{ref}\right)}{\dot{Q}_S} = \frac{3.6 \cdot P_Z - \dot{Q}_b \cdot \eta_{ref}}{\dot{Q}_S} \tag{12}$$

$$\dot{Q}_S = A_{field} \cdot \rho_{field} \cdot I \tag{13}$$

The standard coal consumption rate of the power plant is the boiler coal consumption for per kWh electricity output (g/kWh), defined as:

$$b = \frac{1000 \, m_{sc}}{E} \tag{14}$$

Case Study

In this case, a tower solar energy collector-aided a 660-MW supercritical coal-fired power unit in Dunhuang, China (40°N and 94°E), is described and exemplified. Dunhuang is a city with rich solar resources. Figure 6 shows the DNI data of a typical year in Dunhuang.

Size selection of tower solar energy collector system

It is necessary to determine an appropriate size of the heliostat field (molten salt tower size also changed accordingly) in order to match the 660-MW supercritical coal-fired power unit. The direct normal irradiance (DNI) of the sun is the prerequisite for the design of the heliostat field. This paper uses the weather database of TRNSYS and selects

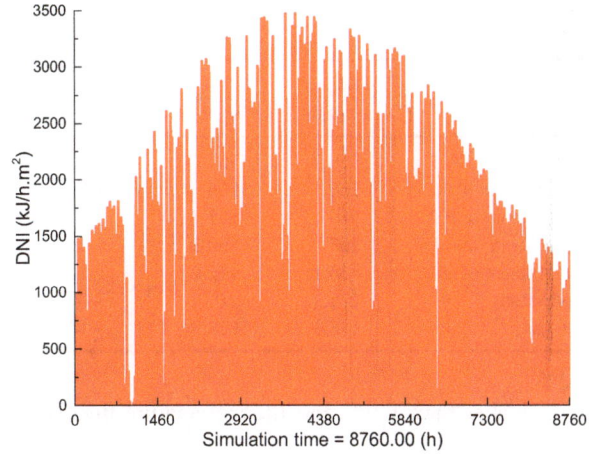

Figure 6. DNI data of typical year in Dunhuang (kJ/h.m²).

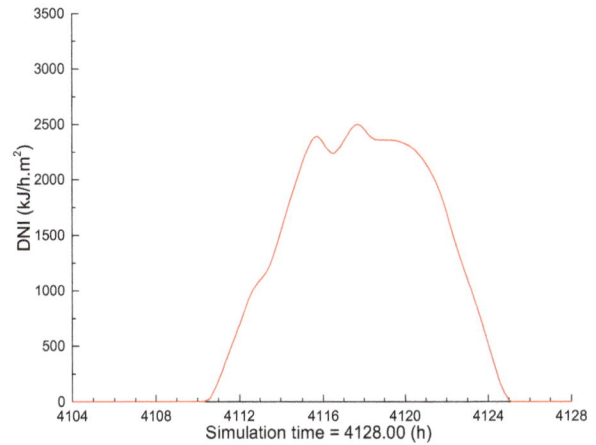

Figure 7. DNI data of the summer solstice in Dunhuang (kJ/h.m²).

the DNI value 2499 kJ/h.m² (equal to 694 W/m²) as the heliostat field design point. It is at 13:00 on 21 June, that is, the summer solstice day (4104–4128 h) in Dunhuang of the typical year data, as shown in Figure 7. According to the characteristics of the scheme, it is preliminarily determined that the heliostat surface area of 120,000 m² and the molten salt mass flow rate of 1.80×10^6 kg/h are set as the reference for the performance analysis. In the later section of heliostat area optimization, the heliostat surface area will be further optimized.

Selection of the boiler extraction steam fraction

As we know, if the flue gas temperature of the power plant is too low, it will cause the serious low-temperature corrosion. And the temperature fluctuation of boiler heat exchanger will be great if the high-extraction steam fraction is selected. In this paper, the flue gas temperature

Table 3. The temperature changes of main parameters with 11% extraction steam.

Temperature (°C)	Original system	After solar-aided	Variation
Inlet steam of HP	566	566	+0
Inlet steam of IP	566	566	+0
Outlet steam of HSh	566	563.7	−4.25
Outlet steam of HRh	566	550.3	−15.7
Outlet steam of LRh	438.1	426.0	−12.1
Feedwater	275.3	275.3	+0
Flue gas	142.8	128.3	−14.5

being higher than 120°C and the temperature fluctuation of high-temperature superheater (HSh) being lower than 5°C are used as the constraint conditions to select the largest extraction steam fraction with the change in the heliostat field area. Under the heliostat surface design area of 120,000 m², the extraction steam fraction of the boiler is set at 11%. And Table 3 shows the changes in temperature with the extraction steam fraction of 11%.

Performance analysis

The performance of the TSCACPG system is analyzed at the full load condition with the coal-saving operating mode (the boiler feedwater mass flow rate is constant and the coal consumption rate is reduced).

The logic flowchart for TRNSYS to carry out the analysis is shown in Figure 8. The loop algorithm is the iterative method. When the error is <0.01°C, the iteration terminates.

Figure 9 shows the thermal load variations in the boiler heat exchangers. After adopting the coal-saving operating mode, the thermal load of each boiler heat exchanger declines from 1.2% of Eco to 15.1% of LSh. And the total decline is 40.8 MW with the reduction rate of 3.2%.

Figure 10 shows the molten salt flow distributions of #1 and #2 solar heater. In Figure 10, L1 is the mass flow rate of molten salt for #1 solar heater ($\dot{m}_{ms1}$); L2 is the mass flow rate of molten salt for #2 solar heater ($\dot{m}_{ms2}$); L_Z is the total mass flow rate of molten salt ($\dot{m}_{salt}$). This

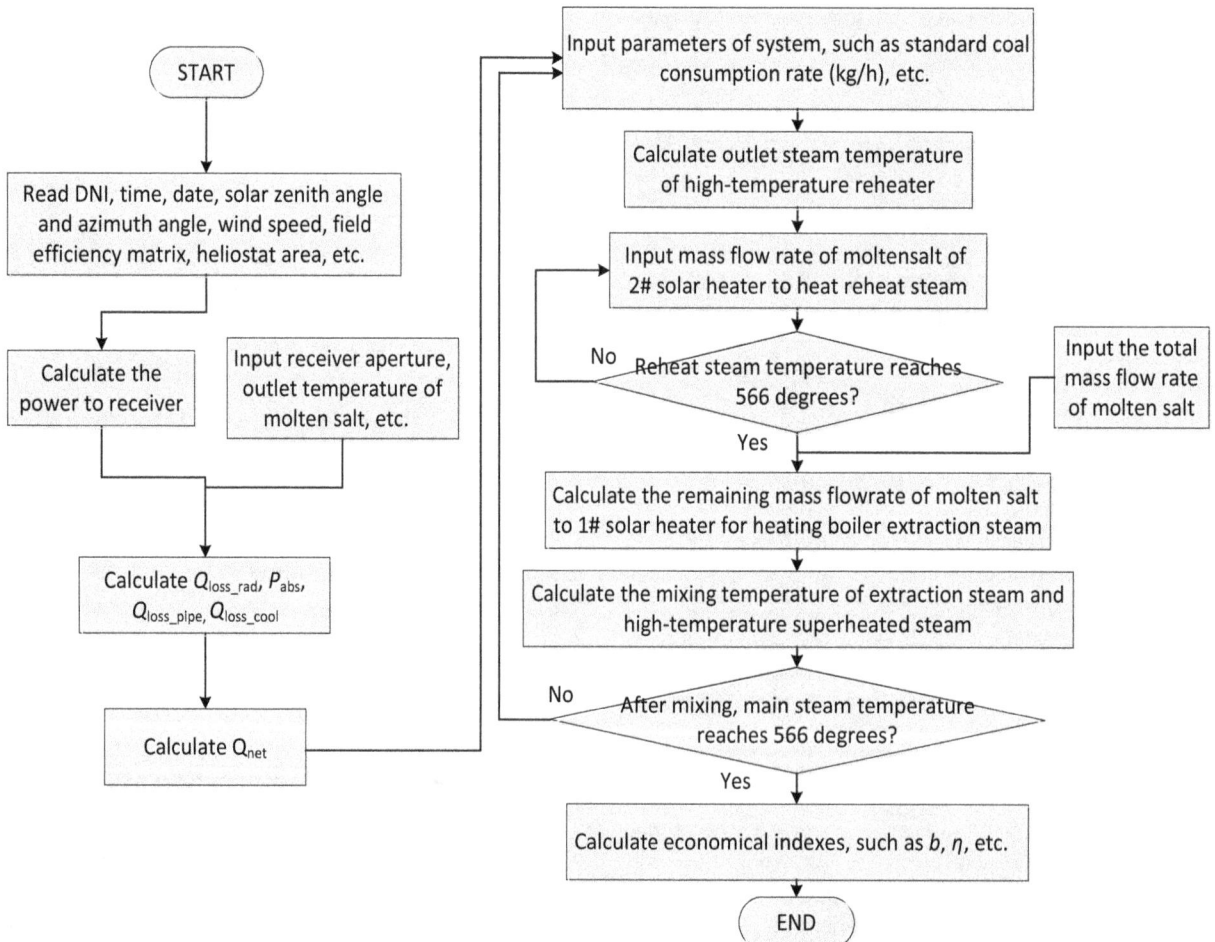

Figure 8. Logic flowchart for the integrated system performance analysis.

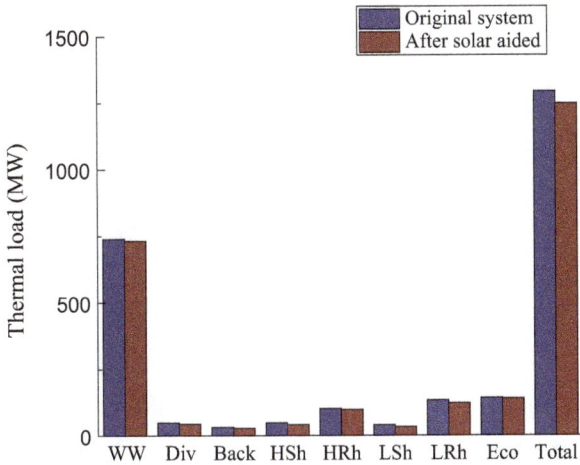

Figure 9. The variations in thermal load of boiler heat exchangers.

figure also reveals the scheme of start-up and shutdown. When the DNI value is greater than 1405 kJ/h.m^2 (equal to 390 W/m^2), the solar power generation system will be put into the conventional coal-fired power system; when the DNI value is less than 1405 kJ/h.m^2, the solar power generation system will be removed. It should be noted that the outlet temperature and outlet flow rate of the molten salt of receiver remain unchanged during the stable operation. This means a portion of solar energy is not used when the DNI value exceeds the value of 2030 kJ/h. m^2. The main reason is for the consideration of the stable and safe operation of the overall power plant system. In this paper, for the economic reason, the TSCACPG system without the energy storage unit is established. In addition, on the basis of the data from the literature [24], the

additional electrical consumption (losses) to prevent the crystallization of the heat transfer medium in our study is assumed as 1% of the solar power output, which is considered in the simulation.

Figure 11 shows both the efficiency and power of the tower solar energy system. In Figure 11, the field power is the heliostat field power ($\dot{Q}_{rec}$); the rec_power is the molten salt tower power ($\dot{Q}_{net}$); ms_abs_power is the molten salt tower thermal power (molten salt absorbed power) ($\dot{Q}_{th}$); the field_efficiency is the heliostat field efficiency (η_{field}); the rec_efficiency is the molten salt tower efficiency (η_{rec}). The molten salt tower power is the result of DNI and the heliostat field efficiency factors. And the efficiency of molten salt tower is close to 90%. The thermal power of the solar power tower is 1.49×10^8 kJ/h (equal to 41.39 MW). The solar power generation system runs for 8.7 h on the summer solstice day.

The simulation results are listed in Table 4. From Table 4, it can be seen that the net plant efficiency is increased by 1.07% and the standard coal consumption declines to 7.7 g/kWh correspondingly due to the aid of solar energy.

In addition to the performance of the summer solstice day, the annual performance is also very important for the TSCACPG system. Therefore, the annual performance of the TSCACPG has been calculated using the DNI data of typical year in Dunhuang. Figure 12 shows the monthly utilization hours of the solar energy, and the annual performances are shown in Table 5. It can be seen that the annual utilization hours of solar energy reaches 2047 h and the annual solar-to-electric efficiency reaches to 15.14%.

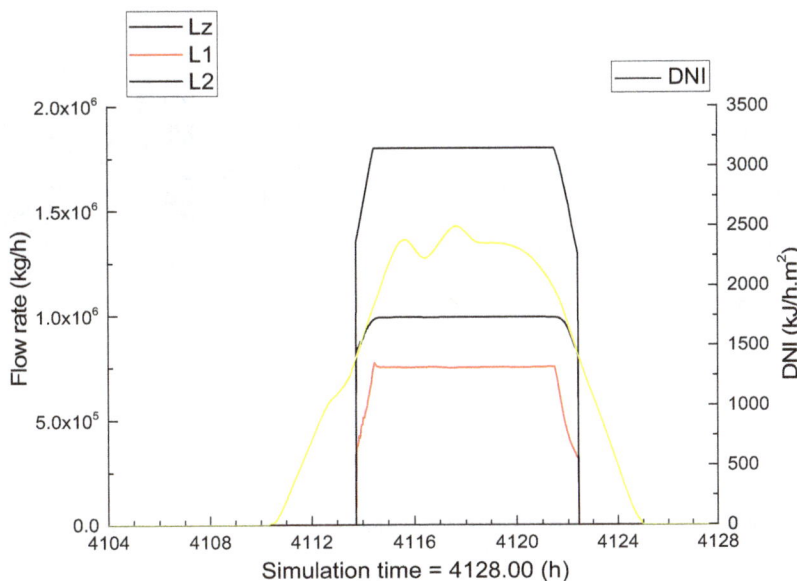

Figure 10. Molten salt flow distributions of #1 and #2 solar heater.

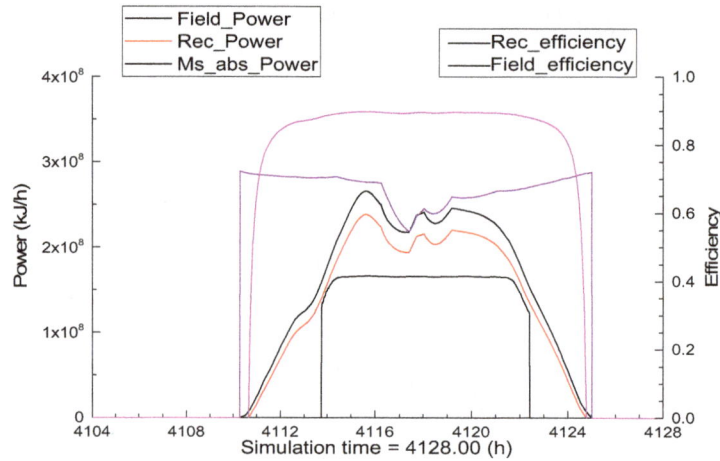

Figure 11. Efficiency and power of the solar power generation system.

Heliostat area optimization

The DNI of each day in a year is different, so the heliostat surface area mentioned above is not the optimal value which is designed on the summer solstice day. In order to get the optimal value, it is necessary to analyze the annual performance (such as the annual power generation, annual solar-to-electric efficiency) with different heliostat surface areas. In consideration of the economic cost, the area with

the maximum annual solar-to-electric efficiency is selected as the optimal value. Figure 13 shows the variation in the boiler extraction steam fraction under different heliostat surface areas. Then, the correlation between the annual performance and the heliostat surface area is analyzed. And the analysis results are shown in Figures 14–16.

Figure 14 shows that the annual power generation from the solar energy increases quickly at the beginning, and then the growth rate slows down with the increase in the heliostat surface area. Figure 15 shows that the standard coal consumption rate decreases quickly at first, and

Table 4. Performance of the TSCACPG system at the design point.

Items	Unit	Value
Power output of original unit	MW	660
Power output of solar-aided unit (TSCACPG system)	MW	660
Boiler feedwater mass flow rate	t/h	2150.6
Inlet temperature of molten salt for #1 solar heater	°C	620
Outlet temperature of molten salt for #1 solar heater	°C	516.9
Inlet temperature of molten salt for #2 solar heater	°C	620
Outlet temperature of molten salt for #2 solar heater	°C	572.8
Outlet temperature of boiler extraction steam for #1 solar heater	°C	600.4
Outlet temperature of reheat steam for #2 solar heater	°C	566
Heliostat field area	m²	121,700
Thermal power of solar power tower	MW	41.39
Solar-to-electric efficiency	%	22.92
Standard coal consumption of original unit	g/kWh	301.5
Standard coal consumption of solar-aided plant	g/kWh	293.8
Standard coal consumption decline	g/kWh	7.7
Net efficiency of original plant	%	40.80
Net efficiency of solar-aided plant (TSCACPG system)	%	41.87

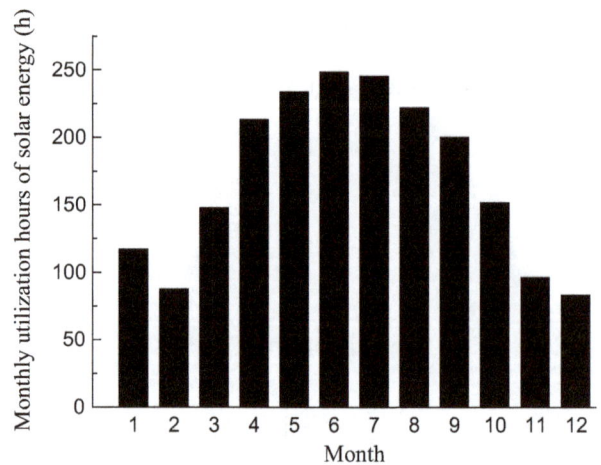

Figure 12. Monthly utilization hours of the solar energy.

Table 5. The annual performance of the TSCACPG system.

Item	Unit	Value
Annual utilization hours of solar energy	h	2047
Annual power generation of solar energy	kWh	3.72×10^7
Annual solar-to-electric efficiency	%	15.14
Annual standard coal saved	t	10,414

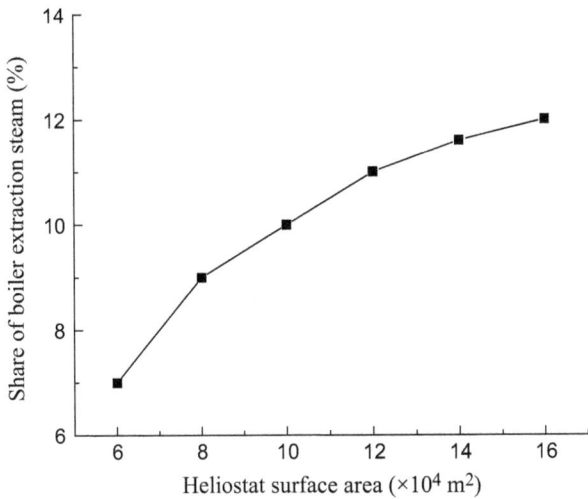

Figure 13. Variation in the boiler extraction steam fraction with the heliostat surface area.

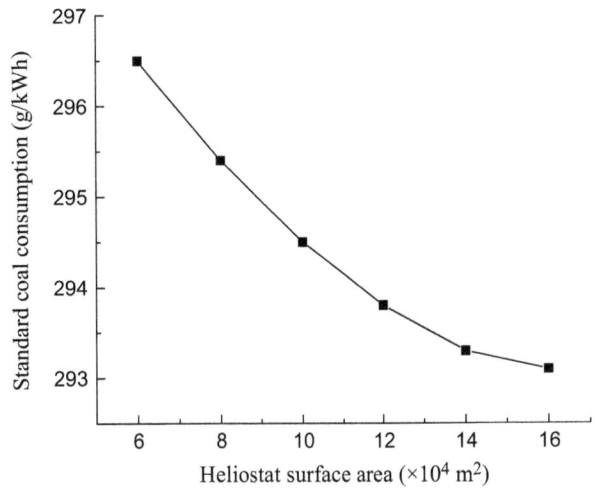

Figure 15. Variation in standard coal consumption with the heliostat surface area.

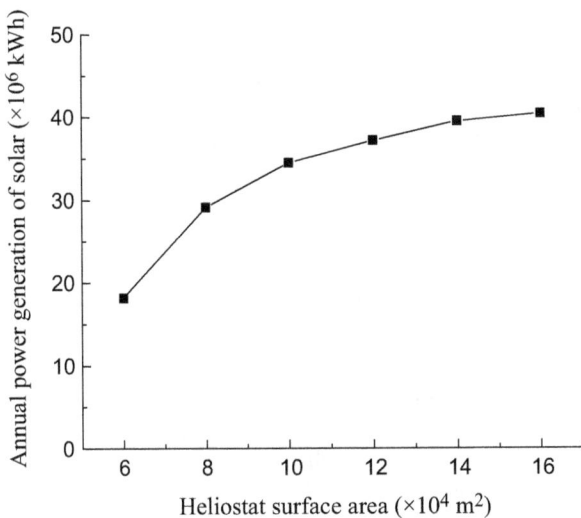

Figure 14. Variation in annual power generation of solar with the heliostat surface area.

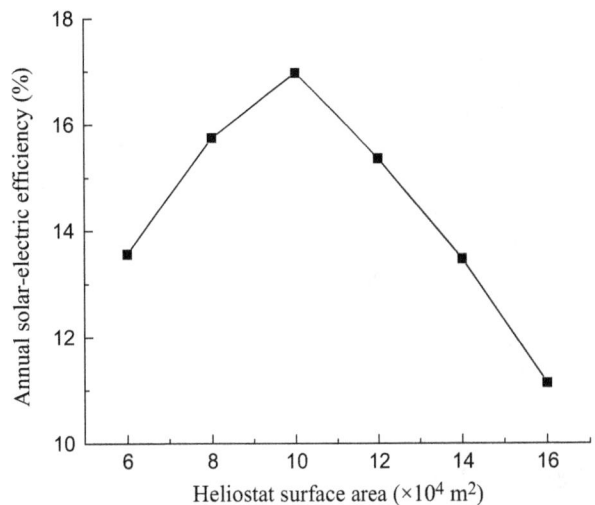

Figure 16. Variation in annual solar-to-electric efficiency with the heliostat surface area.

then the decline rate slows down with the increase in the heliostat surface area.

As shown in Figure 16, the annual solar-to-electric efficiency increases at first and then decreases with the increase in the heliostat surface area. It has a maximum value when the heliostat surface area is 100,000 m². The reason is that as the area increases, more coal will be saved and more solar energy will be used. But if the area continues to increase, the annual solar-to-electric efficiency will decrease yet due to the increase in the unused solar energy. In other words, the growth rate of the unused solar energy is greater than the growth rate of the used solar energy. The maximum annual solar-to-electric efficiency is 16.74%. Obviously, the value of 16.74% is higher than that listed in Table 4 (15.14%)

of the design condition. Moreover, the performances are shown in Table 6 under the optimal situation.

Conclusions

For the economic and security reasons, the TSCACPG system with the coal-saving operating mode without the energy storage is designed on the summer solstice day. The system performances of 1 day and 1 year are all carried out by using the DNI data of typical year in Dunhuang by using the simulation platform TRNSYS.

The research results show that as the heliostat area increases, the annual power generation from the solar

Table 6. Performance of the TSCACPG system under the optimal situation.

Item	Unit	Value
Heliostat field area	m^2	101,400
Net efficiency of original plant	%	40.80
Net efficiency of solar-aided plant	%	41.77
Standard coal consumption of original unit	g/kWh	301.5
Standard coal consumption of solar-aided plant	g/kWh	294.5
Standard coal consumption decline	g/kWh	7.0
Annual utilization hours of solar energy	h	2231
Annual power generation of solar energy	kWh	3.45 × 10^7
Annual solar-to-electric efficiency	%	16.74
Annual standard coal saved	t	9655

energy increases, while the standard coal consumption rate declines. In addition, there is a maximum value of the annual solar-to-electric efficiency. In consideration of the economic costs, the heliostat field area with the maximum annual solar-to-electric efficiency is selected as the optimal value. The results show that, for the case studied, the optimal heliostat surface area is 100,000 m^2, and the maximum annual solar-to-electric efficiency is 16.97%. And under the optimal situation, the standard coal consumption rate is reduced from 301.5 to 294.5 g/kWh when the solar power tower aids the power plant. This article aimed to provide the integration method and ideas for engineering projects of the tower solar collector-aided coal-fired power generation system. The achievement obtained from this paper can guide the choice of solar island scale in a specific project and provide a method of further reducing coal consumption of coal-fired power plants.

Acknowledgments

This study has been supported by the National Nature Science Foundation Project (No. 51576062 and No. 51276063) and the National Basic Research Program of China (No. 2015CB251505).

Nomenclature

A_{field}　total heliostat surface area, m^2;

A_p　receiver aperture area, m^2;

A_{pipe}　surface area of piping, m^2;

b　standard coal consumption rate of power supply, g/kWh;

Back　back platen superheater;

c_{salt}　specific heat capacity of molten salt, kJ/kg. °C;

Div　division wall;

E　electrical energy output of the plant, kWh;

Eco　economizer;

eps_{abs}　emissivity of absorber;

eps_{pipe}　emissivity of piping;

eta_{opt}　optical efficiency;

$f_{cooling\_loss}$　receiver cooling loss factor;

$h_{i,ms1}$　enthalpy of molten salt for #1 solar heater, kJ/kg;

$h_{o,ms1}$　outlet enthalpy of molten salt for #1 solar heater, kJ/kg;

$h_{i,ms2}$　inlet enthalpy of molten salt for #2 solar heater, kJ/kg;

$h_{0,ms2}$　outlet enthalpy of molten salt for #2 solar heater, kJ/kg;

$h_{i,es}$　inlet enthalpy of extraction steam, kJ/kg;

$h_{o,es}$　outlet enthalpy of extraction steam, kJ/kg;

$h_{i,rs}$　inlet enthalpy of reheat steam, kJ/kg;

$h_{o,rs}$　outlet enthalpy of reheat steam, kJ/kg;

HRh　high-temperature reheater;

HSh　high-temperature superheater;

I　direct normal irradiance, kJ/h.m^2;

k_{pipe}　convective loss coefficient of piping, kJ/h.m^2.K;

LRh　Low-temperature reheater;

LSh　Low-temperature superheater;

$\dot{m}_{es}$　mass flow rate of extraction steam from boiler steam-water separator, kg/h;

$\dot{m}_{ms1}$　mass flow rate of molten salt for #1 solar heater, kg/h;

$\dot{m}_{ms2}$　mass flow rate of molten salt for #2 solar heater, kg/h;

$\dot{m}_{rs}$　mass flow rate of reheat steam, kg/h;

$\dot{m}_{salt}$　total mass flow rate of molten salt, kg/h;

m_{sc}　coal consumption converted to the standard coal, kg;

P_S　solar power output, W;

P_Z　total output of the TSCACPG, W;

$\dot{Q}_{abs}$　solar radiation absorbed by the receiver, kJ/h;

$\dot{Q}_b$　heat load of the boiler, kJ/h;

$\dot{Q}_{loss\_cool}$　receiver cooling losses, kJ/h;

$\dot{Q}_{loss\_pipe}$　piping losses, kJ/h;

$\dot{Q}_{loss\_rad}$　receiver radiation losses, kJ/h;

$\dot{Q}_{net}$　thermal output of the receiver, kJ/h;

$\dot{Q}_{rec}$　the power to the receiver, kJ/h;

$\dot{Q}_S$　total solar projection energy, kJ/h;

$\dot{Q}_{th}$　thermal power of solar power tower, kJ/h;

$\overline{T}_{abs}$　average temperature of absorber, °C;

t_{amb}　ambient temperature, °C;

t_{in}　inlet temperature of molten salt for receiver, °C;

t_{out}　outlet temperature of molten salt for receiver, °C;

WW　water wall;

Greek symbols

Γ　a control parameter describing the fraction of the field in track;

σ　Stefan constant, $\sigma = 5.67 \times 10^{-8}$W/(m^2.K^4);

ρ_{field} heliostat field reflectivity;
η_{field} the heliostat field efficiency;
η_{rec} the receiver efficiency;
η_{ref} the thermal efficiency of the original coal-fired power unit;
η_{se} the solar-to-electric efficiency;

Conflict of Interest

None declared.

References

1. Jianli, Y., L. Rumou, J. Hongguang, H. wei, and H. hui 2007. Classification of solar thermal power system (1). Sol. Energy 4:30–33.
2. Jianli, Y., L. Rumou, J. Hongguang, H. wei, and H. hui 2007. Classification of solar thermal power system (2). Sol. Energy 5:29–32.
3. Chuanqiang, Z., H. Hui, and J. Hongguang. 2010. Development situation of power generation technology using heat of light-concentrating solar energy. Therm. Power Generation 39:5–9. 13.
4. Jing, C., L. Jianzhong, S. Wangjun, Z. Junhu, and C. Kefa. 2012. Status quo in research of solar energy thermal power generation system. Therm. Power Generation 41:17–22.
5. Shu-qing, C., C. Juan, and Y. Su-rong. 2009. Analysis of the prospects for solar power generation in China. Inner Mongolia Sci. Technol. Econ. 17:77.
6. Junsheng, Z.. 2003. Development situation of new & renew energy in China. Renew. Energy 2:3–10.
7. Yinghong, C., and C. Juan. 2007. Review on hybrid solar thermal generating system. Modern Electric Power 24:24–28.
8. Zinian, H.. 2009. Solar thermal. University of Science and Technology of China Press, Hefei, 7.
9. Mills, D. 2004. Advances in solar thermal electricity technology. Sol. Energy 76:19–31.
10. Yongping, Y.. 2009. Integrative power generation technology and economical analysis of solar aided coal-fired power plant. China Sci. E Sci. Technol. 39:673–2009.
11. Eric, H., Y. P. Yang, A. Nishimura, and F. Yilmaz. 2010. Solar-aided power generation. Appl. Energy 87:2881–2885.
12. Yang, Y., Q. Yan, R. Zhai, A. Kouzani, and E. Hu 2011. An efficient way to use medium-or-low temperature solar heat for power generation-integration into conventional power plant. Appl. Therm. Eng. 31:157–162.
13. Hong, H., Y. Zhao, and H. Jin. 2011. Proposed partial repowering of a coal-fired power plant using low-grade solar thermal energy. Int. J. Thermodyn. 14:21–28.
14. Hongjuan, H., W. Mengjiao, and Y. Yongping. 2015. Exergy evaluation of solar aided coal-fired power generation system. Proc. CSEE 35:119–125.
15. Peng, S., Z. Wang, H. Hong, D. Xu, and H. Jin 2014. Exergy evaluation of a typical 330 MW solar-hybrid coal-fired power plant in China. Energy Convers. Manage. 85:848–855.
16. Gupta, M. K., and S. C. Kaushik. 2009. Exergetic utilization of solar energy for feedwater preheating in a conventional thermal power plant. Int. J. Energy Res. 33:593–604.
17. Eck, M., E. Zarza, M. Eickhoff, J. Rheinlander, and L. Valenzuela. 2003. Applied research concerning the direct steam generation in parabolic troughs. Sol. Energy 74:341–351.
18. Dimityr, P.. 2011. An option for solar thermal repowering of fossil fuel fired power plants. Sol. Energy 85:344–349.
19. Jian, M., Y. Yongping, H. Hongjuan, and Z. Nan. 2015. Techno-economic analysis of solar thermal aided coal-fired power plants. Proc. CSEE 35:1406–1412.
20. Larrain, T. J., and R. A. Escobar. 2010. Solar thermal power plant performance model to determine fossil fuel backup consumption for different locations in northern Chile. Renew. Energy 35:1632–1643.
21. Yawen, Z., H. Hui, and J. Hongguan. 2014. Mid and low-temperature solar–coal hybridization mechanism and validation. Energy 74:78–87.
22. Schwarzbozl, P., U. Eiden, R. Pitz-Paal, D. Zentrum, für Luft und Raumfahrt e.V. (DLR) and D-51170 Köln. 2002. A TRNSYS Model Library for Solar Thermal Electric Components (STEC). A Reference Manual Release 2.2, 35–37.
23. Schwarzbozl, P., U. Eiden, R. Pitz-Paal, D. Zentrum, für Luft und Raumfahrt e.V. (DLR) and D-51170 Köln. 2002. A TRNSYS Model Library for Solar Thermal Electric Components (STEC). A Reference Manual Release 2.2, 34.
24. Alice. Mojave solar power plant will use the method of electric tracer heating system to prevent the crystallization of the conduction oil. Available at: http://www.cspplaza.com/article-2130-1.html (accessed 02 August 2013).

16

Comprehensive study of performance degradation of field-mounted photovoltaic modules in India

Rajiv Dubey[1] ⓘ, Shashwata Chattopadhyay[1], Vivek Kuthanazhi[1], Anil Kottantharayil[1], Chetan Singh Solanki[1], Brij M. Arora[1], Krishnamachari L. Narasimhan[1], Juzer Vasi[1], Birinchi Bora[2], Yogesh Kumar Singh[2] & Oruganti S. Sastry[2]

[1]National Centre for Photovoltaic Research and Education, Indian Institute of Technology Bombay, Powai, Mumbai 400076, India
[2]National Institute for Solar Energy, Ministry of New and Renewable Energy, New Delhi 110003, India

Keywords
Degradation, photovoltaic modules, PV defects, silicon

Correspondence
Rajiv Dubey, National Centre for Photovoltaic Research and Education, Indian Institute of Technology Bombay, Powai, Mumbai 400076, India. E-mail: dubey.rajeev.iitb@gmail.com

Funding Information
Ministry of New and Renewable Energy India, Solar Energy Research Institute for India and the United States (SERIIUS) (DE-AC36-08GO28308).

Abstract

The All India Survey of Photovoltaic Module Reliability 2014 is an enhanced version of the survey conducted in the previous year, with detailed characterization of PV modules including current-voltage, infrared and electroluminescence imaging, visual inspection, insulation resistance test and interconnect breakage test. More than a thousand modules were inspected in the field and the main results of the survey are presented in this paper. The average P_{max} degradation rate for the so-called 'good' modules (Group X) is 1.33%/year which is higher than that commonly projected by manufacturers, and widely employed in financial calculations. Modules falling in the 'not-so-good' category (Group Y) show even higher degradation rates, and it is at least partly due to higher number of micro-cracks in the modules, and increased degradation of the packaging materials like encapsulant, backsheet, etc. Modules in 'Hot' climates degrade faster than modules in the 'Non-Hot' climates. Degradation in fill factor is the primary cause for performance degradation in the young modules (ages <5 years), whereas short-circuit current degradation is the main contributor to power degradation in the older modules. Small installations (<100 kW_p capacity) show higher degradation than large systems, which may be partly due to lack of proper due diligence by the owner at the time of procurement and installation.

Introduction

India has embarked on an ambitious plan to install 100 GW of solar power by 2022, under the Jawaharlal Nehru National Solar Mission (JNNSM), in order to reduce reliance on fossil fuels and help ameliorate global warming. Consequently, large solar power plants are being set up very rapidly, mostly in the western and southern states which are rich in solar resource. The present day installed solar power capacity in India stands at 8.1 GW_p [1], out of which about 44% capacity has been added in the last 1 year. There is a disparity in the distribution of the solar power projects, with some of the largest power consuming states like Karnataka, Maharashtra, and Uttar Pradesh contributing <20% of the installed capacity [1]. Considering the enormous investments being made in solar photovoltaic (PV) power plants, it is important to understand the long-term performance of the PV modules in the Indian climatic conditions, since the energy yield would decide the ROI (return on investment) for the investors. The climatic conditions in India are very different from those of Europe, where solar panels have traditionally been installed in large capacities. In particular, the regions with high solar resource also tend to have hot climate, which may accelerate the degradation mechanisms due to the higher temperatures. It is necessary to evaluate the effects of the hotter Indian climate on the performance degradation of the solar panels, so as to enable installers and investors to take informed decisions on the investments they are making in this sector. Field surveys need to be conducted periodically to assess the actual on-the-ground realities of the solar PV technology. Some data are available from such surveys in other parts of the world. Jordan et al. [2, 3] reported that the average degradation rate for crystalline silicon modules is 0.8–0.9%/year whereas it is around 1%/year for thin film modules. They have indicated that hotter climates and mounting configurations may lead to higher degradation in some, but not all, products. Suleske et al. [4] have investigated the degradation of modules installed in a grid-tied power plant in Arizona, for 10–17 years, and has reported degradation rates ranging from 0.9%/year to 1.9%/year for nonhot spot modules, while for modules with hot spots, the degradation rate was found to be as high as 5%/year. Sastry et al. [5] have also reported that the degradation rates of crystalline silicon PV modules, monitored at their test bed at the National Institute of Solar Energy (NISE) in Gurgaon (near New Delhi) over a time of ten years, are more than the expected level [6]. Studies done by NISE on the performance analysis of different photovoltaic technologies (a-Si, multi c-Si and HIT) found that HIT and a-Si have performed better than multi c-Si [7, 8]. Furthermore, studies show that in the hot and dry type of climate, there is significant increases in the number of modules suffering from solder joint failure. Tamizhmani [9] have shown that modules in the hot climates of USA show a higher degradation rate than in other climatic zones. Kuitche et al. [10] have mentioned that for Hot and Dry type of climatic condition of Arizona, solder bond failure and discoloration are the two dominant failure modes. A comparative study of modules in different climatic zones in India and USA also indicates that modules degrade faster in hot climates [11]. It can be readily appreciated that comprehensive field surveys can yield a wealth of valuable information. The National Centre for Photovoltaic Research & Education (NCPRE) has been conducting 'All-India Surveys of PV Module Reliability' since 2013 in collaboration with the National Institute of Solar Energy (NISE) in order to evaluate the performance of PV modules in different parts of India. The 2014 edition of the survey has generated valuable data on the degradation of PV modules with respect to different climatic zones, technology, system size and age. In this survey, detailed characterization of the modules has been carried out in situ in the field, which is briefly explained in the next section, along with some of the important statistics which will allow us to understand the results.

Survey Statistics and Methodology

A total of 1148 modules were inspected in the six different climatic zones of India. India has a large geographic stretch, and different areas of the country experience very different climates –hot and dry, hot and humid, composite, moderate, cold and dry, and cold and cloudy [12]. This has been shown in Figure 1, along with the 19 locations marked in black dots, where 51 individual sites were surveyed by us during the 2014 All-India Survey. The details of sites and modules inspected in different climatic zones are given in Table 1. These comprised 983 c-Si modules and 165 thin film modules. Around 70% of the modules were young (0–5 years in age), around 20% were in the 10–20 years age group, and the remaining was older than 20 years. Almost 50% of the inspected modules had rated power >100 W_p, and 80% modules were connected to MPPT inverters.

The methodology adopted for the survey has been explained in detail elsewhere [13, 14], but is being briefly mentioned here. At every site, the first activity was to clean the modules using water, followed by current-voltage (I-V) characterization of individual modules using PVPM and Solmetric I-V tracers. Infrared (IR) thermography was performed next using FLIR E-60 Infrared camera, with the module short-circuited for about 5 min. Interconnect breakage was tested, using Togami Cell Line Checker. Insulation resistance was measured both under dry and wet conditions, by applying a high voltage (1000 V DC) between the shorted module terminals and the module frame. Based on the above measurements, some of the modules were selected for further tests like dark I-V, dark IR, and electroluminescence (EL) measurement. EL was taken in the late afternoon or after sunset, in the presence of ambient light, using the "Image Difference Technique" which has been described elsewhere [15]. Dark I-V was performed with the module covered, using a programmable DC power supply. For dark IR, the module was forward biased at rated short circuit current, and IR images were taken from the back side.

Analysis of Survey Data

I-V data taken at irradiances below 500 Wm^{-2} were removed from the analysis, since translation of these I-V curves to standard test conditions (STC) will introduce very high error. For the rest of the data, taken above 500 Wm^{-2}, the total error (which includes the instrument error and the translation error) is estimated to be <9%. Out of the 1148 modules inspected, the I-V data from 161 modules were discarded due to low irradiance measurement. Of the remaining 987 modules, 22% showed very high degradation rates (exceeding 6%/year), which were termed as "Outliers" and not considered for further analysis, as there appeared to be serious quality issues and other extraneous issues like

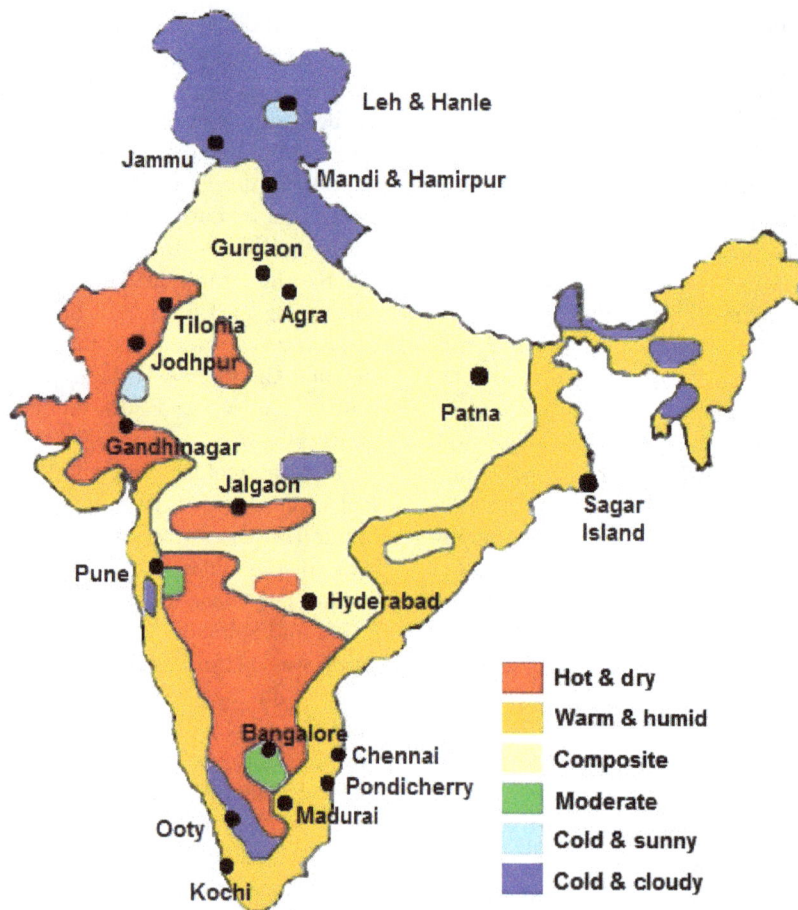

Figure 1. Climatic zones of India as per Bansal and Minke [12]. The black dots indicate the locations surveyed.

Table 1. Site summary.

Climatic zone	Number of sites surveyed	Number of modules surveyed
Hot and Dry (H and D)	15	194
Warm and Humid (W and H)	10	333
Composite (Comp)	9	308
Moderate (Mod)	8	135
Cold and Sunny (C and S)	6	56
Cold and Cloudy (C and C)	3	122
Total	51	1148

over-rating. Figure 2(A) shows the histogram of the P_{max} degradation rate of the remaining 766 modules, and it can be seen that there is a wide variation in the degradation rates, with the median and average of the histogram at 1.96%/year and 2.07%/year, respectively. These values are significantly higher than those reported by Jordan et al. [3], which raises concerns about the quality of the PV modules installed in India. The wide distribution in Figure 2(A) leads us to conclude that the quality of the modules and/or installation procedures

vary significantly. There are 'good' sites, and there are 'not-so-good' ones. (Note that we cannot a priori distinguish between poor quality of modules used and poor transportation/installation procedures, both of which could lead to higher degradation rates). To understand this distribution better, we have plotted the degradation rate separately for each module manufacturer and site, as shown in Figure 3. In this figure, the x-axis represents the manufacturers (code named from A to Z to preserve anonymity) and the different sites for these manufacturers (1, 2, 3, etc.) are also mentioned at the top. For example, modules of manufacturer A were found in 2 sites, manufacturer B in 3 sites, C in 5 sites, and so on. Further, the data are color coded to represent the different climatic zones. It can be seen that some of the sites show tightly bound degradation rates, while others show a highly dispersed rate. Further the average degradation rate of modules at a particular site varies a lot from site to site. The red horizontal bar indicates the average P_{max} degradation rate of each particular site. We therefore decided to divide the sites into two categories – Group X (sites where average

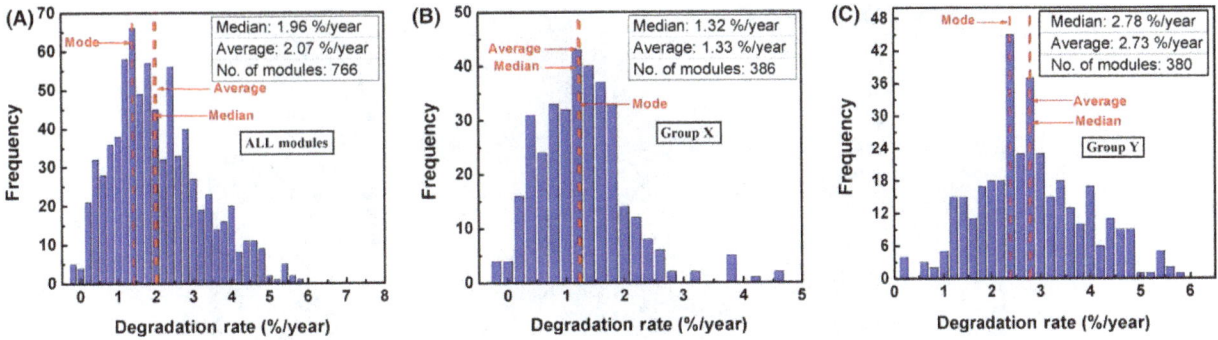

Figure 2. Histogram of Pmax degradation rate of (A) 'All' modules, (B) Group X modules, and (C) Group Y modules.

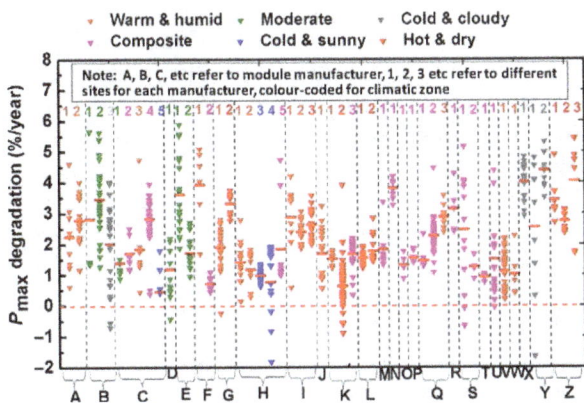

Figure 3. Manufacturer-wise and site-wise Pmax degradation rates. A through Z are the different manufacturers and 1, 2, etc. are the different sites for each manufacturer. Note that site 1 for manufacturer A is not the same as site 1 for other manufacturers, etc.

degradation rate of the modules measured is <2%/year) and Group Y (sites where average degradation rate of the modules is more than 2%/year). The '2%/year' criterion has been chosen somewhat arbitrarily, but was guided by the fact that the average degradation rate of all modules is 2%/year, as seen in Figure 2(A). The demarcation arising from the 2% criterion is purely empirical and therefore unbiased (we have not invoked tier ratings or reputation of the EPC), but, as will be seen later, leads to a good appreciation of the reasons for the wide variance, an understanding of the climatic variation, and receives support from other field observations such as visual, EL, IR, etc. Figure 2B and C show the histograms of the P_{max} degradation rates of Group X and Group Y modules. The average degradation rate of Group X modules is 1.33%/year, which is almost half of the average degradation rate for Group Y modules, which shows that there is significant difference in the performance of these two groups.

Figure 4 shows EL images of some representative modules of Groups X and Y. Group X modules have lesser number of cracks as compared to group Y modules.

According to Kontges et al., cracks can be classified into three categories [16]:

1. Mode A cracks which are basically hair-line cracks, not associated with any dark area,
2. Mode B cracks which are associated with a gray (not very dark) area in the cell,
3. Mode C cracks which are associated with a dark area in the cell.

Table 2 shows the statistics for the 51 modules (consisting of a total of 1416 cells), for which we took the EL measurement. Many of these modules are older than 10 years, and have <60 cells per module. From Table 2, we can see that modules in Group Y have a larger number of cracks as compared to Group X modules. Hence, cracks in the cells are likely to be one of the reasons behind the higher degradation rate observed in group Y modules. Table 3 gives the statistics for modules affected (in percentage) by various types of visual defects, for Groups X and Y in different age groups. The values in brackets indicate the number of samples in the respective category. It can be seen that in most cases, a larger percentage of Group Y modules are affected as compared to Group X modules, which indicates that the material quality and/or manufacturing process is inferior in case of Group Y modules, which has led to higher P_{max} degradation rates.

In the following sub-sections, the effect of system size, climate, technology, and age on the module performance will be discussed in detail.

Module size, system size and installation based variation

The inspected modules varied widely in their rated capacity, ranging from 10 W_p to 386 W_p. The smaller size modules (<100 W_p) are mostly old, while the larger wattages (>100 W_p) are seen only in recently installed systems.

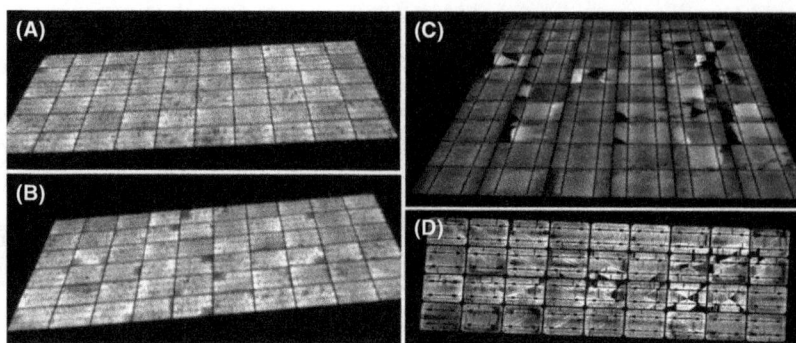

Figure 4. Electroluminescence images of some Group X modules (A), (B) and Group Y modules (C), (D) taken in the field.

Table 2. Percentage of cells affected by different types of cracks in Group X and Group Y modules.

| | Percentage of cells affected by different types of cracks | | | |
	Mode A cracks	Mode B cracks	Mode C cracks	Total no. of cells
Group X	2.4%	14.1%	3.2%	972
Group Y	29.7%	42.3%	14.2%	444

Table 3. Statistics of major visual defects for Group X and Group Y modules (the sample size is given in brackets).

| | Age groups | |
Type of visual defect	Young (<5 years)	Old (>5 years)
Discoloration	X: 9% (82)	X: 88% (148)
	Y: 18% (219)	**Y: 92% (135)**
Front-side delamination	X: 9% (82)	X: 31% (148)
	Y: 16% (219)	**Y: 43% (135)**
Snail trails	**X: 35% (82)**	X: 0% (148)
	Y: 26% (219)	Y: 0% (135)
Metallization discoloration	X: 38% (82)	X: 90% (148)
	Y: 54% (214)	**Y: 99% (135)**
Backsheet degradation	X: 26% (82)	X: 77% (148)
	Y: 31% (219)	**Y: 90% (135)**

Higher value in the category is indicated in bold.

Figure 5 shows the variation in the power degradation rate for two categories of modules, with size <100 W_p and >100 W_p. In this and subsequent figures, each data point refers to an individual module, the number on top indicates the number of modules, the red horizontal bar shows the mean value, the diamond represents the 95% confidence interval, and the error bars refer to error due to name plate tolerance (left) and measurement and translation (right). If relevant, the symbols are color coded to represent climatic zone, and open or filled representing young (<5 years) or old (>5 years) modules. We can see that when considering modules from 'All' sites (refer

Fig. 5A), the smaller capacity modules show less degradation on the average as compared to the larger size modules. However, the same comparison done only considering Group X modules does not show such a wide difference, as seen in Figure 5(B), which indicates that for the good sites, the degradation rate does not depend on module size. On the other hand, for the 'not so good' sites (Group Y), there is a very strong dependence, with the larger size modules (which are Young in age) showing much higher degradation rates.

To understand the effect of the system size (as distinct from module size) on the degradation rate, the surveyed sites have been segregated into two groups– small/medium systems (size ≤100 kW_p), and large systems (size more than 100 kW_p). Figure 6 shows the effect of system size on the performance degradation rate of surveyed modules. It is evident that large systems show lower degradation in performance as compared to smaller systems, the reason for which may be attributed to the greater due diligence and better installation practices adopted for the large systems, which may be lacking in the case of smaller systems. If we compare the mounting location – roof-mounted versus ground-mounted (Figure 7) – we come to the conclusion that roof-mounted systems show higher degradation rates than ground-mounted systems. This may be due to lack of proper due diligence on part of small PV system installers and owners which might result in use of lower grade materials, and also since roof-mounted modules tend to run hotter.

Figure 8 shows the percentage of modules which passed or failed the dry insulation resistance test for small/medium systems and large systems. It is evident that the modules in large installations perform slightly better than small/medium installations, and 100% of the modules in the large installations have passed the dry insulation test.

The severity level of material degradation has been quantified based on the degree (intensity) of the defect and the affected fractional area, estimated by visual inspection. The degree of the discoloration can be light

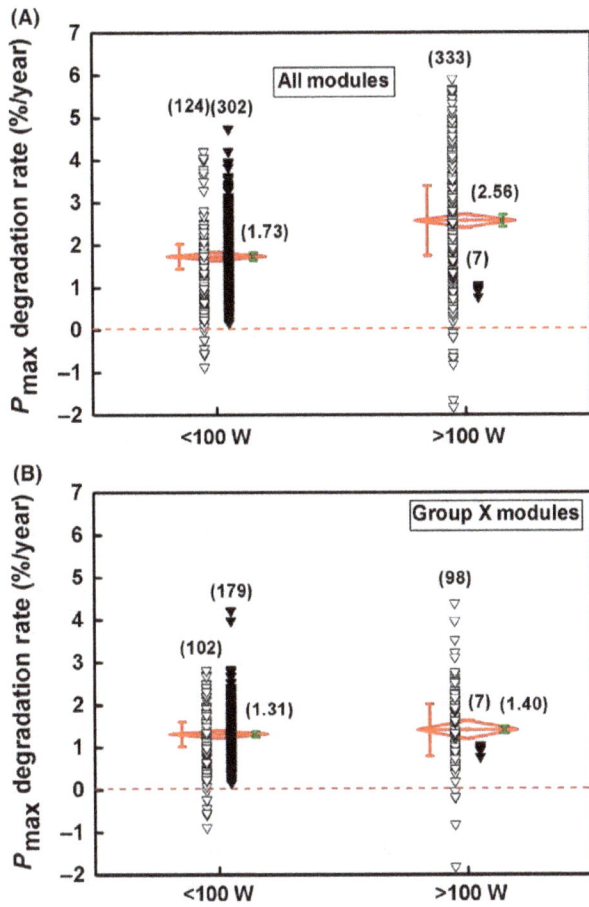

Figure 5. Effect of module size on P_{max} degradation rate for (A) 'All' modules (B) Group X modules.

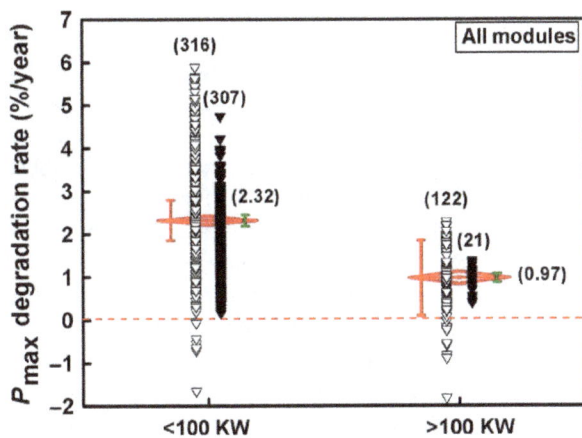

Figure 6. Effect of system size on P_{max} degradation rate for 'All' modules.

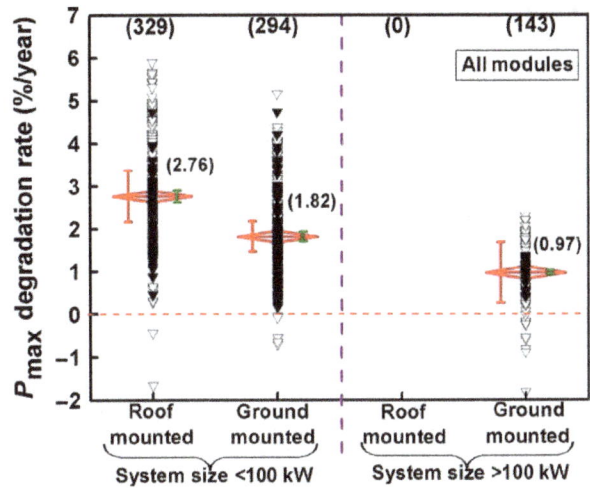

Figure 7. Effect of installation type on P_{max} degradation rate for 'All' modules.

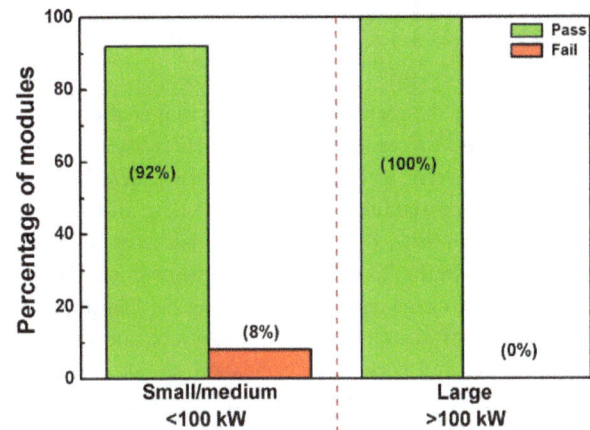

Figure 8. Percentage of modules passed or failed dry insulation resistance test for small/medium and large installations.

$$\text{Discoloration Index (DI)} = \frac{\text{Degree of discoloration} \times \text{Affected area of cell}}{\text{normalizing factor}}, \quad (1)$$

where degree of discoloration = 1 for light/yellow discoloration, = 2 for dark/brown discoloration; normalizing factor = 2 (based on maximum possible value of numerator).

Based on the discoloration index, the modules can be grouped into 5 categories as follows, with the severity of discoloration ranging from 'Nil' to 'Very High' as given in Table 4.

Similarly, modules have also been classified into five severity categories for the other defects like delamination, snail trails, etc. The severity levels in increasing order of magnitude are 'Nil', 'Low', 'Medium', 'High' and 'Very High'. Detailed

(yellow) or dark (brown), and the fractional cell area can range between 0 and 1. The discoloration index for the module is calculated as given below:

Table 4. Discoloration category based on discoloration index.

Discoloration index	0	0–0.25	0.25–0.5	0.5–0.75	0.75–1
Discoloration category	Nil	Low	Medium	High	Very high

Table 5. Severity levels of major visual defects for 'All' modules (the sample size is given in brackets).

Type of visual defect	Size of installation	Age groups	
		Young	Old
Discoloration	Small	Low (519)	Medium (361)[1]
	Large	Low (50)	Nil (5)
Front-side delamination	Small	Low (519)	Low (399)
	Large	Nil (50)	Nil (5)
Snail tracks	Small	Medium (519)	Nil (399)
	Large	Medium (50)	Nil (5)
Metallization discoloration	Small	Low (514)[2]	Medium (399)
	Large	Low (50)	Low (5)
Backsheet degradation	Small	Low (519)	Medium (399)
	Large	Nil (50)	Low (5)

[1]Excluding glass–glass modules which have negligible discoloration.
[2]Excluding All-Back-Contact modules in which metallization is concealed.

explanation of the methodology for arriving at the severity levels is given in a separate publication [17]. Table 5 shows the average severity level of the major visual defects found in the surveyed modules for small and large installations. Severity of front-side delamination and backsheet degradation is higher in the Young modules from small installations as compared to large installations. Among the Old modules, the severity of discoloration, delamination, metallization discoloration, and backsheet degradation are all higher for the modules in small installations, but the number of samples is very small for large installations to give a definitive picture. Overall, it may be said that severity of these defects points toward manufacturing and/or installation related issues with the modules in small installations.

Table 6 shows the result of analysis of infra-red images for 136 modules belonging to Group X, which have been segregated based on the type of mounting and the climatic zone. In the hot zone, modules mounted on the roof have on an average 5°C higher temperature as compared to modules on the ground. It should be noted that the modules on the rooftop were placed on open racks, and not directly on roof. Based on the above discussion, it is evident that the modules on rooftops, usually in small installations, run hotter, besides having material quality and installation related issues, which results in a higher degradation rate.

Climatic zone variation

The effect of climatic zone on the $P_{\max}$ degradation rate is shown in Figure 9. Figure 9(A) shows the plot for all

Table 6. Translated module temperatures for different types of mounting.

Translated module temperature (°C)	Hot zones		Nonhot zones	
	Roof	Ground	Roof	Ground
Median	67.63	60.85	49.20	46.35
Average	65.52	59.75	46.14	45.76
No. of samples	37	37	40	18

system sizes; Figure 9B and C show the plots for small and large systems, respectively. We focus on the climatic variation of the 'good' Group X modules, since the degradation of Group Y modules is likely to be dominated by other factors like quality, installation, etc. When considering all system sizes (Fig. 9A), the average degradation rate for Group X modules is highest in warm and humid zone, closely followed by the hot and dry zone and the composite zone, and is lowest in the cold and sunny zone. This trend is more clearly visible when looking at the degradation rates of only small-sized systems (Fig. 9B) with a minor difference being that the small systems in the hot and dry climate are degrading faster than those in the warm and humid climate. For modules in large systems (Fig. 9C), the trend is almost similar, except for the very low average degradation rate in hot and dry due to the presence of CdTe modules in this category (which are degrading at a much lower rate as compared to all other technologies). Since the average temperatures seen in the first there zones (warm and humid, hot and dry, and composite) are all quite high, these may be clubbed together into the so-called 'Hot' zone, while the moderate, cold and sunny, and cold and cloudy zones may be clubbed together into the 'Non-Hot' zone. Figure 10(A) shows that the $P_{\max}$ degradation rate is about 35% higher in the Hot zone (1.42%/year) as compared to the nonhot zone (1.05%/year), when considering all system sizes. This is understandable since the degradation processes for the various constituent materials of PV modules are accelerated by higher temperatures. Figure 10B and C show the degradation rates of power for small and large systems, respectively. The large systems have almost similar degradation rates in hot and nonhot climates, but this is not the case for the small systems. Thus, small systems are more affected by the climate (hot or nonhot) while the large systems tend to be less affected (which may be because the material quality, installation practices, and maintenance may be better in the large systems).

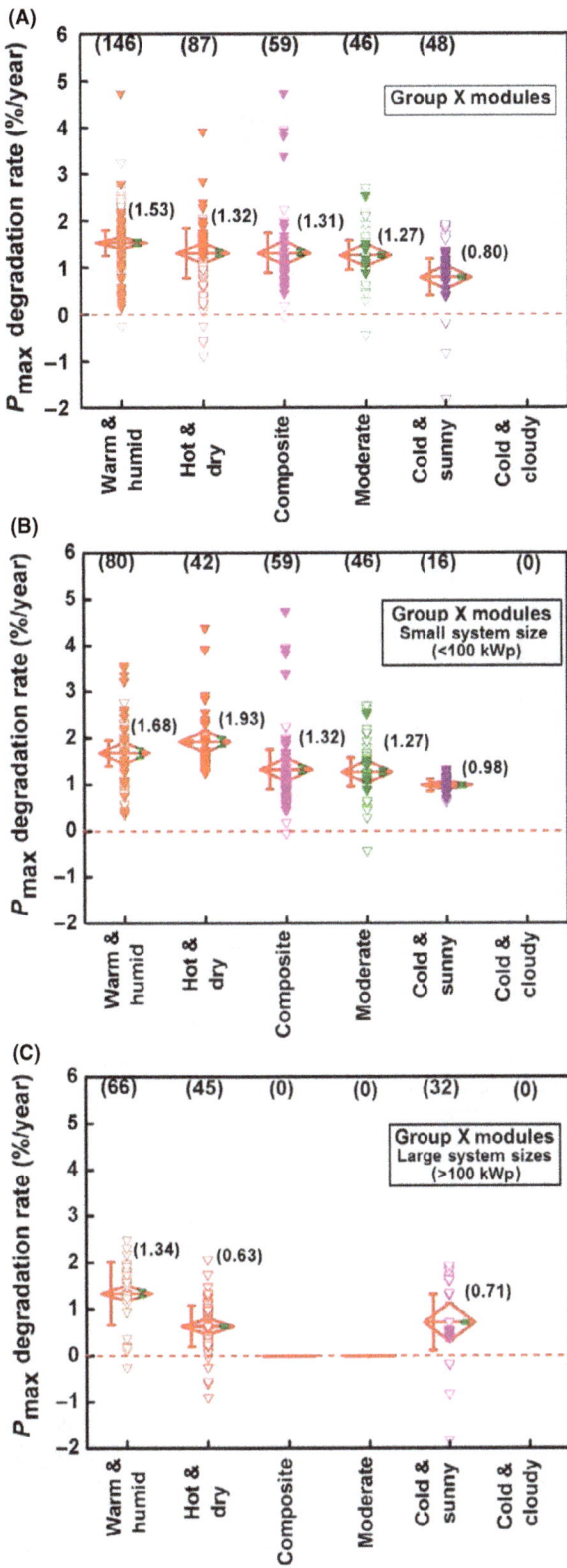

Figure 9. Comparison of P_{max} degradation rate with respect to six-zone classification system for Group X modules, for (A) all system sizes (B) Small system sizes (<100 kWp) (C) Large system sizes (>100 kWp).

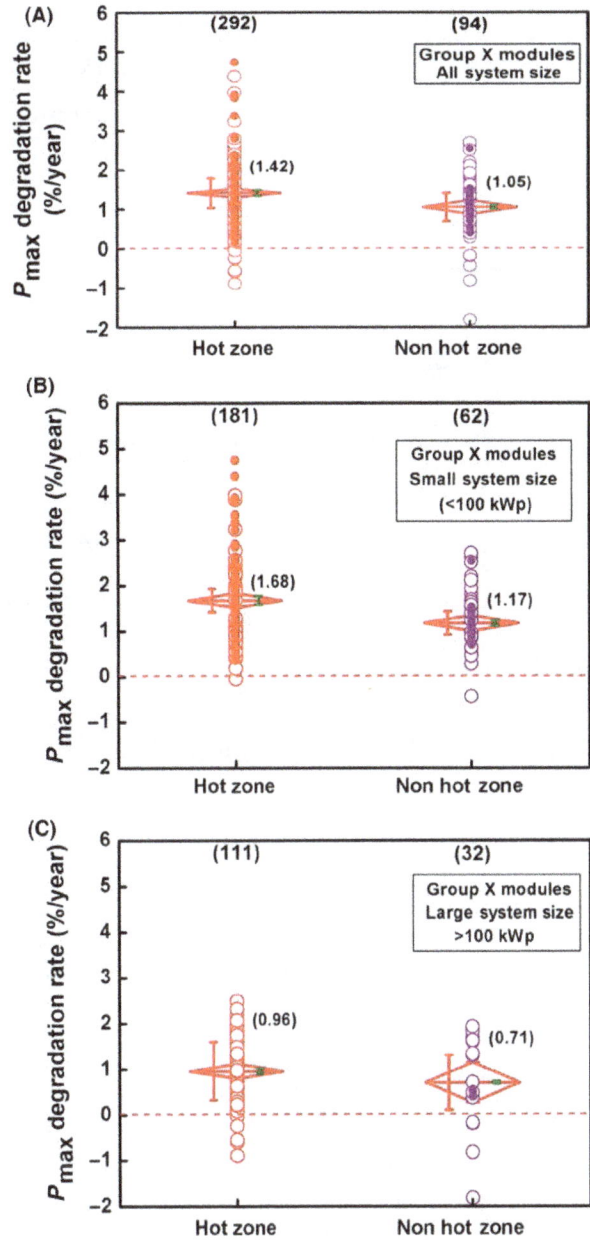

Figure 10. Comparison of P_{max} degradation rate with respect to hot and nonhot zone for Group X modules, for (A) all system sizes (B) Small system sizes (<100 kWp) (C) Large system sizes (>100 kWp).

Table 7 shows the effect of hot cells and module temperature on the module's power degradation rate in the hot and nonhot climatic zones. The mode of the temperature histogram (T_{mode}, as shown in Fig. 11B) of a module has been considered as the module's representative temperature. The difference in the maximum cell temperature of the module (T_{max} in Fig. 11B) from the modal temperature (T_{mode} in Fig. 11B) is referred to as *module* ΔT. For a meaningful comparison, the module temperatures measured in the field need to be translated to reference

Table 7. Effect of Hot Cells on power degradation rate of PV modules.

Parameter (average of many samples)	Hot zone		Nonhot zone	
	Without hot cells	With hot cells	Without hot cells	With hot cells
Power degradation rate (%/year)	1.39	2.02	1.39	1.70
Normalized *module* ΔT (°C)	6.8	24.7	5.7	18.9
Normalized maximum cell temperature (°C)	68.8	86.1	50.3	58.6
No. of samples	10	18	14	17

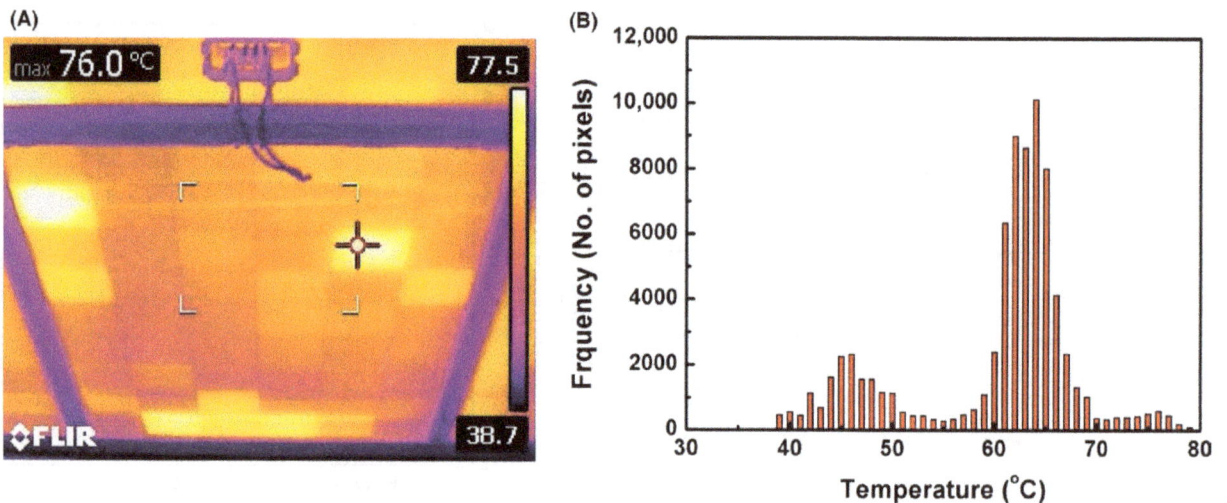

Figure 11. Infrared (IR) image of a PV module, (A), and its temperature histogram extracted from the IR image, (B), showing mode of 63°C and maximum temperature of 79°C, so the module ΔT is 16°C [20].

conditions of 1000 Wm^{-2} irradiance and 40°C ambient temperature [18, 19]. The following relation is used to translate the *module* ΔT to the reference conditions:

$$\text{Translated } module\,\Delta T = \frac{Module\,\Delta T \times 1000}{\text{Measured Irradiance}}. \quad (2)$$

If the translated *module* ΔT is >10°C, we have considered the module to be having Hot Cells [21]. We can see from Table 7 that modules with Hot cells are degrading faster than modules without Hot cells, and this degradation rate is higher in the hot zone as compared to the nonhot zone. This trend agrees well with the normalized (translated) *module* ΔT, which is found to be highest for modules with hot cells in hot zones. The maximum cell temperature is also highest for this category. Modeling and simulation [22] results show that there is very little heat transfer from the Hot cells to the surrounding connected cells. These Hot cells are created under short-circuited condition usually due to mismatch in photo-currents. Figure 12 plots the power degradation rate versus the normalized *module* ΔT, which shows a positive correlation between the two.

Table 8 shows the average severity of visible defects for Group X modules. The severity levels in increasing

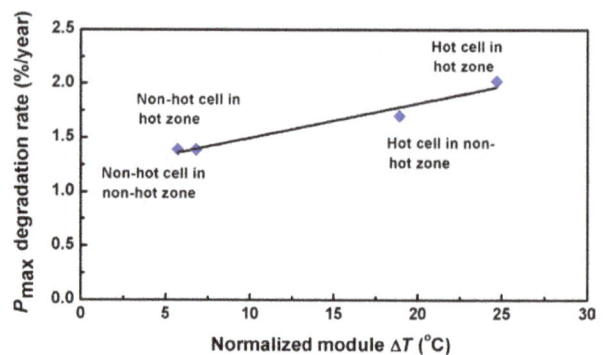

Figure 12. Correlation of power degradation rate with normalized module ΔT.

order of magnitude are 'Nil', 'Low', 'Medium', 'High' and 'Very High'. Detailed explanation of the methodology for arriving at the severity levels is given in a separate publication [17]. Among the Old modules, the average severity of discoloration, delamination, metallization discoloration, and backsheet degradation are all higher for modules in Hot zones as compared to nonhot zones. Among the Young modules, snail tracks are more severe in hot zones than the nonhot zones. Hence, it can be said that, in

Table 8. Average severity of visual degradation in hot and nonhot climates for Group X modules (the sample size is given in brackets).

Type of visual defect	Climatic zone	Age groups	
		Young	Old
Discoloration	Hot	Low (51)	High (155)
	Nonhot	Low (64)	Medium (34)
Front-side delamination	Hot	Nil (51)	Low (193)
	Nonhot	Low (64)	Nil (34)
Snail tracks	Hot	Medium (51)	Nil (193)
	Nonhot	Low (64)	Nil (34)
Metallization discoloration	Hot	Low (46)**	Medium (193)
	Nonhot	Low (64)	Low (43)
Backsheet degradation	Hot	Low (51)	Medium (155)
	Nonhot	Low (64)	Low (34)

Table 9. Percentage of modules affected by interconnect breakage in different climatic zones and age groups (the sample size is given in brackets).

Climatic zone	0–5 years	5–10 years	10–20 years	20+ years
Hot and Dry	NA	NA	100% (20)	2.6% (38)
Warm and Humid	0% (24)	28% (39)	54% (26)	NA
Composite	NA	NA	34% (30)	NA
Moderate	0% (31)	NA	16% (12)	NA
Cold and Sunny	0% (33)	NA	35% (17)	NA
Cold and Cloudy	NA	NA	NA	NA

general, modules in hot zones show more severe material degradation than modules in the nonhot zones.

The percentage of Group X modules affected by interconnects breakage in different climatic zones and age groups are presented in Table 9. Interconnect breakage test has been performed on crystalline silicon modules, using Togami Cell Line Checker. It has not been found in the young modules, and among the old modules, those in the hot and dry climate seems to be affected the most, followed by the warm and humid climate. This is due to stress caused by thermal expansion and contraction as a result of daily thermal cycling. Also moisture induced corrosion causes loss of adhesion strength. Such a trend is also reported in literature for modules installed in the USA [10, 23].

Based on the above discussion, it can be concluded that climate has a significant influence on the performance degradation rates, with modules degrading faster in hot zones as compared to the nonhot zones. Modules in the hot zones suffer from faster packaging material degradation and higher incidence of interconnect breakages as compared to modules in nonhot zones.

Technology based variation

Different technologies of PV modules show different degradation rates as shown in Figure 13. Multicrystalline silicon modules are found to degrade slightly faster than mono

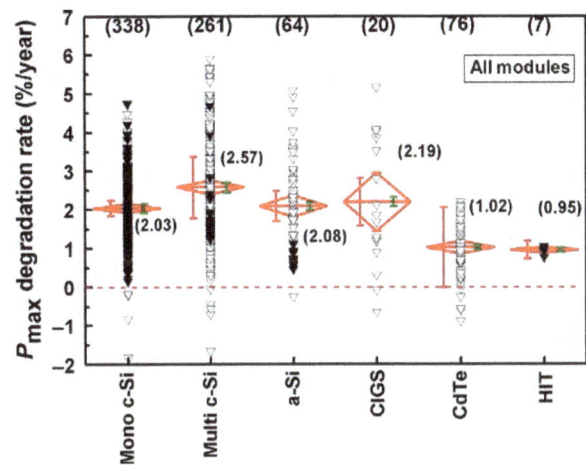

Figure 13. P_{max} degradation rate for different technologies for Group X modules.

c-Si modules. The average degradation rate for some of the thin film technologies like CIGS, CdTe and HIT is found to be lower than the c-Si modules. This may be due to the fact that we encountered modules from only one manufacturer for these thin film technologies, whereas for the c-Si modules there were multiple manufacturers, including reputed experienced manufacturers and also relatively unknown manufacturers. Also, since the number of samples for some of the technologies like HIT and CIGS is very less, no definitive conclusion can be arrived

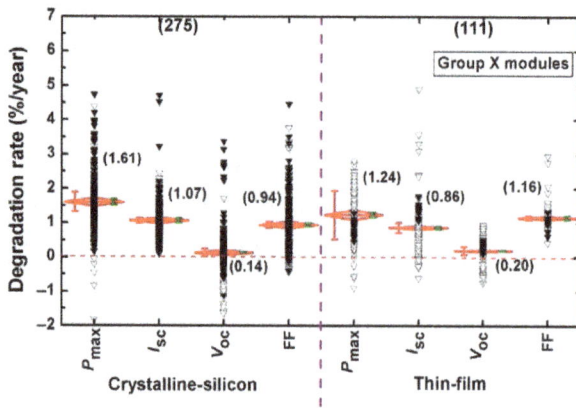

Figure 14. P_{max}, I_{sc}, V_{oc}, and *FF* degradation of different technologies for Group X modules.

at for these technologies. Jordan et al.[3] have indicated that the average degradation rate for CdTe modules has been observed to be around 0.6%/year (based on multiple reports) while the degradation rates for amorphous silicon power plants from various parts of the world has been reported to be within 1%/year in general. Also, the degradation rate for c-Si is commonly reported to be within 1%/year [3]. These values are lower than those seen in our survey, and raises concern about the long term performance of these modules in India. Figure 14 shows the degradation rates for all the *I-V* parameters like short circuit current, open circuit voltage and *FF*, for Group X crystalline silicon and thin film modules. It can be seen that in case of c-Si modules, the degradation in P_{max} is mainly caused due to degradation in I_{sc}, followed by *FF*, and lastly by V_{oc}, whereas for thin film modules, it is mainly the degradation in *FF* which causes this loss of performance, followed by I_{sc} degradation. The degradation in I_{sc} in c-Si is mainly due to discoloration, delamination and cracks in the cell. A small extent of degradation of I_{sc} is due to light-induced degradation and soiling [24, 25].

Corrosion and solder-bond breakage causes the degradation in the *FF* for the c-Si module. In the case of thin films technologies higher value in *FF* degradation is seen as compared to c-Si which is due light-induced degradation of a-Si and an increase in series resistance in CIGS [26].

The predominant defect observed in young crystalline silicon modules is snail trails, which is mostly due to cracks in the solar cells. The cracks tend to increase the series resistance of the modules, and may also cause power loss (depending on the type of crack) [16]. In case of thin film modules, the predominant defect is white spots, which form due to embedded impurities during the manufacturing process owing to improper cleaning of the substrate glass [26]. For the old crystalline silicon modules, the predominant visual defect is encapsulant discoloration, which reduces the short circuit current, and corrosion of metallization, which reduces the fill factor.

Age based variation

The influence of module age on the degradation rate of c-Si modules is shown in Figure 15. In the young modules, *FF* degradation is the primary cause of degradation in P_{max}, whereas in the old modules, I_{sc} degradation is the main reason. P_{max} degradation rate is higher for old modules as compared to young modules in case of Group X but this trend reverses for 'All' c-Si modules. Another interesting fact which emerges is that whereas in old modules P_{max} degradation is dominated by I_{sc} degradation due to encapsulant browning, in young modules it is dominated by *FF* degradation probably due to series resistance caused by more cracks as shown below in Table 11. This indicates that the young modules in Group Y have higher degradation rates than the old modules, which suggests that the raw materials and/or manufacturing process and/or installation practices are not optimized or well-controlled.

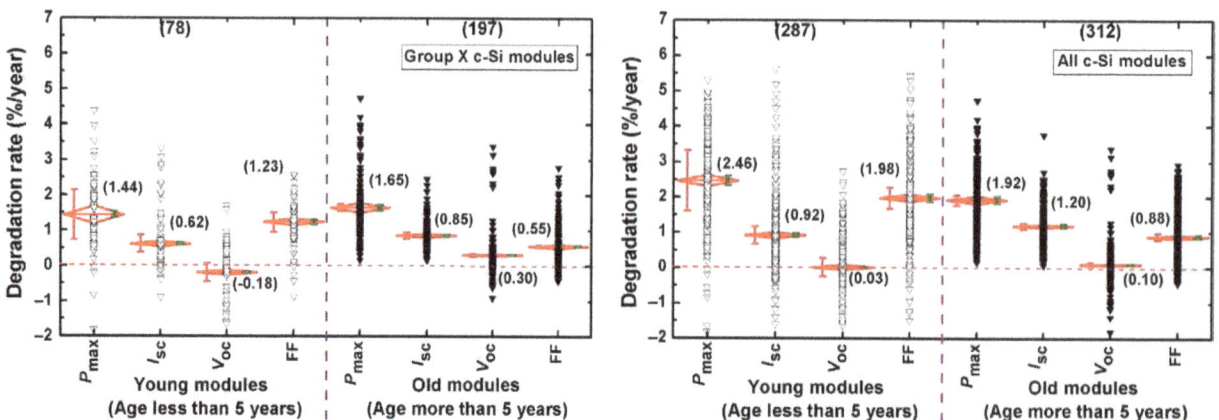

Figure 15. Effect of age on P_{max}, I_{sc}, V_{oc} and *FF* degradation of crystalline silicon modules for (A) Group X, and (B) 'All'.

Table 10. Interconnect failures for Group X modules (the sample size is given in brackets).

Climatic zone	Young modules	Old modules
Hot zones	0% (24)	43% (153)
Nonhot zones	0% (64)	28% (29)

Table 11. Percentage of cells affected by different types of cracks in young and old modules.

	Percentage of cells affected by different types of cracks			
	Mode A cracks	Mode B cracks	Mode C cracks	Total no. of cells
Young modules	37.9%	14.9%	12.2%	408
Old modules	0%	26.2%	4.4%	1008

With aging, the materials in the module degrade due to one or more of the various environmental factors like UV radiation (from sunlight), daily temperature cycling, humidity, etc. The percentage of modules affected by interconnects failure in hot and nonhot zones have been shown in Table 10. The young modules do not have any interconnect breakage, whereas a high percentage of old modules show breakage, particularly in the Hot zones. Temperature cycling (hot in daytime and cooler at night) has been identified as one of the major causes of interconnect failure in the field [10]. Also, with aging, there is increase in encapsulant discoloration, delamination, corrosion and backsheet degradation, as evident from Tables 3, 5 and 8.

From Table 11, it is evident that percentage of cracked cells in young modules is more than that in old modules, which hints that the transportation and/or installation practices for young modules has not been carefully undertaken. Also, the present day modules have thinner cells, which make them more susceptible to cracks as compared to the cells in the older modules.

Conclusions

It is important to understand the performance of the PV modules in actual operating conditions in the field, considering the large investments which are being made for setting up utility scale PV power plants in the near future. The degradation rates have serious implications on the return-on-investment calculations of investors, and can make the difference between a financially successful and unsuccessful project. Field data obtained during the 2014 All-India Survey of Photovoltaic Module Reliability have been analyzed in detail. This shows that there is a wide variation in the annual module degradation rate, and the average degradation rate for the so-called 'good' (Group X) sites is 1.3%/year, though there are also modules which have very low degradation rates of around 0.5%/year. The average power degradation rate for large installations is significantly lower (0.97%/year) than that for small installations (2.3%/year). The performance degradation of young c-Si modules is mainly due to FF degradation, whereas Isc degradation is the main reason for performance degradation in old c-Si modules. For the 'good' sites, the power degradation rate for young modules is lesser than that for the old modules, which is expected considering the continuous improvement in the module manufacturing technology. However, the trend is reversed in the 'not-so-good' sites (Group Y). This suggests that there may be issues with the module quality and/or installation process for these sites. This is supported by the observation that the module packaging materials also degrade at a faster rate in Group Y modules. Micro-cracks are more prevalent in Group Y modules as compared to Group X. Our results suggest that there is a larger degradation in the modules in hot climates. The degradation is further aggravated by the presence of Hot Cells in the modules. Mono c-Si modules are seen to degrade at a slightly lower rate as compared to multi c-Si modules. The results which have emerged from this survey of module reliability in India provide useful indicators for future deployment in terms of location, technology type and diligence required for module selection and handling during installation. Such PV surveys need to be conducted regularly to provide us data on the performance degradation of the modules with age, so that it would be possible to generate the degradation curve over years for each type of effect. With this objective in view, NCPRE and NISE have completed another All-India Survey in 2016 and are planning to undertake more such surveys in the future.

Acknowledgments

This research is based upon work supported in part by (a) the National Centre for Photovoltaic Research and Education funded by Ministry of New and Renewable Energy of the Government of India through the Project No. 31/17/2009-10/PVSC dated 29th September 2010 and (b) the Solar Energy Research Institute for India and the U.S. (SERIIUS) funded jointly by the U.S. Department of Energy subcontract DE AC36-08G028308 (Office of Science, Office of Basic Energy Sciences, and Energy Efficiency and Renewable Energy, Solar Energy Technology Program, with support from the Office of International Affairs) and the Government of India subcontract IUSSTF/JCERDC-SERIIUS/2012 dated 22nd

November 2012. The authors acknowledge assistance from the various State Renewable Energy Development Agencies and from S. P. Gonchaudhuri, and Hemant Lamba, who provided encouragement and valuable inputs for the survey.

Conflict of Interest

None declared.

References

1. Sengupta, D. India adds 3.6 GW to solar capacity. Available at: http://economictimes.indiatimes.com/industry/energy/power/india-adds-3-6-gw-to-solar-capacity/articleshow/53823001.cms

2. Jordan, D. C., and S. R. Kurtz. 2013. Photovoltaic degradation rates-an analytical review. Prog. Photovoltaics Res. Appl., 21:12–29.

3. Jordan, D. C., S. R. Kurtz, K. Van Sant, and J. Newmiller. 2016. Compendium of photovoltaic degradation rates. Prog. Photovoltaics Res. Appl. 24:978–989.

4. Suleske, J. S., J. Kuitche, and G. Tamizh-Mani. 2011. Performance degradation of grid-tied PV modules in a hot-dry climatic condition. *SPIE 8112: Reliability of Photovoltaic Cells, Modules, Components, and Systems IV.*

5. Sastry, O. S., S. Saurabh, S. Shil, P. Pant, R. Kumar, A. Kumar et al. 2010. Performance analysis of field exposed single crystalline silicon modules. Sol. Energy Mater. Sol. Cells 94:1463–1468.

6. Magare, D., O. S. Sastry, R. Gupta, and A. Kumar. 2012. Data logging strategy of photovoltaic (PV) module test beds. *27th European Photovoltaic Specialists Conference and Exhibition, Frankfurt*, Pp. 3259–3262.

7. Sharma, V., O. S. Sastry, A. Kumar, B. Bora, and S. Chandel. 2014. Degradation analysis of a-Si, HIT and mono c-Si solar photovoltaic technologies under outdoor conditions. Energy, 72:536–546.

8. Bora, B., O. S. Sastry, A. Kumar, and M. Bangar. 2015. Performance modelling of three PV module technologies based on clearness index and air-mass using contour map. *42nd IEEE Photovoltaic Specialists Conference, New Orleans*, Pp. 1–4.

9. Tamizhmani, G. 2016. Climate dependent degradation rates based on nation-wide onsite I-V measurements. Presented at *2016 PV Module Reliability Workshop, Lakewood, CO.*

10. Kuitche, J., R. Pan, and G. Tamizhmani. 2014. Investigation of dominant failure modes for field-aged c-Si modules in desert climatic conditions. IEEE J. Photovolt. 4:814–826.

11. Tamizhmani, G., S. Tatapudi, R. Dubey, S. Chattopadhyay, C. Solanki, J. Vasi et al. 2016. Comparative study of performance of fielded PV modules in two countries. *26th International Photovoltaic Science and Engineering Conference, Singapore.*

12. Bansal, N. K., and G. Minke. 1995. Climatic zones and rural housing in India. ForschungszentrumJulich GmbH, Julich.

13. Chattopadhyay, S., R. Dubey, K. Vivek, J. John, C. S. Solanki, K. Anil et al. 2015. All India Survey of Photovoltaic Module Degradation 2014: Survey Methodology and Statistics. *Proceedings of the 42nd IEEE Photovoltaics Specialists Conference, New Orleans*, doi: 10.1109/PVSC.2015.7355712.

14. Dubey, R., S. Chattopadhyay, K. Vivek, J. John, C. S. Solanki, A. Kottantharayil et al. 2016. Correlation of electrical and visual degradation seen in field survey in India. *Proceedings of the 43rd IEEE Photovoltaics Specialists Conference, Portland, OR.*

15. Dubey, R., S. Chattopadhyay, K. Vivek, J. John, C. S. Solanki, K. Anil et al. 2014. Daylight electroluminescence imaging of photovoltaic modules by image difference technique. *6th World Conference on Photovoltaic Energy Conversion, Kyoto.*

16. Kontges, M., I. Kunze, S. Kajari-Schroder, X. Breitenmoser, and B. Bjorneklett. 2011. The risk of power loss in crystalline silicon based photovoltaic modules due to micro-cracks. Sol. Energy Mater. Sol. Cells 95:1131–1137.

17. Chattopadhyay, S., R. Dubey, V. Kuthanazhi, C. S. Solanki, A. Kottantharayil, K. L. Narasimhan et al. Correlating visual degradation with electrical performance of PV modules in different climates of India. IEEE J. Photovolt. [E-pub ahead of print].

18. Moreton, R., E. Lorenzo, J. Leloux, and J. M. Carrillo. 2014. Dealing in practice with hot spots. *Proceedings of the 29th European Photovoltaic Solar Energy Conference and Exhibition, Amsterdam*, Pp. 2722–2727, doi: 10.13140/2.1.3066.7528

19. Oh, J., and G. Tamizh Mani. 2010. Temperature testing and analysis of PV MODULES per ANSI/UL 1703 and IEC 61730 standards. *Proceedings of the 35th IEEE Photovoltaic Specialist Conference, Honolulu*, Pp. 984–988, doi: 10.1109/PVSC.2010.5614569

20. Chattopadhyay, S., R. Dubey, V. Kuthanazhi, C. S. Solanki, A. Kottantharayil, K. L. Narasimhan et al. 2016. Effect of hot cells on power degradation rate of PV modules. *Proceedings of the NREL PV Module Reliability Workshop, Denver, Colorado.*

21. Tatapudi, S., C. Libby, C. Raupp, D. Srinivasan, J. Kuitche, B. Bicer et al. 2016. Defect and safety inspection of inspection of 6 PV technologies from 56,000 modules representing 257,000 modules in 4

climatic regions of the United States. *Proceedings of the 43rd IEEE Photovoltaic Specialists Conference, Portland, OR.*

22. Sun, X., R. Dubey, S. Chattopadhyay, M. R. Khan, R. V. Chavali, T. J. Silverman et al. 2016. A novel approach to thermal design of solar modules: selective – spectral and radiative cooling. *Proceedings of the 43rdIEEE Photovoltaics Specialists Conference, Portland, OR.*

23. Jeong, J.-S., N. Park, and C. Han. 2012. Field failure mechanism study of solder interconnection for crystalline silicon photovoltaic module. Microelectron. Reliab. 52:2326–2330.

24. Jordan, D. C., J. H. Wohlgemuth, and S. R. Kurtz. 2012. Technology and climate trends in PV module degradation. *Proceedings of the 27th European Photovoltaic Specialists Conference and Exhibition, Frankfurt,* Pp. 3118–3124.

25. Wohlgemuth, J. H., and R. C. Petersen. 1993. Reliability of EVA modules. *Proceedings of the 23rd Photovoltaic Specialists Conference, Louisville, KY.* doi: 10.1109/PVSC.1993.346972

26. Kontges, M., S. Kurtz, C. Packard, U. Jahn, K. A. Berger, K. Kato. T. Friesen et al. 2014. Review of failures of photovoltaic modules. *International Energy Agency Report No.* IEA-PVPS T13-01:2014.

17

Inverse photovoltaic yield model for global horizontal irradiance reconstruction

Boudewijn Elsinga[1] (iD), Wilfried van Sark[1] (iD) & Lou Ramaekers[2] (iD)

[1]Copernicus Institute of Sustainable Development, Utrecht University, PO Box 80.115, 3508 TC Utrecht, The Netherlands
[2]ECOFYS, PO Box 8408, 3503 RK Utrecht, The Netherlands

Keywords
Decomposition model, diffuse irradiance, distributed sensor network, photovoltaics, PV yield model

Correspondence
Boudewijn Elsinga, Copernicus Institute of Sustainable Development, Utrecht University, PO Box 80.115, 3508 TC Utrecht, The Netherlands.
E-mail: b.elsinga@uu.nl

Funding Information
Netherlands Enterprise Agency (RVO) (Grant/Award Number: TKISG02017).

Abstract

This article describes the method of deriving Global Horizontal Irradiance (GHI) from combining power measurements with static meta data (tilt, orientation, brand/type) of rooftop photovoltaic (PV)-systems. This *inverse* PV model implements a forward yield model that is based on a modified Orgill and Hollands decomposition model and the Perez transposition model for irradiance. The forward as well as the inverse PV model were validated with DC power measurements of four different mono- and polycrystalline modules combined with weather station data (2 minute resolution data over a period ranging from the 11th of June through the 24th of August 2014). The bias-corrected forward PV model shows a best (r)RMSE of 16.0 W (15.1%) with a (r)MBE of −1.67 W (−1.57%) for one of the polycrystalline modules. The bias-corrected inverse PV model shows a best (r)RMSE of 65.6 Wm^{-2} 15.1% with a (r)MBE of 0.994 Wm^{-2} (0.229%) for one of the polycrystalline modules. Similar results were obtained for the three other modules.

Introduction

For the study of spatial effects of irradiance fluctuations in relation to photovoltaic (PV) systems, a dense homogeneous network of irradiance sensors is ideally required for obtaining data, as done by e.g. [1–3]. In the absence of such a network in our region of interest (Province of Utrecht, The Netherlands), we use AC power output measurements of existing distributed rooftop PV-systems. Using PV-systems for the estimation of synchronous total or averaged power output in a region is reported in [4, 5]. In our paper, however, we describe a model that translates measured AC power from monitored small residential rooftop PV-systems (< 5 kWp) into Global Horizontal Irradiance (GHI). This way, monitored PV-systems can be employed as a network of GHI sensors (a.k.a. *PV-sensor field*) to perform a range of studies on variability and short-term forecasting, expanding our previous work [6, 7].

Brief literature overview

There is a long history of PV yield models that use decomposition of GHI into direct and diffuse and transposition to a tilted surface, of which a thorough review was recently published [8]. In this paper we use a modified version of the Orgill and Hollands decomposition model that statistically models the relation between (hourly) direct and diffuse irradiance [9] and the Perez transposition model to relate GHI to the global irradiance on an inclined plane, I_{POA} [10]. The details of these methods, as well as our inversion of it will be discussed in *Forward PV Yield Model*. Recently published work on the same subject, or comparable to it, is shown here for comparison. Details and results will be discussed in *Comparison to other studies*.

In the study of Marion et al. [11], a modified, tilted DIRINT model, which has its origin in the DISC model by Maxwell [12] and which was further developed by Perez et al. [13], was used to convert I_{POA} into GHI based on 5-minute resolution data in three locations (Cocoa, Florida; Eugene, Oregon; Golden, Colorado (USA)).

Yang et al. [14] report on performance of combinations of decomposition and transposition models to reconstruct GHI from I_{POA} measured at inclined silicon photosensors. This comes close to our aim of using (silicon) PV-systems as GHI sensors in the tropical region of Singapore.

In Yang et al. [15], reconstruction of GHI from sensors inclined at various inclination angles is reported using the Perez transposition model.

Furthermore, Killinger et al. [16] present the *power projection* model that aims to describe PV performance relative to other (nearby) PV-systems while incorporating the different orientation and tilt through reverse iteration of the Perez DIRINT model, among several other models. A byproduct of this method is the reconstruction of GHI from measured Power for the 45 PV-systems at 5-min interval over the period of 2010–2014.

Outline

The conversion from power to GHI naturally introduces ambiguity or errors that result from the large amount of variables that influence the power output of (aging) PV modules in an outdoor situation. In this paper, we quantified these uncertainties for the specific model that was implemented as part of the *Solar Forecasting and Smart Grids* (SF&SG) project.[1] First, the data acquisition and treatment will be explained. After that, the forward PV-system yield model will be discussed, as this will be the method that is inverted in this study. Finally a discussion on the uncertainties and applicability of the inverse PV model is provided. Symbols used in this work are listed in Table 1.

Data Acquisition and Treatment

Measurement data from the Utrecht Photovoltaic Outdoor Testing Facility (UPOT) was used to verify the presented models [17–19]. Apart from Global Horizontal Irradiance (GHI, I), Global Array Plane Irradiance (I_{POA}), Direct Normal Irradiance (DNI), and Diffuse Horizontal Irradiance $I_{h,diff.}$, also DC power output (P_{DC}) and module temperature of several PV modules are measured for validation and calibration of the *inverse PV model*, see *Visual validation*. Two identical mono-crystalline modules of both 245 W_p and two identical polycrystalline modules of both 240 W_p are used for validation. All modules, from this point on to be called `mono1`, `mono2`, `poly1`, and `poly2` are part of the same array that is directed due South at fixed 37° tilt angle. GHI and DHI is measured using EKO MS-802 and EKO MS-401 pyranometers respectively. DNI is

Table 1. Overview of the used symbols and their units, in order of appearance.

Symbol	Name	Unit
I	Global Horizontal Irradiance (GHI)	Wm^{-2}
$I_{h,diff.}$	Diffuse Horizontal Irradiance	Wm^{-2}
I_{beam}	Direct Normal Irradiance (DNI)	Wm^{-2}
$I_{h,beam}$	Direct Horizontal Irradiance	Wm^{-2}
I_{POA}	Global Array Plane Irradiance	Wm^{-2}
γ	Solar elevation angle	degree
θ_z	Solar zenith angle $\theta_z = 90° - \gamma$	degree
ψ	Solar azimuth angle	degree
I_0	Extraterrestrial Global Horizontal Irradiance	Wm^{-2}
K	Extraterrestrial Clearness Index	–
K_D	Diffuse fraction of the GHI	–
$R(\gamma)$	De Jong's modification term	–
(r)MBE	(Relative) Mean Bias Error	(%) Wm^{-2}
MAPE	Mean Absolute Percentage Error	%
(r)RMSE	(Relative) Root Mean Square Error	(%) Wm^{-2}
θ	Solar-Array Plane incidence angle	Degree
$F_{xy}', (\bar{F}_{xy}')$	(interpolated) Perez factors	–
ε	Perez factors parameter	–
ρ	Average surface albedo	–
$I_{POA,beam}$	Direct Array Plane Irradiance	Wm^{-2}
α	Array Plane azimuth angle	degree
β	Array Plane tilt angle	degree
$I_{POA,diff.}$	Diffuse Array Plane Irradiance	Wm^{-2}
$I_{POA,refl.}$	Array Plane Irradiance from reflections	Wm^{-2}
f	Forward PV model function	W
P_{DC}	Direct-Current Power	W
P_{AC}	Alternating-Current Power	W
η_{EU}	European Inverter Efficiency	–
$a_1, a_2, \kappa_0, \kappa_1$	Connection parameters of the adapted decomposition model	–
$e_{(P)}/e_{(GHI)}$	Power/GHI error	–
$e_{rel.(P)}/e_{rel.(GHI)}$	Relative Power/GHI error	–
$c_{(P)}/c_{(GHI)}$	Power/GHI correction factors	–

measured by an EKO MS-56 pyrheliometer. The power of the test modules at UPOT is measured using an EKO MP-160 IV-curve tracer, and all the modules are individually optimized via 1st generation Femtogrid power optimizers [20].

Data treatment

For the verification of the decomposition part of the yield model, the QCRad VAP method was implemented, see [21]. This method filters the data for extraordinary occurrences of (GHI, I) divided by the sum[2] of the diffuse horizontal ($I_{h,diff.}$) and direct horizontal ($I_{h,beam}$) irradiance, dependent on solar zenith angle θ_z, which complements the solar elevation angle γ: ($\theta_z = 90° - \gamma$). Values of I and ($I_{h,diff.} + I_{h,beam}$) of <50 Wm^{-2} were also discarded. The standard bounds of the QCRad method were applied as in equation 1, see also [21], p. 4:

$$\begin{cases} 92\% < I/(I_{\text{h,diff.}} + I_{\text{h,beam}}) < 108\% \text{ for } (0° \leq \theta_z < 75°) \\ 85\% < I/(I_{\text{h,diff.}} + I_{\text{h,beam}}) < 115\% \text{ for } (75° \leq \theta_z < 93°) \end{cases} (1)$$

The extraterrestrial clearness index $K(t)$ and the diffuse fraction $K_D(t)$ are defined as in equation 2. Here, I_0 = extraterrestrial horizontal irradiance, with maximal value of 1364 Wm^{-2}.

$$K = I/I_0$$
$$K_D = I_{\text{h,diff.}}/I \qquad (2)$$

Following [8] unrealistic values of $K_D > 1$ were discarded from the validation data as these represent effects that do not benefit the understanding of the tested model: in principle the diffuse fraction should be <1. Additionally, due to some unavoidable topographical shading of both the modules and the pyranometers, morning values with $\theta_z > 62°$ and evening values of $68° < \theta_z < 74°$ were removed. The topographical shading of the modules and irradiance sensors at UPOT in the morning and evening is best visible in subfigures C of Figures 7 and 8 at around 06:00 and 17:00 UTC, respectively, and are most strongly present during clear sky conditions.

When the irradiance changes too much during an IV-curve measurement, the data point is flagged as "not to be trusted", and hence the accompanying GHI value at that time stamp will be discarded as well. In order to compare all four module measurements and the weather station data, all (2 min.) module measurement time series ware resampled using linear interpolation to match the 2 min time stamp of the weather station data synchronously. After QCRad was applied and topographical shading in the morning and evening at specified solar zenith angles was removed, 17,642 2 min data points and 593 hourly data points are available per module per data type (power, irradiance, temperature, solar zenith angle) in the measurement period that covers the 11th of June 2014 through the 24th of August 2014. During this period, the following days were excluded due to large parts of missing data: (the 18th and 19th of June; the 4th, 19th, 21st, and 30th of July and the 1st of August).

Forward PV Yield Model

The Ineichen clear sky model was used for calculations regarding clearness index and clear sky index [22]. Details of its implementation in MATLAB PVLIB can be found in [23].

A yield model of a PV-system calculates the electrical power output of the system from the available measured irradiance components: preferably array plane global irradiance I_{POA}, or its principal components direct, diffuse, and reflected irradiance on the array plane. Global Horizontal Irradiance (GHI) I can be input as well,

which is more readily available, but requires some assumptions on the translation from the horizontal plane to the tilted (module) plane at the location of the PV-system. Our aim is to invert a GHI-to-power yield model. The calculation steps in the yield model that is used in our study are shown in Figure 1.

Sun position

The first step calculates the position of the sun in the sky (solar elevation γ and solar azimuth ψ) from the current date and time and from the location of the system (latitude, longitude, time zone). The standard formulas of this calculation are described in e.g. [24].

DC and AC yield

A standard single diode model is used to calculate the DC power from given module parameters and array plane irradiance. See De Soto et al. [25] for explanation of this procedure.[3] The cell temperature (assumed equal to module temperature) follows from the equilibrium between heat loss through the module material and the heat gain by solar irradiance that is absorbed and not converted into electricity.

The inverter's DC/AC efficiency is characterized by a power transfer function, depending on the instantaneous DC power, see equation 3.

$$P_{\text{AC}}(t) = a \cdot (P_{\text{DC}}(t) - P_{\text{thresh}}) - b \cdot (P_{\text{DC}}(t) - P_{\text{thresh}})^2 (3)$$

The value of this weighted efficiency is obtained by assigning a percentage of time the inverter resides in a given operating range. The efficiency at X% of nominal

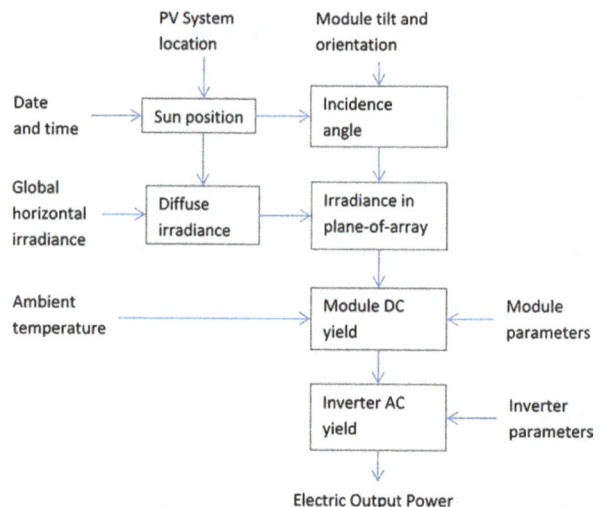

Figure 1. Block diagram of the data processing that is at the basis of the forward PV model.

power is denoted by "η_X, and the weighted average is defined as:

$$\eta_{EU} = 0.03 \cdot \eta_{05} + 0.06 \cdot \eta_{10} + 0.13 \cdot \eta_{20} + \ldots$$
$$\ldots + 0.1 \cdot \eta_{30} + 0.48 \cdot \eta_{50} + 0.2 \cdot \eta_{100} \qquad (4)$$

By setting $\eta_{max} = \eta_{50}$ and by using the expression for η_{EU} as in equation 4, the power efficiency expression can be solved for the parameters a and b. Manufacturers generally provide the power threshold (P_{thresh}), the maximum efficiency (η_{max}) and the (JRC/Ispra) "European Efficiency" (η_{EU}) of the inverter, see [26].

Array plane irradiance

The total irradiance on the tilted PV module surface, I_{POA} is defined as in equation 5. The time variable (t) is omitted for brevity.

$$I_{POA} = I_{POA,beam} + I_{POA,diff.} + I_{POA,refl.} \qquad (5)$$

The direct part $I_{POA,beam}$ is calculated from the I_{POA} that follows from the decomposition model, corrected for by geometric factors, that can be found in many textbooks, e.g. [27].

$$I_{POA,beam} = I_{beam} \cdot \cos\theta = I_{h,beam} \cdot \cos\theta / \sin\gamma \qquad (6)$$

Here, $\cos\theta = \cos\beta \sin\gamma + \sin\beta \cos\gamma \cos(\alpha-\psi)$; and θ is the incidence angle, β the array plane tilt angle, α the azimuth of the array plane, ψ the azimuth of the sun, and γ the solar elevation.

To calculate the (sky-dome) diffuse irradiance in the array plane, $I_{POA,diff.}$ we use the method as described by Perez et al. in [10], see equation 7, where a and c depend on the solid angle of the circumsolar region.

$$I_{POA,diff.} = I_{h,diff.}\left(\frac{1}{2}(1+\cos\beta)(1-F_1'(\varepsilon)) + \cdots \right.$$
$$\left. \cdots + \frac{a}{c}F_1'(\varepsilon) + \sin\beta\, F_2'(\varepsilon)\right) \qquad (7)$$

The Perez method determines the amount of irradiance coming from the part of the sky dome as "seen" by the PV module and relies on the dependence of the so-called empirical Perez factors for circumsolar and horizon-band brightness $F_1'(\varepsilon)$ and $F_2'(\varepsilon)$ that depend on ε:

$$\varepsilon = 1 - (I_{h,beam}/\sin\gamma)/I_{h,diff.} \qquad (8)$$

The details of this method and how the factors $F_{xy}'(\varepsilon)$ are constructed, based on sums of subfactors F_{xy} where x can stand for 1 or 2 and y can stand for 1, 2 or 3, can be found in [10]. Our method makes use of the interpolated values $\tilde{F}_{xy}'(\varepsilon)$ with respect to ε, as visualized in Figure 2.

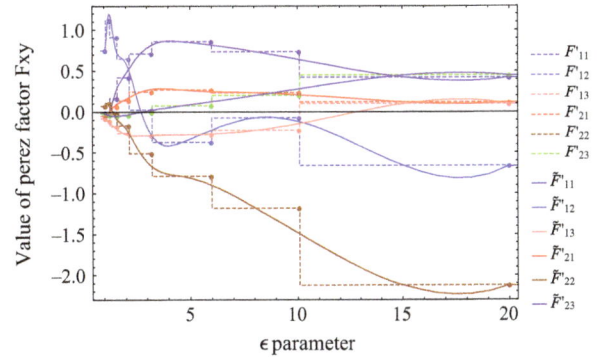

Figure 2. Visualization of the original Perez parameters F_{xy}' (dashed lines) and the interpolated values $\tilde{F}_{xy}'$ (solid lines) as a function of parameter ε. The interpolation (spline) was done through values halfway between the original ε values to minimize overall deviation.

The ground-reflected irradiance on the array plane, $I_{POA,refl.}$ is given by:

$$I_{POA,refl.} = \rho \cdot I \cdot \frac{1}{2}(1-\cos\beta) \qquad (9)$$

The parameter ρ is the albedo: the reflection coefficient of the surface. An average value of $\rho = 0.2$ is taken as it is the standard value used in urban environment, see e.g. [28] for reference.

Basically, I_{POA} is a function with many input parameters and principally depends on the total GHI, decomposed into the direct (beam) part, $I_{h,beam}$ and the diffuse irradiance, $I_{h,diff.}$ via:

$$I = I_{h,diff.} + I_{h,beam}$$
$$= I(K_D + (1-K_D)) = I_0(KK_D + K(1-K_D)) \qquad (10)$$

Ultimately, the forward PV-model can be thought of as a function $f:I \rightarrow I_{POA} \rightarrow P_{AC}$, and this is the function that needs to be inverted to go from P_{AC} to I.

PV-system Model Inversion

Decomposition

The diffuse ratio, K_D, is found to depend on K to some extent via a so-called *decomposition relation*. The starting point is the modified decomposition relation of Orgill and Hollands (O&H) that was originally developed for Toronto, Canada [9], adapted to the climate of The Netherlands via modified threshold values and a solar elevation angle γ dependent $R(\gamma)$, according to the correction of De Jong [29] in [30], p. 106. This will be called the "De Jong" decomposition model, see equation 11.

$$K_D = \begin{cases} 1 & \text{for} \quad K \leq 0.22 \\ 1 - 6.4 \cdot (K - 0.22)^2 & \text{for} \quad 0.22 < K \leq 0.35 \\ 1.47 - 1.66 \cdot K & \text{for} \begin{cases} 0.35 < K \\ \text{and } K \leq \frac{1.47 - R(\gamma)}{1.66} \end{cases} \\ R(\gamma) & \text{for} \quad K > \frac{1.47 - R(\gamma)}{1.66} \end{cases} \quad (11)$$

The function $R(\gamma)$, in which γ is the solar elevation angle, is given by equation 12 and the effect on the decomposition relation is illustrated in Figure 3.

$$R(\gamma) = 0.847 - 1.61 \cdot \sin \gamma + 1.04 \cdot \sin^2 \gamma \quad (12)$$

For values of $K > 0.35$ the diffuse fraction K_D is differentiated according to the solar elevation angle γ: the diffuse fraction is given a higher value when the solar elevation angle γ is relatively low. Especially for superirradiance (or cloud-enhancement) events when the clearness index surpasses the terrestrial[4] clear-sky = 1 value: $K > (1/1.364 \approx 0.73)$, the solar elevation is considered low when $\gamma < 51°$: this value of γ roughly determines the minimum of $R(\gamma)$ and is close to the highest possible $\gamma \approx 62°$ in the Netherlands. This means that for a large part of the year, the diffuse fraction at high K will be higher than with the uncorrected O&H model. The correction of De Jong partially corrects for this underestimation via $R(\gamma)$ that increases rapidly from $\gamma \approx 51°$ toward lower values of γ. For low γ in the context of the associated K values, see the curved bend in the "Original De Jong" 3D-plot in Figure 3 that indicates the inequality $K > (1.47 - R(\gamma))/1.66$ of equation 11, and plots of data versus the decomposition models in Figure 6. In general terms, this means the modified model attributes a higher diffuse fraction to high (extraterrestrial) clearness index

when the sun is relatively low and this correction becomes less pronounced for lower values of K.

The O&H model was developed for hourly irradiance values, and lacks detail where more sophisticated models include cloud enhancement effects at that are mostly visible at subhourly intervals, see [8] for a recent comparison. However, we chose to use the De Jong's decomposition model for the PV measurement time series of approx. 2 min interval due to its simplicity which is important for the computation speed of the inversion of the forward yield model.

Inversion

The translation from power to irradiance is illustrated for a 13.8 kWp (AC) PV-system "Velt & Vecht", for which high resolution data were available prior to the measurements at UPOT. A clear sky day (21st of April 2013) was used for this purpose. From Figure 4A) it can be seen that the inversion process is not unique (blue line). For most values of P, there are two or more solutions as can be seen from the vertical intersections at the chosen power output value of 0.9 kW. The following four steps (I-IV) are taken during each iteration step to make the outcome inverse relation unique, see Figure 4.

At a given moment, the forward PV model is inverted, shown at step I in Figure 4 (A). At steps I and II, the measured power value gives different (not unique) possible values for GHI denoted by the horizontal (orange) intersecting lines in (A) and (B). One of the factors causing the nonuniqueness is the discreteness of the so-called Perez factors that are used in determining the diffuse radiation part on the tilted PV array. The original values

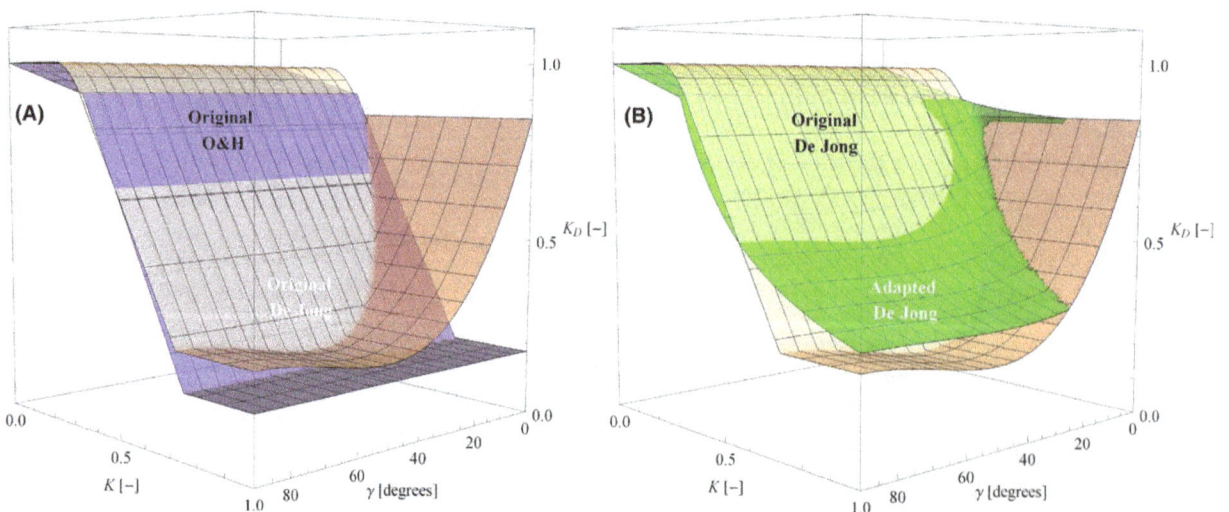

Figure 3. 3D Graphs illustrating the dependence of K_D on K and γ of the original De Jong (orange) compared with (A) the Orgill and Hollands (blue), and (B) the Adapted De Jong (green) decomposition models.

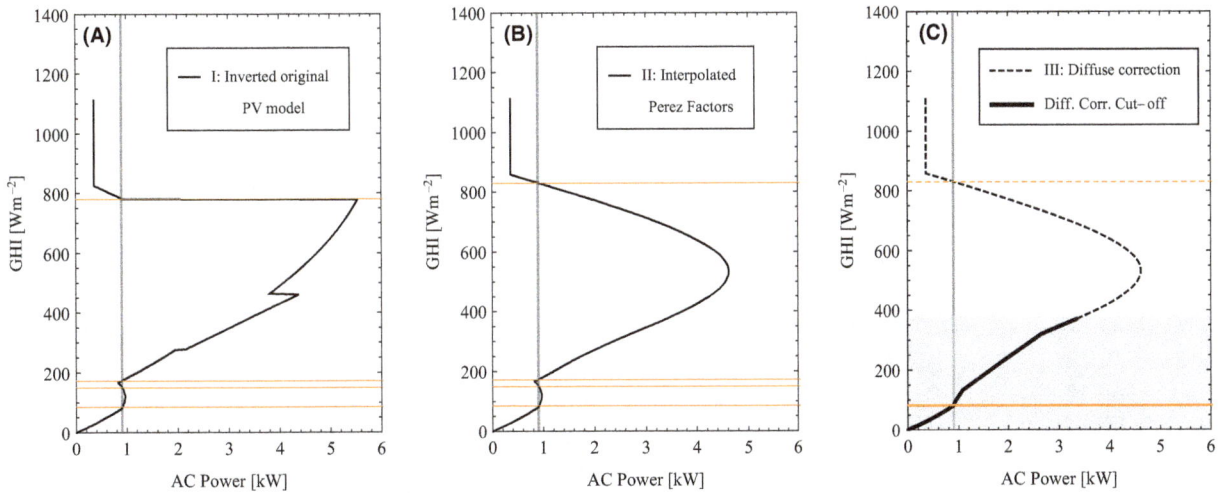

Figure 4. An example of GHI versus measured AC Power of the PV-system "Velt & Vecht". This is the inverse relation of the original GHI to power PV model. At a certain moment, 21-4-2013 (17:00) a power of 0.9 kW was measured (vertical line). The power to GHI relation is cut-off at 370 Wm^{-2} inside the gray area, to avoid final ambiguity in the inverse relation (see the dashed line for the other possible inverse solution). From the clear-sky model it is known that GHI could not reach values higher than this one for this PV-system at this time of day. The combination of solar elevation $\gamma \approx 25°$ and clearness index $K > 0.7$ at this moment is a good example of a situation that benefits from the linearized $K \cdot K_D$ relation, see also Figure 5.

of these factors are shown as dashed lines in Figure 2 as a function of the parameter ε. In the forward PV model these factors are interpolated using spline interpolation (at step II), and are attributed a continuous value (solid lines in Fig. 2). Implementation of the interpolated Perez factors shows as the black line in Figure 4(B) and (C).

The modification of the original O&H decomposition model by $R(\gamma)$ reduces the ambiguity of $0.2 < K < 0.6$, especially at relatively low $\gamma < 20°$, see the blue and orange dashed line corresponding to the values of $\gamma = 25°$ and $\gamma = 10°$ in Figure 5, however, the general relation $K \cdot K_D(K)$, given γ, is not fully one-to-one yet.

Inversion uniqueness

Finding an unique value of GHI at a known P requires a one-to-one invertible relation of $P_{AC}(I)$. From equation 10 we can see that this requires that the product of $K \cdot K_D$ as a function of K is invertible. The problem with this lies in the steeply decreasing slope of the $K_D(K)$ relation in the range $0.22 < K \leq (1.47 - R(\gamma))/1.66$: the product $K \cdot K_D$ will have a decreasing, concave parabolic shape over that range, that will result in an overall surjective function (that maps multiple values to the same result value, and is not uniquely invertible); this is visible in Figure 5. Additional to De Jong's modification of the O&H model, we therefore propose the following adaptation in order to turn $K \cdot K_D$ into a bijective (one-to-one) relation. First, the constant and the quadratic part (for $0 < K \leq 0.35$) of equation 11 are replaced by a single linear function of which the slope is sufficiently shallow.

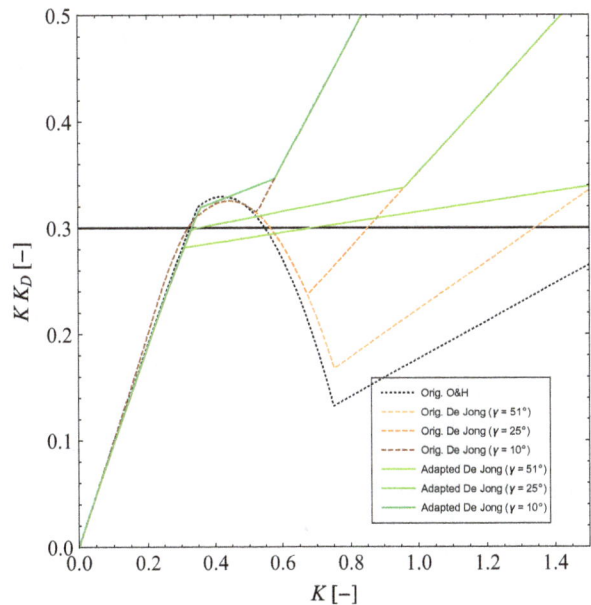

Figure 5. Example of how the adapted decomposition model results in a unique relation between KK_D and K (green, solid lines). The lines associated to three values of solar elevation angle γ are shown and intersect uniquely with an arbitrary value of KK_D (the thick horizontal line), whereas the original De Jong decomposition model (orange, dashed lines) and the Orgill and Hollands (blue, dotted) model would have multiple values of K associated to the shown $K_D = 0.3$.

Secondly, the strongly linear decreasing part (for $0.35 < K \leq (1.47 - R(\gamma))/1.66$) is turned into a decreasing convex function $\propto (a_1(\gamma) + a_2(\gamma)K)/K$ in order to have a *linear* relation when multiplied by K. This leads to the following adapted decomposition relation, see equation 13, and is visualized in Figures 3 and 6.

$$K_D = \begin{cases} 1 - 0.290K & \text{for} \quad 0 < K \le \kappa_0(\gamma) \\ \dfrac{a_1(\gamma) + a_2(\gamma)K}{K} & \text{for} \quad \kappa_0(\gamma) < K \le \kappa_1(\gamma) \\ R(\gamma) & \text{for} \quad K > \kappa_1(\gamma) \end{cases} \quad (13)$$

The numerical values and *connection* parameters $a_1(\gamma)$, $a_2(\gamma)$, $\kappa_0(\gamma)$, and $\kappa_1(\gamma)$ are determined such that the (now one-to-one) linearized part of the $K \cdot K_D$ to K relation connects to the other parts for every choice of γ, see Appendix. Also such that it retains the same shape beyond the surjective region and that there is minimal overall deviation with respect to the measured values. This is visible in Figure 5 where the straightened parts of the green, solid line segments continue as the straight parts of the blue, dashed lines where they intersect.

The resulting graph is shown as the black, dashed line in Figure 4C: this is considered step III. For this particular day and time it is known from the Ineichen clear sky model that the GHI can never be larger than roughly 370 Wm^{-2} (indicated by the gray area in Figure 4C), such that the ambiguous value at around 830 Wm^{-2} can be dismissed (step IV). The thick black solid line in Figure 4C represents the final one-to-one relation: the horizontal, orange line shows the unique value of GHI = 80 Wm^{-2} that is related to the measured power of 0.9 kW for this PV-system at that moment. This value of the GHI is found by starting off at estimating the value of $I^{i=0}$ such that $f(I^{i=0}) = P$ from the forward PV-model. When the initial value $I^{i=0}$ is too far off the inverted model line, its value is varied, keeping all the other variables constant until the intersection of the line $P_{AC} = P_{AC}(t)$ meets the inverted model line (within machine precision of 10^{-14}). The iteration procedure was implemented in MATLAB and follows the Levenberg–Marquardt algorithm. Consequent application of the inversion procedure then transforms a $P_{AC}(t)$ time series into a GHI(t) time series for a given PV-system so the PV-system can be used as a GHI sensor.

Validation

The following error metrics are used to numerically describe the validity of the decomposition model in terms of modeled (K_D) and measured values $(\tilde{K}_D)$, similar as used by [8]:

$$\text{rMBE} = 100\% \times \frac{1}{N} \sum_{i=1}^{N} \left(\frac{K_{D_i} - \tilde{K}_{D_i}}{\overline{\tilde{K}_D}} \right) \quad (14)$$

$$\text{MAPE} = 100\% \times \frac{1}{N} \sum_{i=1}^{N} \frac{|K_{D_i} - \tilde{K}_{D_i}|}{\tilde{K}_{D_i}} \quad (15)$$

$$\text{rRMSE} = 100\% \times \sqrt{\frac{1}{N} \sum_{i=1}^{N} \left(\frac{K_{D_i} - \tilde{K}_{D_i}}{\overline{\tilde{K}_D}} \right)^2} \quad (16)$$

$$R^2 = 1 - \frac{\sum_{i=1}^{N} \left(K_{D_i} - \tilde{K}_{D_i} \right)^2}{\sum_{i=1}^{N} \left(\tilde{K}_{D_i} - \overline{\tilde{K}_D} \right)^2} \quad (17)$$

where $\overline{\tilde{K}_D}$ denotes the mean of the measured values. The similarity of the error metrics for both hourly and 2 min values, as in Table 2 shows that the application to sub-hourly resolution is approximately as valid as for hourly values. Although the values of rRMSE of around 20% are considered high [8], it should be stressed that the choice for the adapted De Jong model is based on the existence of the unique inverse of the $K \cdot K_D$. Furthermore, the bias (rMBE) of the adapted De Jong model is much lower than the other two.

As can be seen from Figure 6, clearly both the QCRad procedure as well as the sun elevation correction $R(\gamma)$ influence the 2-min data more strongly than the hourly data. The rRMSE of the QCRad treated data is somewhat higher than for the raw data, but that is mainly due to the removal of mainly low-irradiance K values that are usually quite well fit by the decomposition model.

The inverse PV-system model was further validated at UPOT in the measurement period. Numerical results are shown for all four modules, but graphical results are only shown for module `mono1` as they are similar for the four modules. Overall correction factors are discussed, followed by the statistics on the performance of the forward and inverse PV model.

At UPOT, the DC output power is measured directly and AC power measurements are not available. To use the output power in the inverse PV-system model calculation, an ideal inverter is assumed. In this situation we always have $P_{DC} = P_{AC}$. In the application of the inverse PV model to real PV-systems, the inverter types are known and the α and β values are implemented accordingly. The benefit of verifying the model at UPOT is the

Table 2. Error Metrics of the three decomposition models at both time resolutions for QCRad data.

Data		rMBE	MAPE	rRMSE	R^2
Hourly data	O&H	−10.9%	21.8%	21.4%	0.817%
Hourly data	DJ	−3.82%	22.3%	19.3%	0.850%
Hourly data	ADJ	−0.23%	36.4%	23.9%	0.772%
2-min. data	O&H	−12.5%	23.4%	23.6%	0.778%
2-min. data	DJ	−4.7%	22.4%	20.1%	0.840%
2-min. data	ADJ	−0.78%	36.4%	24.4%	0.763%

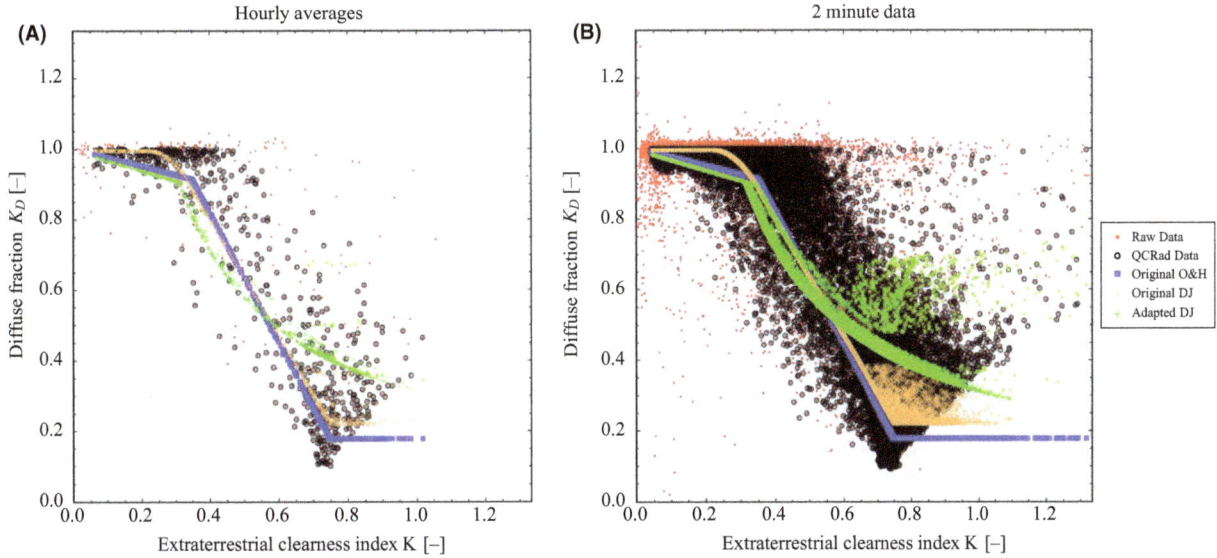

Figure 6. Visualization of the decomposition models over the measurement period. In both (A) and (B), the red dots show the uncorrected data; the black circles show the values that were corrected for QCRad and shading. The orange (×) and green (+) markers indicate the original De Jong and Adapted De Jong decomposition models on the corrected values. The O&H model is shown in blue (□) for reference. (A) is for hourly averaged values; (B) is for 2 min values. The notable absence of modified markers around $K_D = 0.4$ is due to omission of shaded measurement values at the specified values of θ_z. These plots can be seen as a side-view of Figure 3 as viewed in the γ direction.

availability of on-site registered irradiance (GHI, I_{POA}, DNI and diffuse) and ambient temperature.

Numerical Results of the Inverse PV Model

Correction factors

To be able to correct for an overall bias in the model calculations in a pragmatic way, we have added an overall correction factor to the PV model output. For each module the correction factor was calculated by linear regression with 0 constant bias: the slope of the fitted line through the $\{P_{calc.}, P_{meas.}\}$-tuples defines the correction factor through $c_P \equiv 1/slope$. The calculated GHI is slightly larger than the measured GHI. This is in line with the power bias found in the forward PV model: the P_{DC} calculated from the measured GHI was a little lower than the measured P_{DC}. The forward PV model underestimates the output power and thus the calculated GHI has to be increased to compensate for this underestimation.

The GHI correction factors c_{GHI}, are found similar to c_P through linear fit of $\{GHI_{calc.}, GHI_{meas.}\}$-tuples, and the correction factor is then defined as: $c_{GHI} \equiv 1/slope$. The values of the factors c_{GHI} and c_P, are presented in Table 3. When the forward and inverse factors are multiplied per module, they match unity within 1.5%.

Visual validation

In the Figures 7 and 8 calculated and measured time series are graphed for 3 days in the measurement period. The modeled values follow the measured values very well, except for a few instances. In Figure 9 data from the both models spanning the full measurement period is plotted.

Error metrics

The following definitions for absolute error $e_{(X)}$ and relative Error $e_{rel.(X)}$ are used to quantify the statistical accuracy of the forward and inverse PV model (where X can stand for P_{DC} or GHI and the overline denotes the average value during the measurement period); see equations 18 and 19 respectively.

$$e_{(P_{DC})}(t) = P_{DCmeas.}(t) - c_P \cdot P_{DCcalc.}(t)$$
$$e_{(GHI)}(t) = GHI_{meas.}(t) - c_{GHI} \cdot GHI_{calc.}(t) \qquad (18)$$

Table 3. Used correction factors c_X for DC Power and GHI.

Module	c_P	c_{GHI}
mono1	1.067	0.944
mono2	1.064	0.945
poly1	1.084	0.929
poly2	1.080	0.928

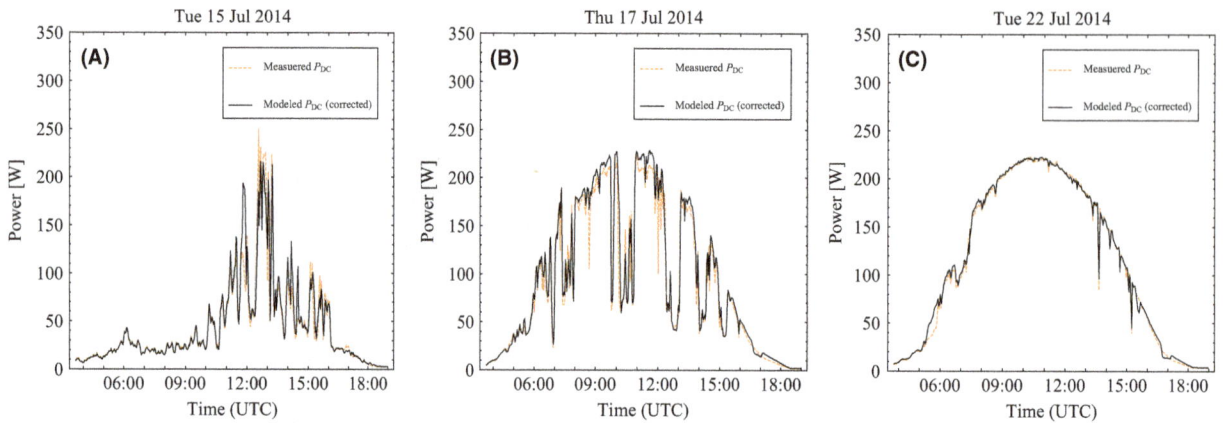

Figure 7. Forward PV Model. Three example days that show the result of the forward PV model at UPOT: measured P_{DC} (orange, dashed line) and modeled, bias-corrected P_{DC} of module mono1 (black line): (A) partly overcast, (B) variable, (C) mostly clear.

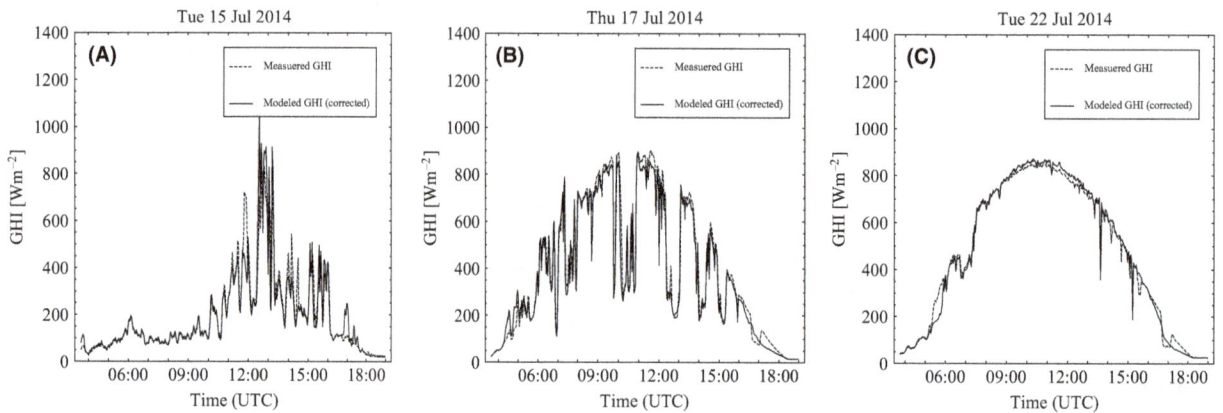

Figure 8. Inverse PV Model. Three example days that show the result of the inverse PV model at UPOT: measured GHI (blue, dashed line) and modeled, bias-corrected GHI of module mono1 (black line): (A) partly overcast, (B) variable, (C) mostly clear.

$$e_{rel.(X)}(t) = \frac{e_{(X)}(t)}{\overline{X}_{meas.}} \qquad (19)$$

The comparison of the calculated versus measured P_{DC} and GHI in Figure 9(A) and (B), respectively, shows that the calculated and measured values correspond well, but there is considerable spread in the values, which is quantified by the root mean square of the (relative) error RMSE, see Figure 10. This spread is caused mainly by the inherently statistical property of the transposition model that assumes a simple relationship through scattered data points relating the diffuse fraction to clearness index, see equation 11. The numerical error metrics for both models and all four modules show similar results.

The mean measured values of P_{DC} and GHI in the measurement period were $\overline{P_{DC}} = 107$ W and $\overline{GHI} = 435$ Wm^{-2}, respectively, for panel mono1. These values were used to determine the relative errors $e_{rel.(X)}$, as shown in Tables 4 and 5. For the four modules, slightly different values for

average GHI were found. This is due to the fact that the UPOT system registers the measurements from the modules individually through sequential IV-curve measurements. When the irradiance changes too much during an IV-curve measurement, the datapoint is flagged as "not to be trusted", and hence the accompanying GHI value at that time stamp will be discarded as well. By normalizing each module to its own measured time series and GHI, the results can still be compared well.

Discussion

A general-purpose power-to-irradiance model should in principle be verified for a large range of tilt and azimuth combinations [11, 16]. However, the rooftop PV-systems for which our inverse PV model is developed, are generally speaking oriented due south at a tilt between 25° and 45°, depending on the roof being flat or sloped. For this reason the verification using the UPOT modules as

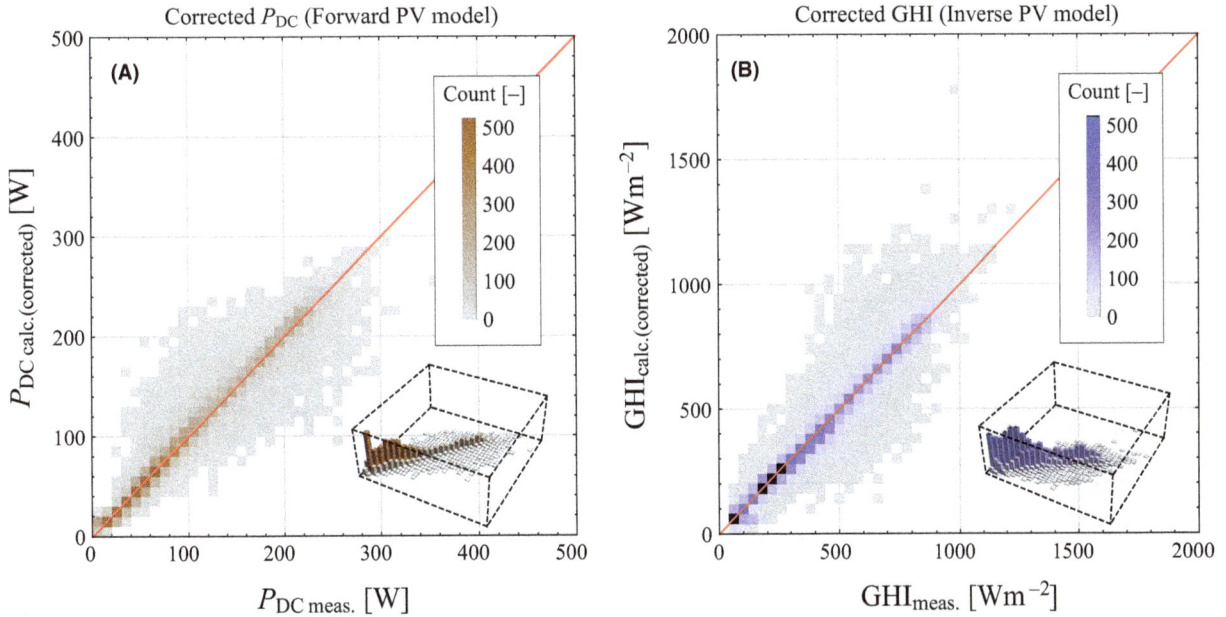

Figure 9. Scatter plot histograms of the calculated versus the measured values of module `mono1`; similar graphs were found for all four test modules. (A) shows the P_{DC} from the forward PV model. (B) shows the GHI from the inverse PV model. The insets show the same data in 3D for additional visualization.

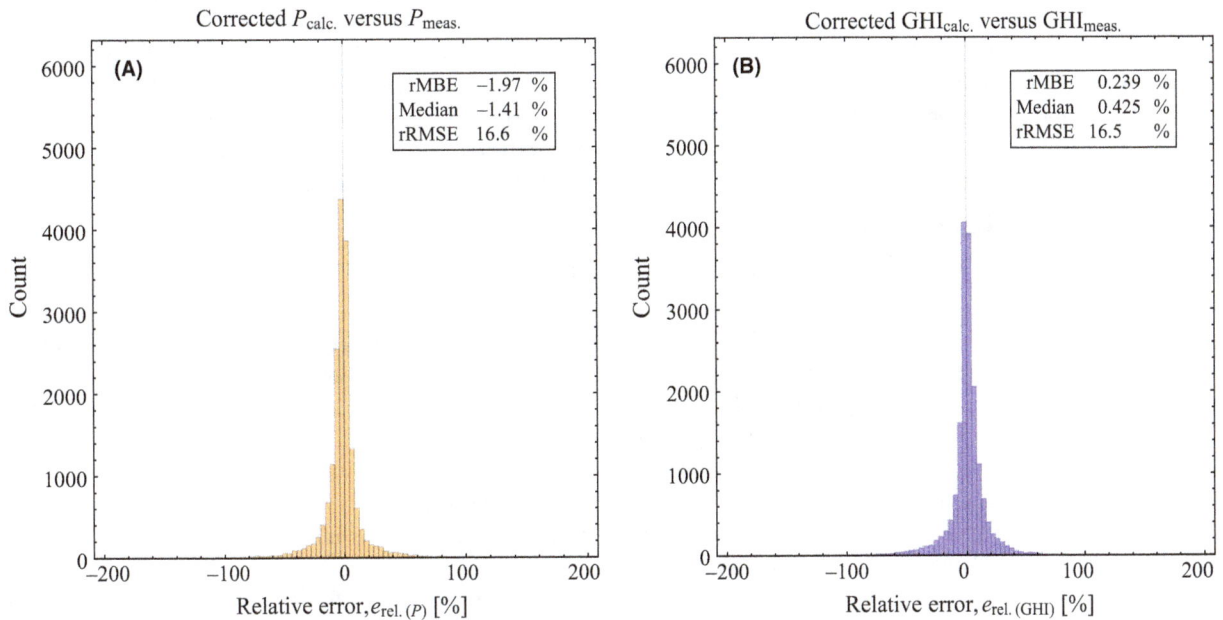

Figure 10. Distribution of relative errors, $e_{(x)}$ of calculated DC power via the forward PV model (A), and GHI via the inverse PV model (B). Data of module `mono1`; similar graphs were found for all four test modules. The error distributions [in W or Wm^{-2}] are equally shaped.

introduced in *Data Acquisition and Treatment* is sufficient.

When the RMSE values of the proposed inverse model are specified by daily values (i.e. calculated over the (corrected) time series ranging over each day instead of the entire measurement period at once), an interesting observation could be made: the RMSE that the model has is proportional to the variability in the fluctuations of the variable of interest, by a factor of 0.74 on $\Delta t = 2$ min basis. For the inverse model this resulted in a linear fit with $R^2 \approx 0.97$ for all four modules together whereas the dependence of MBE to variability is negligible, see

Table 4. Error statistics for the Forward PV model (DC Power) shown for all four modules. The prefix "r-" denotes the error metrics relative to the mean value of P_{DC}. Results are shown after correction by c_p.

Module	$\overline{P_{DC}}$ (W)	MBE (W)	RMSE (W)	rMBE (%)	rRMSE (%)
mono1	107	−2.11	17.7	−1.97	16.6
mono2	107	−1.99	16.4	−1.86	15.4
poly1	107	−2.05	17.5	−1.93	16.4
poly2	106	−1.67	16.0	−1.57	15.1

Table 5. Statistics on bias and relative errors for the Inverse PV model (GHI) shown for all four modules. The prefix "r-" denotes the error metrics relative to the mean value of GHI. Results are shown after correction by c_{GHI}.

Module	$\overline{GHI}$ (Wm⁻²)	MBE (Wm⁻²)	RMSE (Wm⁻²)	rMBE (%)	rRMSE (%)
mono1	435	1.04	71.6	0.239	16.5
mono2	434	1.67	66.7	0.383	15.4
poly1	435	1.05	73.0	0.243	16.8
poly2	434	0.994	65.6	0.229	15.1

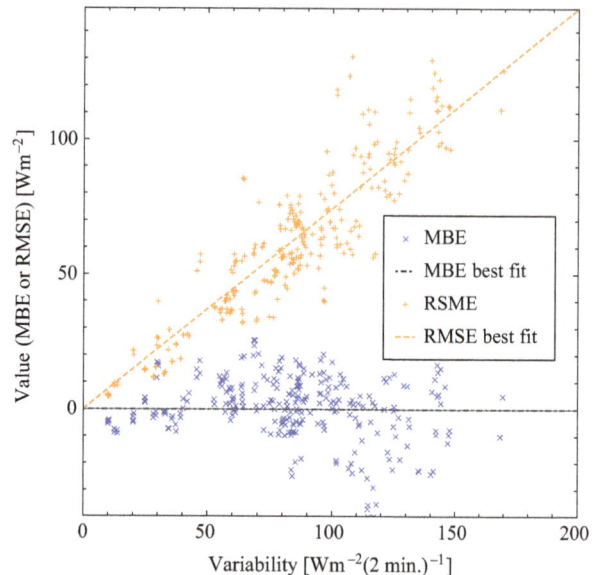

Figure 11. The daily MBE (×) and RMSE (+) of the inverse PV model (for all four modules together) versus the daily GHI variability, along with the best fit lines.

Figure 11. For the forward model, similar relations were obtained but are not additionally shown.

These relations could be of relevance when using the actual output of this model in e.g. quantifying uncertainty for irradiance forecasting on days characterized by their irradiance variability.[5]

Comparison to other studies

As stated in the introduction, the results from our inversion model are compared to recent and comparable studies.

Although the results of Marion et al. [11] are based on different input data-type (I_{POA} instead of Power; 5-min instead of 2-min) and different locations –from which Eugene, Oregon seems the most akin to our location in terms of latitude and climate– their results are shown here for comparison: relative Mean Bias Error: rMBE = −1.1% and relative Root Mean Squared Error rRMSE = 6.4%.

Yang et al. [14] compare different decomposition and transposition combinations and report on the specific combination of anisotropic O&H decomposition + Perez decomposition model that comes close to our approach. They report bias-reduced relative RMSE values ranging from approx. 12% to 15% for an East-facing sensor (at inclination 18.3°) and approx. 6% to 15% for a West-facing sensor (at inclination 6.1°) over 12 months of 5-minute aggregate irradiance data in Singapore. Similar results were found for different

combinations of uni- and bivariate decomposition models. It must be noted that at the latitude of Singapore (1.3° N), PV-modules are usually less steeply inclined (5° − 20°) [14], which makes measured I_{POA} more close to GHI than at our studied latitude (Utrecht, the Netherlands, 52°N), where PV-systems on roofs are usually installed at around 40° tilt, or 25° when on a flat roof.

In [15], reconstruction of GHI from sensors at various inclination angles shows that at measurement inclinations of 30°, and 40°, the inverse Perez model results in RMSE of 16% and 23%, respectively, over 1-min data ranging from January 2014 through May 2014. These results are comparable to our results and boundary conditions (UPOT modules are inclined at 37°).

Furthermore, in the study of Killinger et al. [16], mean values of MBE of −4.8% and RMSE of approx. 40% were reported, compared to measured GHI values at the PV locations.

The results of these aforementioned studies thus show mixed better or worse performance in calculated GHI than our inverse PV model, albeit under different climatic conditions and using different methods. In terms of used method and results, we would like to substantiate the choice for our method by one of the conclusions in [15] (#5: "Elaborate decomposition models do not out-perform simpler models") as it seems to align quite well with our method and results. For our purpose of using the existing PV-fleet as a sensor network, the thus shown uncertainties are acceptable.

Outlook

The here presented inverse PV model is intended to enable the use of 202 PV-systems in the Province of Utrecht (NL) as (proxies for) GHI sensors. P_{AC} (and derived GHI) data from these PV-systems are recorded at $\approx$2 sec resolution using a commercial but calibrated combined power measurement/data logger device [31] for the use within several fields of research: short-term Solar Forecasting using the peer-to-peer (P2P) method [6, 32], verification of total-sky images assisted nowcasting of GHI values and research into spatio–temporal variability in clear-sky index fluctuations [33].

The bias-correction factors are based on validation with the test-modules. However, when the Inverse PV model is applied to different type of module in the PV sensor field, or local effects such as age or soiling, systematic errors may arise. Bias correction of modules of the measured PV-systems in this PV sensor field are done by aligning parts of clear sky day measurements with low variability to clear sky estimates from the forward PV model.

Conclusion

Building on a modified Orgill and Hollands decomposition model in combination with a modified Perez transposition model, we have developed a practical Global Horizontal Irradiance (GHI)-to-Power model (forward PV model) that is also *inverted* to calculate GHI from measured AC PV output power (inverse PV model). This inverse PV model is useful for converting data on measured power from distributed PV-systems into an effective GHI sensor network. Validation of the forward and inverse PV model was done with data measured at an outdoor testing PV facility (UPOT), that measures DC power of individual modules from different types and manufacturers as well as GHI, plane-of-array global irradiance and module temperature. Two monocrystalline and two polycrystalline modules were used to correlate the modeled values of P and GHI to measured values for the forward and inverse PV model respectively. The bias-corrected forward PV model shows a best (r)RMSE of 16.0 W (15.1%) with a (r)MBE of −1.67 W (−1.57%) for one of the polycrystalline modules. The bias-corrected inverse PV model shows a best (r)RMSE of 65.6 Wm^{-2} (15.1%) with a (r)MBE of 0.994 Wm^{-2} (0.229%) for one of the polycrystalline modules. Similar results were obtained for the three other modules. Overall, the benefits of having a de facto distributed GHI sensor network outweigh the introduced uncertainties in our opinion.

Acknowledgments

The authors thank Paul Raats, Bas Vet, Santiago Peñate Vera (DNV-GL), and Ruut Brandsma (ECOFYS) for fruitful discussions and support. Furthermore, thanks to Atse Louwen and Arjen de Waal (UU) for the UPOT data and Geert Litjens (UU) for discussion on the manuscript. This study is financially supported by the Netherlands Enterprise Agency (RVO), through funding the project *Solar Forecasting & Smart Grids* (SF&SG) within the framework of the Dutch Topsector Energy, "TKI Switch2SmartGrids" under TKISG02017.

Conflict of Interest

None declared.

Notes

[1] See http://www.solarforecasting.nl (In Dutch).
[2] Direct horizontal irradiance is calculated from (DNI, I_{beam}) by multiplication with the cosine of the solar zenith angle: $I_{h,beam} = I_{beam} \cos(90 - \gamma)$.
[3] A constant air mass modifier of 1 is used in our model, as the original solar elevation dependent air mass modifier diverged for solar elevations below $\gamma < 4°$.
[4] Clear-sky index is defined as GHI(t) normalized by GHI(t) under clear sky conditions, in this case proportional to 1/1.364 of the maximal extraterrestrial irradiance value.
[5] Irradiance variability is here defined in absolute sense as the Root Mean Square of the GHI fluctuations: RMS($\Delta I(t)$)

References

1. Lonij, V. P., A. E. Brooks, A. D. Cronin, M. Leuthold, and K. Koch. 2013. Intra-hour forecasts of solar power production using measurements from a network of irradiance sensors. Sol. Energy 97:58–66.

2. Lorenzo, A. T., W. F. Holmgren, and A. D. Cronin. 2015. Irradiance forecasts based on an irra-diance monitoring network, cloud motion, and spatial averaging. Sol. Energy 122:1158–1169.

3. Aryaputera, A. W., D. Yang, L. Zhao, and W. M. Walsh. 2015. Very short-term irradi-ance forecasting at unobserved locations using spatio-temporal kriging. Sol. Energy 122:1266–1278.

4. Golnas, A., J. Bryan, R. Wimbrow, C. Hansen, S. Voss, and S. Clemente. 2011. Performance assessment without pyranometers: predicting energy output based on historical Correlation. Pp. 2006–2010 *in* 37th IEEE Photovoltaic Specialists Conference (PVSC).

5. Engerer, N. A., and F. P. Mills. 2014. K PV: a clear-sky index for photovoltaics. Sol. Energy 105:679–693.

6. Elsinga, B., and W. G. J. H. M. Van Sark. 2014. Inter-system time lag due to clouds in an urban PV ensemble. Pp. 754–758 *in* Conference Record ofthe IEEE Photovoltaic Specialists Conference. https://doi.org/10.1109/pvsc.2014.6925029.

7. Elsinga, B., and W. G. J. H. M. Van Sark. 2015. Spatial power fluctuation correlations in urban rooftop photovoltaic systems. Prog. Photovoltaics Res. Appl., 23:1390–1397.

8. Engerer, N. A. 2015. Minute resolution estimates of the diffuse fraction of global irradiance for southeastern Australia. Sol. Energy 116:215–237.

9. Orgill, J. F., and K. G. T. Hollands. 1977. Correlation equation for hourly diffuse radiation on a horizontal surface. Sol. Energy 19:357–359.

10. Perez, R. R., P. Ineichen, R. Steward, and D. Menicucci. 1987. A new simplified version of the Perez diffuse irradiance model for tilted surfaces. Sol. Energy 39:221–231.

11. Marion, B. 2015. A model forderiving the directnormal and diffuse horizontal irradiance from the global tilted irra-diance. Sol. Energy 122:1037–1046.

12. Maxwell, E. L. 1987. A Quasi-physical model for converting hourly global to direct insolation. SERITR-215 - 3087.

13. Perez, R., P. Ineichen, E. Maxwell, R. Seals, and A. Ze-lenka. 1992. Dynamic global-to-direct irradiance conversion models. Pp. 354–369 *in* ASHRAE Transactions - Research Series.

14. Yang, D., Z. Dong, A. Nobre, Y. S. Khoo, P. Jirutitijaroen, and W. M. Walsh. 2013. Evaluation of transposition and decomposition models for converting global solar irradiance from tilted surface to horizontal in tropical regions. Sol. Energy 97:369–387.

15. Yang, D., Z. Ye, A. M. Nobre, H. Du, W. M. Walsh, L. I. Lim et al. 2014. Bidirectional irradiance transposition based on the Perez model. Sol. Energy 110:768–780.

16. Killinger, S., F. Braam, B. Moller, B.-H. Wille-Haussmann, and R. McKenna. 2016. Projection of power generation between differently-oriented PV systems. Sol. Energy 136:153–165.

17. UPOT 2016. Utrecht photovoltaic outdoor testing facility. Available at http://www.upot.nl/system.html (accessed 20 December 2016).

18. Van Sark, W. G. J. H. M., A. Louwen, de Waal A. C., B. Elsinga, and R. E. I. Schropp. 2012. UPOT: The Utrecht Photovoltaic Outdoor Test Facility. Pp. 3247–3249 *in* 27th EUPVSEC.

19. Louwen, A., A. C. de Waal, R. E. I. Schropp, A. P. C. Faaij, and W. G. J. H. M. van Sark. 2016. Comprehensive characterisation and analysis of PV module performance under real operating conditions. Prog. Photovoltaics Res. Appl. 25:218–232.

20. Femtogrid Energy Solutions. Available at http://www.femtogrid.com/products/power-optimizer/solar-power-optimizer/

21. Long, C. N., and Y. Shi. 2006. The QCRad value added product: surface radiation measurement quality control testing, including climatology configurable limits. PNNL; Richland, WA, United States. doi:10.2172/1019540. http://www.osti.gov/scitech/servlets/purl/1019540.

22. Ineichen, P., and R. Perez. 2002. A new airmass independent formulation for the linke turbidity coefficient. Sol. Energy 73:151–157.

23. Photovoltaic Performance Modelling Consortium (PVPMC). PVLIB-toolbox. Available at https://pvpmc.sandia.gov/applications/pv_lib-toolbox/

24. Michalsky, J. J. 1988. The Astronomical Almanac's algorithm for approximate solar position (1950–2050). Sol. Energy 40:227–235.

25. De Soto, W., W. Klein, and W. A. Beckman. 2006. Improvement and validation of a model for photovoltaic array performance. Sol. Energy 80:78–88.

26. Haeberlin, H., F. Kaeser, C. Liebi, and C. Beutler. 1995. Results of recent performance and reliability tests of the most popular inverters for grid connected PV systems in Switzerland. *Proc. 13th EU PV Conference, Nice.*

27. Duffie, J. A., and W. A. Beckman. 1991. Solar engineering of thermal processes. John Wiley & Sons Inc., Hoboken, New Jersey.

28. Liu, B. Y. H., and R. C. Jordan. 1963. The long-term average performance of flat-plate solar-energy collectors: with design data for the U.S., its outlying possessions and Canada. Sol. Energy 7:53–74.

29. De Jong, J. B. R. M. 1980. Een karakterisering van de zonnes-traling in Nederland. Rapportnr. WPS-3-80.05. R306. Technical report, Techn. Univ. Eindhoven.

30. Velds, C. A. 1992. Zonnestraling in Nederland (Dutch, English Summary). Technical report, Thieme/KNMI, Royal Netherlands Meteorological Institute. Available at http://bibliotheek.knmi.nl/knmipubDIV/Zonnestraling_in_Nederland.pdf

31. Upp Energy (part of Upp Smart). 2015. Last referenced on 20 Dec 2016. Available at http://uppenergy.com/product/details/1f

32. Elsinga, B., and W. G. J. H. M. Van Sark. 2017. Short-term peer-to-peer solar forecasting in a network of photovoltaic systems. *in revision.*

33. Elsinga, B., and W. G. J. H. M. Van Sark. 2017. Analytic model for correlations of cloud induced fluctuations of clearness index. Sol. Energy 155:985–1001.

Appendix

Connection of the convex part of the adapted decomposition model

The following steps were taken to ensure that the linearized part of the $K \cdot K_D$ relation connects to the other two parts of the adapted decomposition model:

Based on the notion that a function $\kappa f(\kappa)$ is one-to-one if its derivative is always positive or always negative. Our function should be increasing, so we have the condition:

$$\frac{\partial(\kappa f(\kappa))}{\partial \kappa} > 0 \tag{A1}$$

or, more specifically:

$$f(\kappa) + \kappa \frac{\partial(f(\kappa))}{\partial \kappa} = c \tag{A2}$$

For some $c > 0$, this differential equation yields, with extra introduced constant c_0:

$$f(\kappa) = c + \frac{c_0}{\kappa} \tag{A3}$$

In terms of the studied model, we will write:

$$K_D(K) = a_2 + \frac{a_1}{K} \tag{A4}$$

Now only the values of a_1 and a_2 need to be determined, along with the range K over which this substitution will be applied.

A straight line can be parametrized by two arbitrary points $\{x_0, y_0\}$ and $\{x_1, y_1\}$, such that the linearized part can be written as the straight line $K \cdot K_D = a_1(\gamma) + a_2(\gamma)K$, with parameters:

$$a_1(\gamma) = \frac{x_1 - x_0}{x_0 y_1 - x_1 y_0}$$
$$a_2(\gamma) = \frac{y_1 - y_0}{x_1 - x_0} \tag{A5}$$

All of these depend on γ, but the dependences of the x_0, x_1, y_0, and y_1 parameters on γ is left implicit here. The fixed parameters that guarantee connection to the

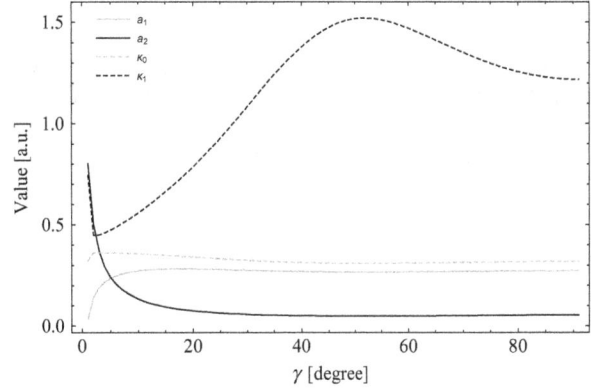

Figure A1. The dependence of the parameters a_1, a_2, κ_0, and κ_1 on γ.

other parts of the decomposition model are: $m = 0.290$, $\Delta = 0.02$ and $\tau = 0.75$.

$$y_0(\gamma) = \tau \kappa_A (1 - m\kappa_A) + (1 - \tau)\kappa_B(\gamma)R(\gamma)$$
$$x_0(\gamma) = \frac{1 - \sqrt{1 - 4my_0(\gamma)}}{2m}$$
$$y_1(\gamma) = y_0 + \Delta \tag{A6}$$
$$x_1(\gamma) = \kappa_B(\gamma)$$

with:

$$\kappa_A = \frac{557}{40(25m - 46)}$$
$$\kappa_B(\gamma) = \frac{1.557 - R(\gamma)}{1.84} \tag{A7}$$

Finally, the convex relation as in equation A4 has the right properties for the range $\kappa_0(\gamma) < K \leq \kappa_1(\gamma)$, which follow from intersection with the first (quadratic) part of $K \cdot K_D$ and the third (linear) part of $K \cdot K_D$ (with K_D as in equation 13):

$$\kappa_0(\gamma) = \frac{\sqrt{X^2 - 4m(x_0 - x_1)(x_0 y_1 - x_1 y_0)} + X}{2m(x_0 - x_1)}$$
$$\kappa_1(\gamma) = \frac{x_0 y_1 - x_1 y_0}{R(\gamma)(x_0 - x_1) - y_0 + y_1} \tag{A8}$$

Here, $X = (x_0 - x_1 - y_0 + y_1)$. The parameters a_1, a_2, κ_0, and κ_1 are visualized in Figure A1.

18

High efficiency thin-film amorphous silicon solar cells

Ahmadreza Ghahremani & Aly E. Fathy

Department of Electrical Engineering and Computer Science, University of Tennessee, Knoxville, Tennessee

Keywords

Amorphous silicon, defects, metallic nanoparticles, solar cell, thin film

Correspondence

Ahmadreza Ghahremandi, Department of Electrical Engineering and Computer Science, University of Tennessee, Knoxville, TN. E-mail: aghahrem@utk.edu. Mailing Address: 1520 Middle Drive Knoxville, TN 37996-2250

Funding Information

National Science Foundation (Grant/Award Number: 'EPS-1004083').

Abstract

Enhancing light absorption within thin film amorphous silicon (a-Si) solar cells should lead to higher efficiency. This improvement is typically done using various light trapping techniques such as utilizing textured back reflectors for pronounced light scattering within the cell thus achieving higher absorption. It is believed that embedding metallic nanoparticles (MNPs) inside the structure could increase light scattering. However, embedding MNPs can also cause significant structure defects and pronounced efficiency drop as well – it has been indicated by many experiments that disproved this belief. In search of ways to improve efficiency, we have investigated the impact of MNP's size, and location within the solar cell, in addition to the effect of defects, and doping levels on the overall efficiency. On the basis of our 3D multiphysics (optical-electric) modeling, we developed a design guideline for embedding these MNPs and reducing the impact of defects created in the embedding process. The results of simulations were compared to relevant measured data, and it showed a good agreement. Subsequently, models were used to predict performance, and over 30% improvement in solar cell efficiency (~13% is predicted); which is beyond the state of the art. This was predicted by optimizing the size and location of the MNPs and tailoring the doping levels to have better forward light trapping and absorption.

Introduction

Remarkable manufacturing cost reduction in solar cells can be achieved using thin film hydrogenate amorphous silicon (A-Si:H) instead of bulk silicon. However, a pronounced efficiency drop could be incurred by utilizing thin film silicon [1]. Typically, any thin film solar cells suffer from a huge reduction in light absorption within absorber layers (semiconductors), and that can cause efficiency drop due to inherent surface reflection.

To achieve higher efficiency, some boosting techniques have been developed for better light absorption. These enhancement methods are based on increasing the optical path length and embedding scatterers within cells. They are designed for constructive interference. Random textured or corrugated external/internal interfaces are used to improve scattering [2–8], while transparent conductive oxide (TCO) layers are utilized to minimize reflections at interfaces, additionally highly reflective surfaces are used to enhance back reflections. Figure 1 illustrates such enhancing techniques.

Alternatively, MNPs are intentionally placed within solar cells. MNPs (few nanometers in diameter) can scatter a wide range of visible light, and also can create high intensity near-fields in their vicinity [10]. However, light can face optical losses for small (few nanometer) MNPs that can supersede scattering. Therefore methods need to be developed to enhance scattering and to improve absorption if possible.

Typically, the optical properties of MNPs are highly controlled by changing size [9], density [10, 15], conductivity [9], location [11, 16], and shape [12–14]. These MNPs can be made out of gold or silver, and both could exhibit great metal/plasmon behavior at optical frequencies and consequently would impact on amorphous silicon thin film solar cell's performance [11]. Several studies have utilized nanotechnology to fabricate embedded MNPs within solar cells. Unfortunately, serious parasitic losses and structure defects were incurred and had been associated with

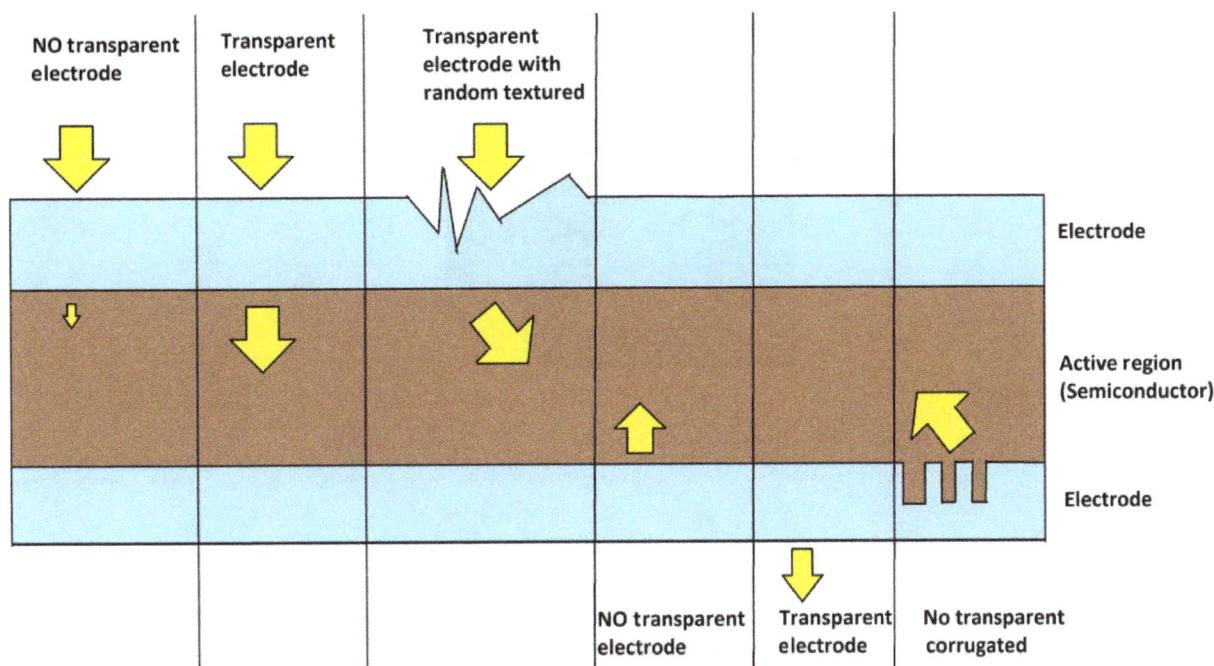

Figure 1. Optical paths inside solar cells for different type of electrodes.

these MNPs that led to significant overall solar cell efficiency degradation. So studies are still going on to explore the feasibility of finding an efficient light scattering scheme whenever MNPs are carefully embedded within the structure to increase optical path length and to provide better absorption for light, while minimizing energy loss.

To combat such efficiency drop, we need to address some challenging issues like: optical losses within MNPs, and those due to fabrication defects. The optical loss is manifested by a large fraction of the impinging light energy absorbed by MNPs and converted to phonons, thus reducing the overall efficiency. Add to that, during the fabrication process, gross material defects can occur. Defects would spread around embedded MNPs causing loss that would increase even further with higher defect density. So these issues associated with the design and fabrication, need to be resolved to enhance efficiency. Several investigations have been carried out to understand the role of these embedded nanoparticles and potential to improve performance [9–11]. Further studies are still needed, given that the impact of MNPs has not been experimentally materialized yet and the need has significantly increased to reveal a successful design recipe.

Along these lines, we carried out a detailed 3D multiphysics modeling of plasmon solar cells [17], and we studied the effect of MNPs on performance in search for efficiency enhancement. In this paper, new design rules for embedding MNPs inside thin film amorphous silicon solar cells will be presented that would lead to solar cell

efficiency enhancement. A modeling toolbox was successfully developed for 3D solar cells performance analysis [17], and it was validated by previously published experimental data carried out by Ref. [11].

The effect of placing MNPs at alternative locations (front, middle, and back of the P-I-N solar cell) to maximize the photocurrent generation will be discussed in section II. A new design methodology will be recommended in sections III and IV. Conclusion will be given in section V. Finally a methodology for a robust simulation will be presented in the Appendix.

Simulation Analysis

A 3D model of a thin film amorphous silicon solar cell has been developed which accounts for surface roughness as well. A view of the structure is shown in Figure 2A. The surface roughness would impact on the overall performance. Typically, the amount of surface roughness is related to transparent conductive oxide (TCO) type. For instance, using TCO film with large grains would increase the surface roughness [24–26]. At the same time, the size of the grains is correlated with the thickness of the thin film TCO layer, where a thinner film may have less surface roughness [24–26]. For instance, the thickness of the thin film TCO, considered for this model, is 75 nm, and its surface roughness is estimated to be <10 nm.

In our investigation, to model the solar cell and to take the effect of surface roughness into account, a 3D

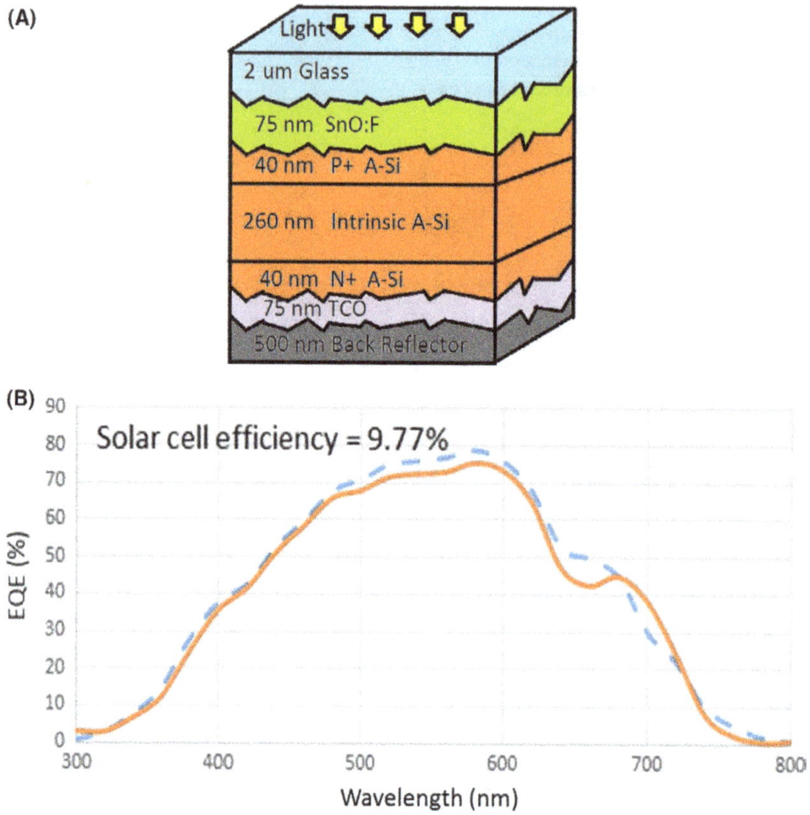

(A)

(B)

Figure 2. (A) schematic of the P-I-N device; (B) EQE curve (dash line represents measurement [11]; solid line represents simulation result) [17, 34].

device model like a trapezoidal grating is assumed [17]. A periodic structure of a trapezoidal shape (like that of [27]) is designed and implemented to the 3D model of [11]. Utilizing our model, a comparison between our results and that measured External Quantum Efficiency (EQE) by [11] is shown in Figure 2B. Only a slight discrepancy is seen – thus validating our models. This success establishes the logistics to extrapolate models that include MNPs effects and the impact of their size, shape, and location of the device layers on solar cell efficiency.

To understand the effect of existing silver nanoparticles, we studied solar cell's performance after embedding these MNPs at different layers, one layer at a time. In our model, silver nanoparticles are designed as spheres with 18 nm diameter and placed in a random 2D array with a maximum center-to-center spacing of 36 nm. First, we embedded MNPs inside the absorber region (t = 50 nm as seen in Fig. 3A), and our simulation results (Fig. 3B) rather huge drop in conversion efficiency which is once more consistent with Ref. [11] observations.

The agreement seen in Figure 3B between simulation and measurement is good, and it validates our model again. Although surprisingly, the efficiency has dropped to 3.5% in contradiction to the common belief that it should be

enhanced upon using MNPs – (however, if no defects exist, the efficiency though would be 9.77% as indicated in Fig. 2B). We have done several modeling for cases of MNPs that were moved from the top of the intrinsic layer to its bottom, however, the solar cell efficiency still dropped even further to 2.55% [17], which is related to effect of defects. In our previous publication [17], we showed that how EQEs can be changed by placing the MNPs at different positions within the layer of the absorber. We also calculated the amount of efficiency, FF, Voc, and Jsc for various scenarios.

Impact of Defects

One of the disadvantages of embedding MNPs inside a semiconductor is increasing the density of defects especially around the MNPs. Presence of defects can cause optical losses. This extra optical loss is due to a large Shockley Read Hall recombination rate – which would mean a huge efficiency drop. For instance, to consider the effect of defects around MNPs (inside the intrinsic region), the recombination rate was considered 100 times higher than normal value estimated when inside intrinsic region.

To demonstrate this effect, a P-I-N structure was analyzed before and after embedding the MNPs [17], and a

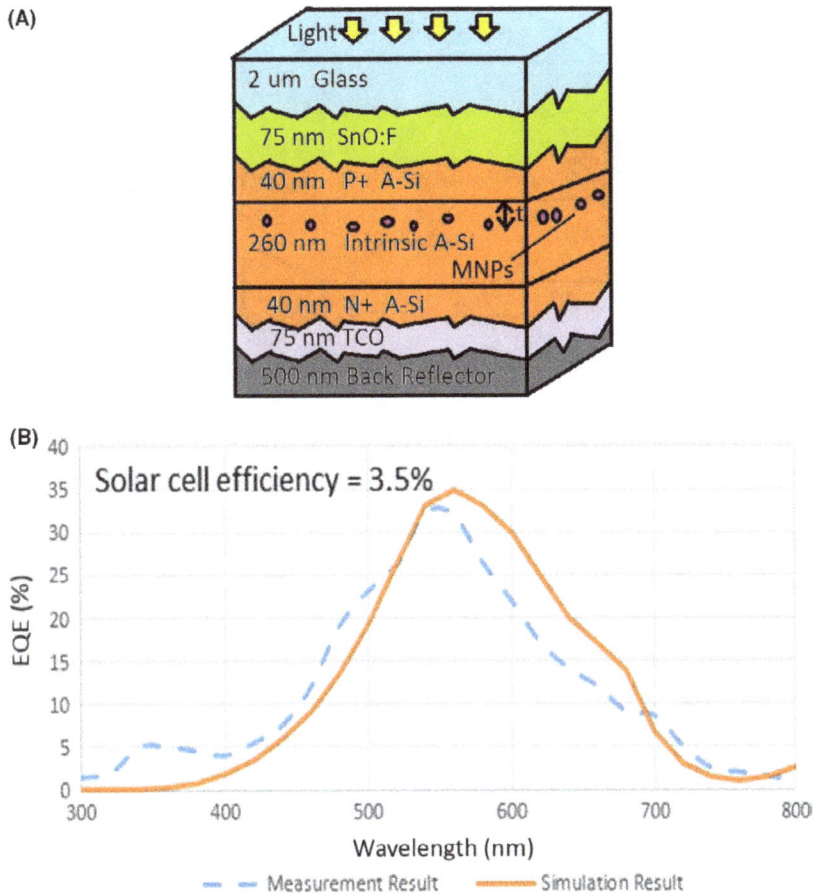

Figure 3. (A) Schematic of the P-I-N device; (B) EQE curve (dash line represents measurement [11]; solid line represents simulation result), after considering of defects [34].

huge difference between the results with and without accounting for the presence of the defects was seen in our first experiment (efficiency of 9.8% without considering defects, and 3.5% with as seen in Fig. 3).

To avoid such a problem, MNPs should be placed in a proper location so they would cause minimal impact on efficiency reduction. It turns out that if defects are placed in a highly doped region, they would not impact on recombination rate in this region, on the other hand the recombination rate would relatively increase if MNPs are placed in a lightly doped region; hence, it is better off placing the MNPs in a highly doped region. In other words, the impact of placing MNPs on recombination rate can be very high if they are placed inside the intrinsic layer compared to being in a highly doped region (P+, or N+) as shown in Figure 4A. Subsequently, the recombination rate should not change much compared to the MNP-free case. [17]. Simulation result shown in Figure 4B, is validating our observation, and indicates pronounced efficiency improvement when placing MNPs only inside the top P+ layer as expected. The improvement happens

due to a relatively strong light intensity propagating through the top layer where strong localized fields exist around MNPs close to depleted region of P-I-N.

To improve the performance even further, the size, and location of MNPs should be optimized as well. Intuitively, the high-frequency spectrum of light is mostly absorbed within the top layers. Hence, if small size MNPs with diameters in the range of 18 nm (i.e., resonating at these high frequencies) are used in the top P+ layer, they would enhance the scattering and absorption of this spectrum. Meanwhile, use of such small MNPs would still allow the relatively low-frequency spectrum to travel through the top layers and reach the bottom ones. However, to have appreciable absorption for the spectrum at low frequencies large MNPs (size around 200 nm in diameter) resonate and enhance absorption. Large MNPs should be placed at the bottom layer (i.e., inside TCO – next to the N+ region, see Fig. 5A). At this point, the intensity of light for the ultraviolet (UV) rays (high frequencies) close to the N-type region (at the back) is very weak, since most of their energies have already been absorbed by the top

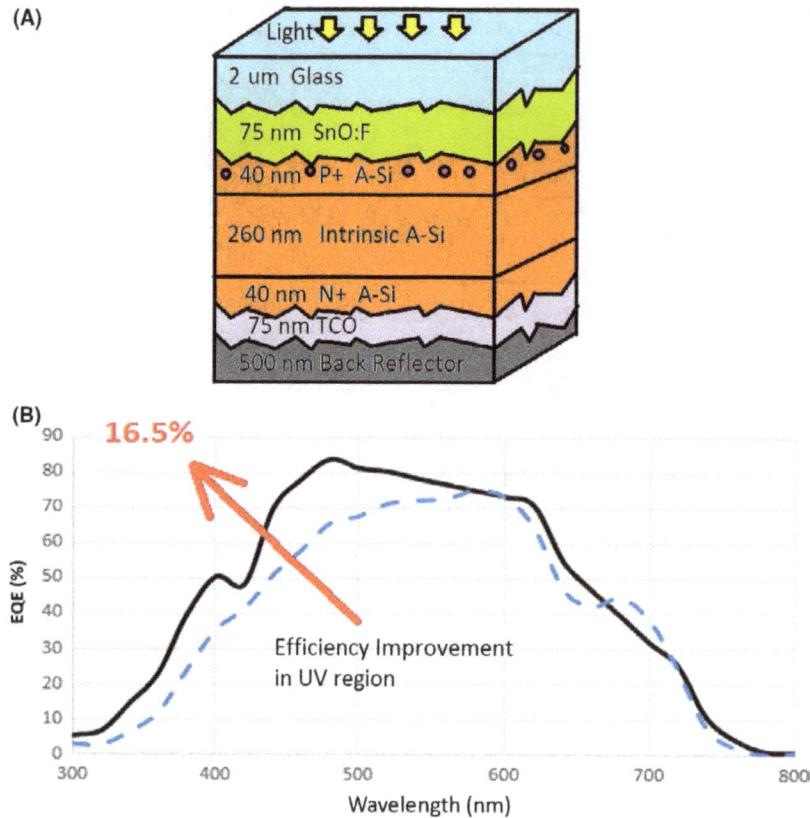

Figure 4. (A) Schematic of the P-I-N device; (B) Comparison of EQE with and without MNPs; solid line represents MNPs inside the P+ top layer of the a-Si; dash line represents the case with No MNPs [34].

layers of the absorber (i.e., inside the P+, and intrinsic region), and mostly Infrared (IR) rays exist. In case of placing small MNPs on the top and large MNPs on the bottom, simulation result shows remarkable improvement for solar cell efficiency by roughly 30%, as seen in Figure 5B. This would simply mean an overall efficiency of 13%.

Optimization of Highly Doped Region

Doped layer thickness and doping level can impact the efficiency of thin film solar cell. At high frequencies (UV region), most of the impinging solar energy on the cell is absorbed at the top of the semiconductor close to the surface (here P+ region), and it happens before approaching UV rays to the junction or depleted region. This is obviously translated to an energy loss [18–21]. Figure 6 shows that the extinction coefficient of amorphous silicon which has exponential growth rate in UV region. It turns out pronounced light absorption happens with very low EQE. In other word UV rays are absorbed dramatically very close to the surface of the semiconductor (free charges will recombine together, because there is no electric field force for separation), and they may not be capable of reaching

to the sweet spot inside the device (which is close to the junction called depleted region). So increasing the thickness of intrinsic layer of the semiconductor is not suggested for efficiency improvement at UV range, unless the absorber has low light absorption at this frequency range. However, if UV rays instead of being absorbed mostly in P+ layer (close to the surface) reach to the depleted region, and propagate in a longer path inside the P-I-N device, then more electron–hole pairs will be generated and will be separated (having more separated charges can be expected as more electricity). This would require: first, optimizing the thickness of the highly doped layer; and second, optimizing the level of dopant. Hence, the thickness of P+ layer should be thinned and the level of dopant (here P+) needs to be decreased enough, thus pushing the depleted region closer to the surface of the semiconductor on the top. Therefore, the probability of generating separated charges (electron–hole pairs) by UV rays will increase. The main reason that light trapping of MNPs for a-Si solar cells (in the state of the art) only occurs in long wavelengths (above 500 nm) is that the observation of UV rays inside intrinsic layer or close to the device junction normally happens with very low intensity (regarding to the huge absorption occurs

(A)

(B)

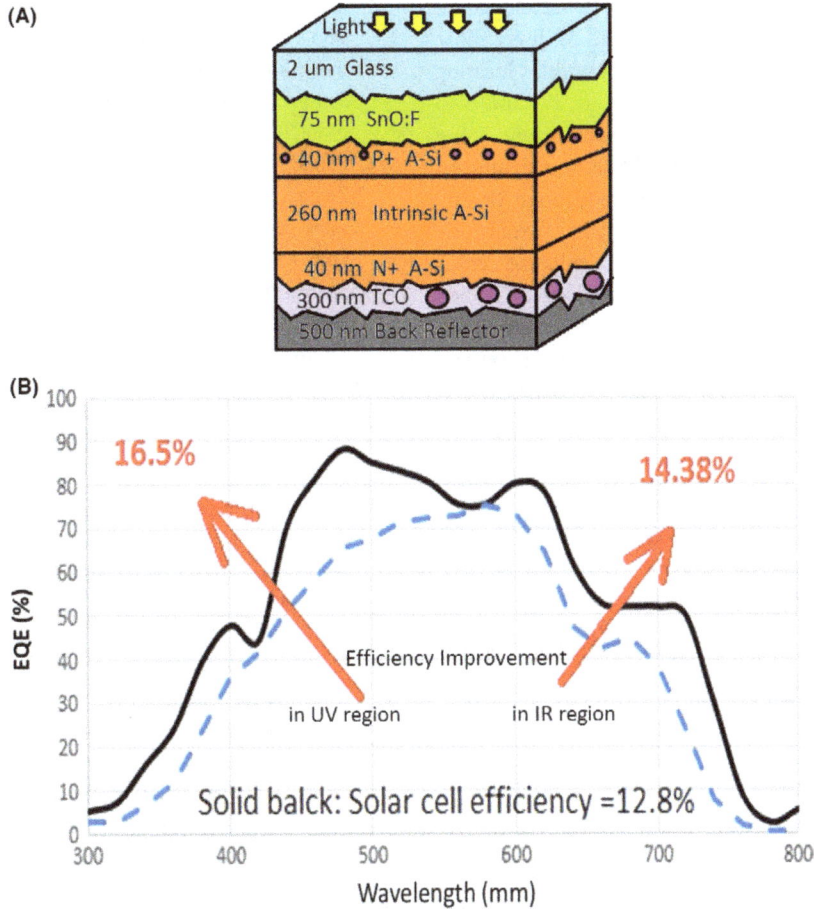

Figure 5. (A) Schematic of the P-I-N device; (B) Using different size of MNPs, in two different locations EQE curve for the simulation (solid line represents after optimization-black; dash line represents No MNPs blue) [34].

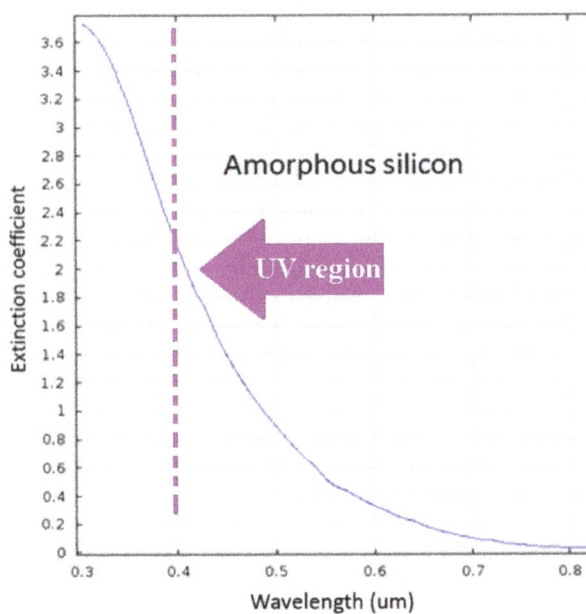

Figure 6. Extinction coefficient of amorphous silicon.

at the surface of the absorber). Also the chance of existing MNPs at the right place (for resonance) is low. It is suggested that small MNPs to be placed between the transparent electrode and the highly doped semiconductor at the top layer side, instead of inside the P+ region for ease of fabrication process. In this case, the near field of those nanoparticles (at resonance) would still have some effects inside the depleted region and could generate more free electron–hole pairs compared to embedding MNPs on the P+ layer.

Conclusion

Random embedding of MNPs has resulted in a drop of solar cell efficiency. Extensive simulation, based on our 3D combined optical-electric modeling toolbox has led to very promising results for ways to achieve higher efficiency. First, a significant efficiency drop detected after adding the MNPs (related to the substantial number of defects left). Presence of defects has resulted in a considerable optical loss around the MNPs. Second, increasing the recombination rate would reduce the conversion of optical energy to electricity. Hence, better manufacturing techniques need to be developed to reduce both defects and recombination, and to direct more light to the depleted region. Such as strategically locating small MNPs at the highly doped regions (i.e., P+ and N+) rather than inside the intrinsic layer. Additionally, using small MNPs at the top (P+) layer should allow a significant portion of the optical energy to propagate through (it acts like a transparent layer for long wavelengths), meanwhile larger MNPs are placed at the bottom to enhance reflection/scattering. It is certainly not recommended to embed large MNPs inside the active region, because it can cause a large amount of optical loss for the whole system. Optimizing the thickness of the highly doped layers and level of dopant can have a huge effect on solar cells' performance as well. Finally simulation results indicate an impressive efficiency enhancement of up to ~30% which amounts to 13% overall efficiency.

Acknowledgments

This work was supported by the grant from the National Science Foundation of USA (Grant No. NSF EPS-1004083).

Conflict of Interest

None declared.

References

1. Zeman, M. 2006. Advanced amorphous silicon solar cell technology. Pp. 173–236 in J. Poortmans and V. Archipov, eds. Thin film solar cells: fabrication, characterization and applications. Wiley, Wiley, Chichester. ISBN: 978-0-470-09126-5.

2. Deckman, H. W., C. B. Roxlo, and E. Yablonovitch. 1983. Maximum statistical increase of optical absorption in textured semiconductor films. Opt. Lett. 8:491–493.

3. Isabella, O., F. Moll, J. KrC, and M. Zeman. 2010. Modulated surface textures using zinc-oxide films for solar cells application. Phys. Status Solid. A 207:642–646. doi:10.1002/pssa.200982828.

4. Kambe, M., A. Takahashi, N. Taneda, K. Masumo, T. Oyama, and K. Sato. 2008. Proceedings of the 33rd IEEE PVSC: San Diego. IEEE, New York, Pp. 1–4, doi: 10.1109/PVSC.2008.4922507

5. Borri, C., and M. Paggi. 2015. Topological characterization of antireflective and hydrophobic rough surfaces: are random process theory and fractal modeling applicable? J. Phys. D: Appl. Phys. 48:045301.

6. Haug, F.-J., T. Söderström, O. Cubero, V. Terrazzoni-Daudrix, X. Niquille, S. Perregeaux, et al. 2008. Materials Research Society Symposium Proceedings 1101: KK13-KK. doi: 10.1557/PROC-1101-KK13-01.

7. Krc, J., M. Zeman, A. C˘ampa, F. Smole, and M. Topic˘. 2006. Novel approaches of light management in thin-film silicon solar cells. Materials Research Society Symposium Proceedings 910: A25-A; doi: 10.1557/PROC-0910-A25-01.

8. Pahud, C., O. Isabella, A. Naqavi, F.-J. Haug, M. Zeman, H. P. Herzig, et al. 2013. Plasmonic silicon solar cells: impact of material quality and geometry. Opt. Express 21:A786–A797. doi:10.1364/OE.21.00A786.

9. Stuart, H. R., and D. G. Hall. 1998. Island size effects in nanoparticle-enhanced photodetectors. Appl. Phys. Lett. 73:3815.

10. Toroghi, S., and P. G. Kik. 2012. Cascaded plasmonic metamaterials for phase-controlled enhancement of nonlinear absorption and refraction. Phys. Rev. B 85:045432.

11. Santbergen, R., R. Liang, and M. Zeman. 2010. A-Si:H Solar Cells with embedded silver Nanoparticles. Photovoltaic Specialists Conference PVSC, IEEE, doi: 10.1109/PVSC.2010.5617095.

12. Chen, Y.-S., W. Frey, S. Kim, K. Homan, P. Kruizinga, K. Sokolov, et al. 2010. 1. Enhanced thermal stability of silica-coated gold nanorods for photoacoustic imaging and image-guided therapy. Opt. Express 18:8867–8878. doi:10.1364/OE.18.008867.

13. Lu, X., M. Rycenga, S. E. Skrabalak, B. Wiley, and Y. Xia. 2009. Chemical synthesis of novel plasmonic nanoparticles Annu. Rev. Phys. Chem. 60:167–192.

14. Mock, J. J., M. Barbic, D. R. Smith, D. A. Schultz, and S. J. Schultz. 2002. Shape effects in plasmon resonance of individual colloidal silver nanoparticles. Chem. Phys. 116:6755–6759.

15. Zhang, W., Q. Li, and M. Qiu. 2013. A plasmon ruler based on nanoscale photothermal effect. Opt. Express 21:172–181. doi:10.1364/OE.21.000172.

16. Sha, W. E. I., H. L. Zhu, L. Chen, W. C. Chew, and W. C. H. Choy. 2015. A general design rule to manipulate photocarrier transport path in solar cells and its realization by the plasmonic-electrical effect. Sci. Rep. 5:8525. doi:10.1038/srep08525.

17. Ghahremani, A., and A. E. Fathy. 2015. A three-dimensional multiphysics modeling of thin-film amorphous silicon solar cells. Energy Sci. Eng. 3(6):520–534. doi:10.1002/ese3.100A.

18. Balanis, C. A. 1989. Advanced engineering electromagnetics. Wiley, Hoboken, New Jersey, USA.

19. Neamen, D. A. 2003. Semiconductor physics and devices: basic principles, 3rd ed. McGraw Hill, New York.

20. Sze, S. 1981. Physics of semiconductor devices, 2nd ed. Interscience, New York.

21. Madelung, O. 2012. Semiconductors: data handbook. Springer, Verlag Berlin Heidelberg.

22. Hall, R. N. 1952. Electron-hole recombination in silicon. Phys. Rev. 87:387.

23. Shockley, W., and W. T. Read. 1952. Statistics of the recombinations of electrons and holes. Phys. Rev. 87:835–842.

24. Gracia, M., F. Rojas, and G. Gordillo. Morphological and optical characterization of SnO2:F thin films deposited by spray pyrolysis. 20th European Photovoltaic Solar Energy Conference, 6–10 June 2005, Barcelona, Spain.

25. Chaaya, A. A., R. Viter, M. Bechelany, Z. Alute, D. Erts, A. Zalesskaya, et al. 2013. Evolution of microstructure and related optical properties of ZnO grown by atomic layer deposition. Beilstein J. Nanotechnol. 4:690–698. doi:10.3762/bjnano.4.78.

26. Kumar, V., N. Singh, R. M. Mehra, A. Kapoor, L. P. Purohit, and H. C. Swart. 2013. Role of film thickness on the properties of ZnO thin films grown by sol-gel method. Thin Solid Films 539:161–165.

27. Isabella, O., S. Solntsev, and M. Zeman. 2013. 3-D optical modeling of thin-film silicon solar cells on diffraction gratings. Prog. Photovolt. Res. Appl. 21:94108.

28. Sakata, I., and Y. Hayashi. 1985. Theoretical analysis of trapping and recombination of photo generated carriers in amorphous silicon solar cells. Appl. Phys. A 37:153–164.

29. Ihalane, E., M. Meddah, A. Elfanaoui, L. Boulkaddat, E. El hamri, X. Portier, et al. 2011. Numerical simulation of photocurrent in a solar cell based amorphous silicon. Moroccan J. Condens. Matter. 13:83–87.

30. Piprek, J. 2003. Semiconductor optoelectronic devices introduction to physics and simulation. Academic Press; 1 edition (January 21, 2003). ISBN-13: 978-0125571906.

31. Weiland, T. 1977. A discretization method for the solution of Maxwell's equations for six-component fields. Electron. Commun. 31:116–120.

32. Moharam, M. G., and T. K. Gaylord. 1981. Rigorous coupled-wave analysis of planar-grating diffraction. J. Opt. Soc. Am. 71:811–818.

33. Bakr, N. A., A. M. Funde, V. S. Waman, M. M. Kamble, R. R. Hawaldar, D. P. Amalnerkar, et al. 2010. Determination of the optical parameters of a-Si:H thin films deposited by hot wire–chemical vapor deposition technique using transmission spectrum only. Pramana J Physics. 76:519–531.

34. Ghahremani, A., and A. E. Fathy. Strategies for designing high efficient thin-film amorphous silicon solar cells. Photovoltaic Specialist Conference (PVSC), 2015 IEEE 42nd.

Appendix

A1. Developing a Methodology for a Robust Simulation

The flowchart in below shows how our 3D model of a solar cell works.

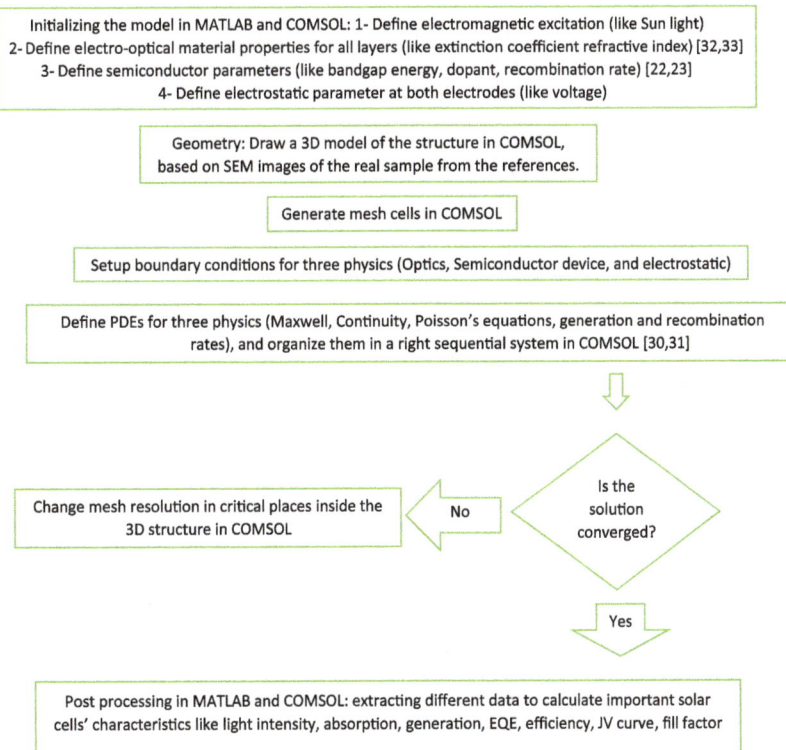

The flowchart in below shows how our 3D model of a solar cell works.

Initializing the model in MATLAB and COMSOL: 1- Define electromagnetic excitation (like Sun light)
2- Define electro-optical material properties for all layers (like extinction coefficient refractive index) [32,33]
3- Define semiconductor parameters (like bandgap energy, dopant, recombination rate) [22,23]
4- Define electrostatic parameter at both electrodes (like voltage)

Geometry: Draw a 3D model of the structure in COMSOL, based on SEM images of the real sample from the references.

Generate mesh cells in COMSOL

Setup boundary conditions for three physics (Optics, Semiconductor device, and electrostatic)

Define PDEs for three physics (Maxwell, Continuity, Poisson's equations, generation and recombination rates), and organize them in a right sequential system in COMSOL [30,31]

Change mesh resolution in critical places inside the 3D structure in COMSOL ← No — Is the solution converged?

Yes

Post processing in MATLAB and COMSOL: extracting different data to calculate important solar cells' characteristics like light intensity, absorption, generation, EQE, efficiency, JV curve, fill factor

A2. Initialization, Mesh Generation, and Boundary Conditions

Initialization of the input data is the crucial part of this work. The accuracy of the results is directly related to the input data. For instance using the accurate solar spectrum of energy as an excitation for electromagnetic propagation, applying the right values for electro-optical material properties to solve light intensity inside the structure, and also initializing semiconductor with right amount of carrier density inside P-I-N, and recombination rate to solve continuity equation (in physics device) are some basic steps to start modeling for solar cells. Table 1 shows the list of parameters that are used for initialization of the 3D model. Selecting right type of mesh with right size for each cell is very important challenge (in numerical methods) for converging the matrices for a nonlinear system of Partial Differential Equations (PDEs). Since three physics are solved sequentially with the same structure of mesh cells, converging the solver with accurate result in minimum time of processing is our target. Free Tetrahedral has been selected as a type of mesh. The minimum size of element for critical regions like the junctions inside the semiconductor, electrodes, and around metallic nanoparticles changes between 0.1 and 1 nm. The maximum size of element depends on: 1-the longest side of each single layer 2-operation wavelength. It changes between 10 nm to 100 nm. This cad tool used Finite Element Method (FEM) as a numerical method to solve the nonlinear system of PDEs. Figure 7 shows the geometry of the whole structure in 3D with considering boundary conditions as well.

Table 1. The utilized value of each parameter for the utilized validation example [17].

Parameter –Name	Value
T(temperature)	300 [K]
Ni -Ref[28]	0.949×10^6 [cm^{-3}]
Doping (N+)	1×10^{20} [cm^{-3}]
Doping (P+)	10^{21} [cm^{-3}]
Thickness (N+, a-Si)-Ref [9]	40 [nm]
Thickness (intrinsic, a-Si)- Ref [9]	260 [nm]
Thickness (P+, a-Si)- Ref [9]	40 [nm]
Thickness (AZO) -(TCO in front)- Ref [9]	75 [nm]
Thickness (glass)- Ref [9]	200 [μm]
Thickness (Air)- Ref [9]	20 [μm]
Thickness -(TCO BACK)- Ref [9]	75 [nm](AZO)
Thickness (Silver) Reflector back- Ref [9]	500 [nm]
Electron mobility, a-Si -intrinsic-Ref [12]	20 [cm^2/(V sec)]
Hole mobility, a-Si -intrinsic-Ref [12]	2 [cm^2/(V sec)]
Electron mobility, a-Si N+-Ref [12]	20 [cm^2/(V sec)]
Hole mobility, a-Si N+-Ref [12]	2 [cm^2/(V sec)]
Electron mobility, a-Si P+-Ref [12]	20 [cm^2/(V sec)]
Hole mobility, a-Si P+-Ref[12]	2 [cm^2/(V sec)]
Electron Life time, a-Si -intrinsic-Ref [27, 28]	20 [nsec]
Hole Life time, a-Si -intrinsic-Ref [27, 28]	20 [nsec]
Electron Life time, a-Si N+-Ref [27, 28]	0.0001 [nsec]
Hole Life time, a-Si N+-Ref [27, 28]	10 [nsec]
Electron Life time, a-Si P+-Ref [20,21]	10 [nsec]
Hole Life time, a-Si P+-Ref [20,21]	0.0001 [nsec]
Density Of State Valence band, a-Si –Ref [25]	2.5×10^{20} [cm^{-3}]
Density Of State Conduction band, a-Si –Ref [25]	2.5×10^{20} [cm^{-3}]
Difference between Defect level and intrinsic level N+,P+-Ref [29]	0.7
Difference between Defect level and intrinsic level intrinsic-Ref [29]	0.3
Energy Band gap a-Si-Ref [28]	1.74
Diameter silver NPs-Ref [9]	20 [nm]
Affinity, a-Si (electro affinity)-Ref [12]	4.00 eV
Incident Light Angle	0 [deg]

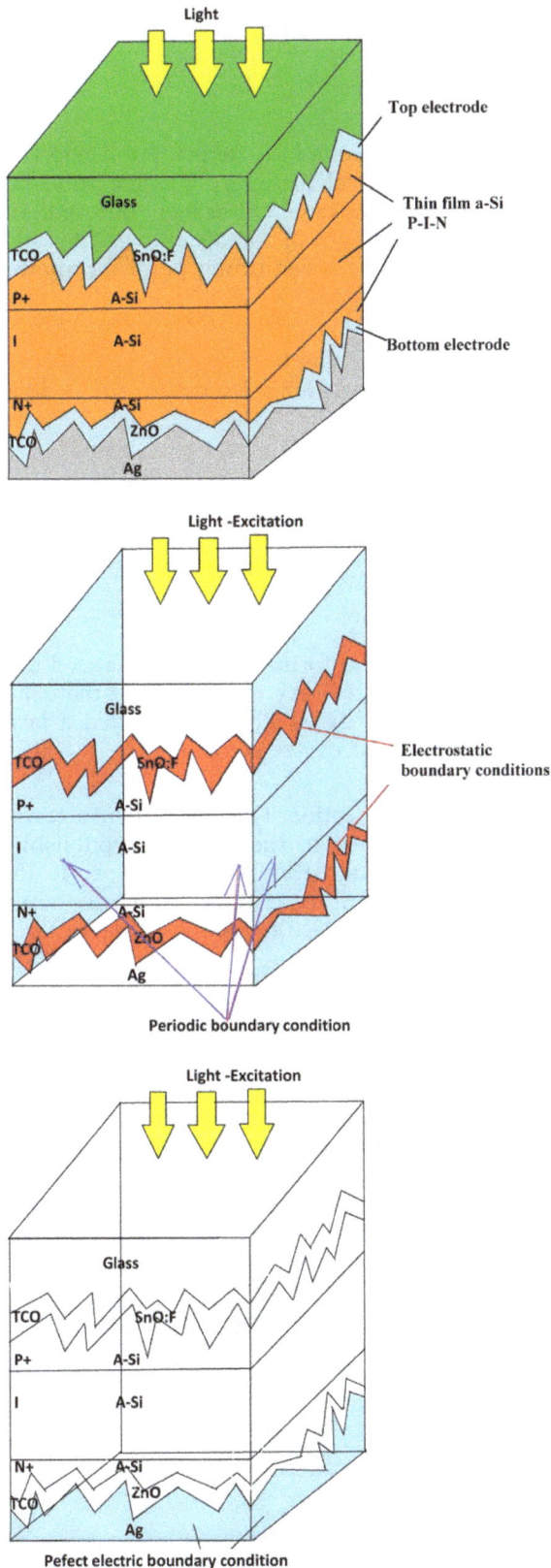

Figure 7. The boundary conditions, and the excitation in a 3D structure of the solar cell.

A3. Optimization Method

For a common problem in solar cells (no plasmon), some degree of freedoms for optimization are presented in bellow to improve the conversion efficiency.

1. Extinction coefficient of semiconductor
2. Refractive index of materials (attached to the top and the bottom of the absorber)
3. Thickness of each layers
4. Doping level of highly doped region
5. Surface roughness of the electrodes
6. Grating at the back reflectors
7. Conductivity of the electrodes

Although the spotlight of this study is based on using plasmon layers (MNPs) to improve the efficiency of thin film amorphous silicon solar cells. For this reason the degree of freedoms to optimize the model are limited by the type and dimension of the original sample has already defined (in case of no MNPs). It means that the only variables that need to be addressed to reach the goal are: 1-shape of MNP 2- size of MNP 3-location of MNP. Since during the fabrication process there is no control on the shape of silver NPs, then only two variables left for tuning thin film plasmon solar cells. Finally we are able to find the sweet spot by tuning the two variables. Although in some cases like placing small MNPs close to the junction inside the semiconductor, it is not easy to get a quick convergence. Therefore, it is a major requirement to define very fine mesh especially around the critical regions. Lack of any control on mesh generation can cause long time of processing, or out of memory due to aggressive computations.

Permissions

List of Contributors

Jehad Abed, Meera Almheiri, Afra Alketbi and Camilia Aokal
Sustainable and Renewable Energy Engineering Department, University of Sharjah, United Arab Emirate

Abdul Hai Alami
Center for Advanced Materials Research, University of Sharjah, United Arab Emirates
Sustainable and Renewable Energy Engineering Department, University of Sharjah, United Arab Emirate

Juan P. Babaro
Instituto Nacional de Tecnología Industrial-Física y Metrología, San Martín, Buenos Aires, Argentina

Kevin G. West
SensorMetrix, San Diego, California, 92126

Behrang H. Hamadani
National Institute of Standards and Technology, Engineering Laboratory, Gaithersburg, Maryland 20899

Johannes Hepp
Bavarian Center for Applied Energy Research (ZAE Bayern), Haberstraße 2a, 91058 Erlangen, Germany
Erlangen Graduate School in Advanced Optical Technologies (SAOT), Friedrich Alexander University Erlangen-Nuremberg (FAU), Paul-Gordan-Str. 6, 91052 Erlangen, Germany

Florian Machui and Hans-J. Egelhaaf
Bavarian Center for Applied Energy Research (ZAE Bayern), Haberstraße 2a, 91058 Erlangen, Germany

Christoph J. Brabec and Andreas Vetter
Bavarian Center for Applied Energy Research (ZAE Bayern), Haberstraße 2a, 91058 Erlangen, Germany
Materials for Electronics and Energy Technology (iMEET), Friedrich Alexander University Erlangen-Nuremberg (FAU), Energie Campus Nürnberg (EnCN), 90429 Nürnberg, Germany

Johannes Hofer, Prageeth Jayathissa, Zoltan Nagy and Arno Schlueter
Architecture and Building Systems, Institute of Technology in Architecture, ETH Zurich, John-von-Neumann Weg 9, 8093 Zürich, Switzerland

Abel Groenewolt
Institute for Computational Design, University of Stuttgart, Keplerstrasse 11, 70174 Stuttgart, Germany

Hiroyuki Kanda, Abdullah Uzum, Norihisa Harano and Seigo Ito
Department of Materials and Synchrotron Radiation Engineering, Graduate School of Engineering, University of Hyogo, 2167 Shosha, Himeji, Hyogo 671-2280, Japan

Seiya Yoshinaga, Yasuaki Ishikawa and Yukiharu Uraoka
Graduate School of Materials Science, Nara-Institute of Science and Technology, 8916-5 Takayama, Ikoma, Nara 630-0192, Japan

Hidehito Fukui and Tomitaro Harada
Daiwa Sangyo Co. Ltd., 3-4-11, Nakayasui, Sakai, Osaka, Japan

Michael D. Kempe, David C. Miller, John H. Wohlgemuth, Sarah R. Kurtz, John M. Moseley and Dylan L. Nobles
National Renewable Energy Laboratory, 1617 Cole Boulevard, Golden, Colorado 80401

Katherine M. Stika, Yefim Brun and Sam L. Samuels
DuPont Company, 200 Powder Mill Road, Wilmington, Delaware 19803

Qurat (Annie) Shah and Govindasamy Tamizhmani
Polytechnic Campus, Arizona State University, 7349 East Unity Avenue, Mesa, Arizona

Keiichiro Sakurai, Masanao Inoue, Takuya Doi and Atsushi Masuda
National Institute of Advanced Industrial Science and Technology, 1-1-1 Umezono, Tsukuba, Ibaraki 305-8568, Japan

Crystal E. Vanderpan
Underwriters Laboratories, 455 East Trimble Road, San Jose, California

Igor Konovalov
Ernst Abbe University of Applied Science in Jena, Carl Zeiss Promenade 2, 07745 Jena, Germany

Vitali Emelianov
Ernst Abbe University of Applied Science in Jena, Carl Zeiss Promenade 2, 07745 Jena, Germany
Thuringian Postgraduate School of Photovoltaics "Photograd", Institute of Physics, Technical University Ilmenau, 98684 Ilmenau, Germany

Thomas Rachow, Stefan Reber, Stefan Janz, Marius Knapp and Nena Milenkovic
Materials – Solar Cells and Technologies, Fraunhofer ISE, Heidenhofstrasse 2, 79110 Freiburg, Baden-Württemberg, Germany

Bruno Soria, Eric Gerritsen and Paul Lefillastre
Photovoltaic Modules Laboratory, CEA-INES, 50 avenue du Lac Léman, Le Bourget-du-Lac F-73375, France

Jean-Emmanuel Broquin
CNRS, IMEP-LACH, Grenoble Alpes University, Grenoble F-38000, France

Vladimir Švrček, Mickael Lozac'h, Takeshi Tayagaki, Tetsuhiko Miyadera and Koji Matsubara
Research Center for Photovoltaics, National Institute of Advanced Industrial Science and Technology (AIST), Central 2, Umezono 1-1-1, Tsukuba, 305- 8568, Japan

Calum McDonald
Research Center for Photovoltaics, National Institute of Advanced Industrial Science and Technology (AIST), Central 2, Umezono 1-1-1, Tsukuba, 305-8568, Japan
Nanotechnology and Integrated Bio-Engineering Centre (NIBEC), Ulster University, Coleraine, UK

Tomoyuki Koganezawa
Japan Synchrotron Radiation Research Institute (JASRI), 1-1-1, Kouto, Sayo-cho, Sayo-gun, Hyogo 679-5198, Japan

Davide Mariotti
Nanotechnology and Integrated Bio-Engineering Centre (NIBEC), Ulster University, Coleraine, UK

Jun Wang, Song Yang, Chuan Jiang and Yaoming Zhang
Key Laboratory of Solar Energy Science and Technology in Jiangsu Province, School of Energy and Environment, Southeast University, No. 2 Si Pai Lou, Nanjing 210096, China

Peter D. Lund
Key Laboratory of Solar Energy Science and Technology in Jiangsu Province, School of Energy and Environment, Southeast University, No. 2 Si Pai Lou, Nanjing 210096, China
School of Science, Aalto University, FI-00076 Aalto (Espoo), Finland

Emily C. Warmann
California Institute of Technology, 1200 E California Blvd, Pasadena, California 91125-0002

Harry A. Atwater
Kavli Nanosciences Institute, California Institute of Technology, Pasadena, California

Emily C. Warmann, Cristofer Flowers, John Lloyd and Harry A. Atwater
California Institute of Technology, Pasadena, California, USA

Carissa N. Eisler
E O Lawrence Berkeley National Laboratory, Berkeley, California, USA

Matthew D. Escarra
Tulane University, New Orleans, Louisiana, USA

Soohyun Bae, Kyung Dong Lee, Seongtak Kim, Hyunho Kim, Sungeun Park, Hae-Seok Lee and Donghwan Kim
Department of Materials Science and Engineering, Korea University, 145, Anam-ro, Seongbuk-gu, Seoul, Korea

Wonwook Oh, Nochang Park and Sung-Il Chan
Electronic Convergence Material and Device Research Center, Korea Electronic Technology Institute, Seongnam, Korea

Yoonmook Kang
KU•KIST Green School, Graduate School of Energy and Environment, Korea University, 145, Anam-ro, Seongbuk-gu, Seoul, Korea

Liqiang Duan, Xiaohui Yu, Shilun Jia and Buyun Wang
School of Energy, Power and Mechanical Engineering, National Thermal Power Engineering and Technology Research Center, Key Laboratory of Condition Monitoring and Control for Power Plant Equipment of Ministry of Education, Beijing Key Laboratory of Emission Surveillance and Control for Thermal Power Generation, North China Electric Power University, Beijing 102206, China

Jinsheng Zhang
Shenhua Guohua (Beijing) Electric Power Research Institute Co., Ltd., Beijing 100025, China

Rajiv Dubey, Shashwata Chattopadhyay, Vivek Kuthanazhi, Anil Kottantharayil, Chetan Singh Solanki, Brij M. Arora, Krishnamachari L. Narasimhan and Juzer Vasi
National Centre for Photovoltaic Research and Education, Indian Institute of Technology Bombay, Powai, Mumbai 400076, India

Birinchi Bora, Yogesh Kumar Singh and Oruganti S. Sastry
National Institute for Solar Energy, Ministry of New and Renewable Energy, New Delhi 110003, India

Boudewijn Elsinga and Wilfried van Sark
Copernicus Institute of Sustainable Development, Utrecht University, 3508 TC Utrecht, The Netherlands

Lou Ramaekers
ECOFYS, 3503 RK Utrecht, The Netherlands

Ahmadreza Ghahremani and Aly E. Fathy
Department of Electrical Engineering and Computer Science, University of Tennessee, Knoxville, Tennessee

Index